THE LOEB CLASSICAL LIBRARY

FOUNDED BY JAMES LOEB

HIPPOCRATES

II

LCL 148

HIPPOCRATES

VOLUME II

WITH AN ENGLISH TRANSLATION BY

W. H. S. JONES

HARVARD UNIVERSITY PRESS
CAMBRIDGE, MASSACHUSETTS
LONDON, ENGLAND

First published 1923
Reprinted 1943, 1952, 1959, 1967, 1981, 1992, 1998

ISBN 0-674-99164-8

Printed in Great Britain by St Edmundsbury Press Ltd,
Bury St Edmunds, Suffolk, on acid-free paper.
Bound by Hunter & Foulis Ltd, Edinburgh, Scotland.

CONTENTS

PREFACE

In this, the second volume of Hippocrates in the Loeb series, it has been found useful to go more fully into textual questions than was necessary when preparing Vol. I. Critical scholars have cleared away most of the blemishes that disfigured the text of *Airs Waters Places* and of *Epidemics* I. and III., but the text of many of the treatises in the present volume is still in places uncertain.

Many kind helpers have made the task of preparing the text easier that it would otherwise have been. The Earl of Leicester and Mr. C. W. James have given me the opportunity of consulting Holkhamensis 282 at my leisure. Dr. Karl Mras, Professor in Vienna, has sent me a photograph of a part of θ, and the Librarians of S. Mark's Library, Venice, and of the Vatican Library, have in a similar way helped me to collate M and V. The Curators of the Bodleian were kind enough to allow me to inspect Baroccian 204. The Librarians of the Cambridge University Library have helped me in various ways, and Dr. Minns has given me the benefit of his expert advice in deciphering places that presented special difficulty.

My colleague the Rev. H. J. Chaytor continues to lend me his invaluable services, and I must thank Sir Clifford Allbutt for a most searching criticism of the first volume.

PREFACE

Dr. E. T. Withington has helped me so much that not a few parts of this book might rightly be described as his, and I am glad to say that he will be the translator of the third volume, which will contain the surgical treatises.

In the Postscript I have gathered together a few notes which I could not put at the foot of the text.

INTRODUCTORY ESSAYS

I

PROGNOSIS

A MODERN doctor, when called to a case of illness, is always careful to diagnose it, that is, to put it in its proper place in the catalogue of diseases. It may be infectious and so need isolation; it may be dangerous and require special nursing. Precautions which are essential in a case of influenza are not so necessary in a common cold. Treatment, too, varies considerably according to diagnosis; diseases may be similar in symptoms and yet call for different medicines.

It is remarkable, and at first rather puzzling, that Hippocrates[1] attached no great value to diagnosis. Although in the works I have called Hippocratic many diseases are referred to by their names, their classification and diagnosis are always in the background. The chief division is into "acute" and "chronic" illnesses, and Hippocrates is mainly concerned with the former. For practical purposes he appears to have divided acute diseases into two main classes: (*a*) chest complaints and (*b*) those

[1] I mean by "Hippocrates" the writer of *Epidemics I.* and *III.*, *Prognostic* and *Regimen in Acute Diseases.*

fevers which we now call malarial. Further than this, at least as far as treatment is concerned, he did not think it necessary to go.[1]

Hippocrates held that it was impossible to decide with certainty when a variation in the symptoms constituted a different disease, and he blamed the Cnidian physicians for multiplying types by assigning essential importance to accidental details. He attached far less value to diagnosis than he did to what may perhaps be called general pathology of morbid conditions, in particular of acute diseases. In all these diseases, according to Hippocrates, there are symptoms, or combinations of symptoms, which point to certain consequences in either the near or the remote future. In other words there is a common element, of which can be written a common medical history. Such a medical history for acute diseases is the work *Prognostic*.

Prognosis, as the knowledge of this general pathology was called, Hippocrates valued for three reasons:

(1) A physician might win the confidence of a patient by describing the symptoms that occurred before he was called in.

(2) He could foretell the final issue with approximate certainty.

(3) A knowledge of dangers ahead might enable him to meet them, or even to prevent them.

Besides these utilitarian reasons, we cannot doubt

[1] In the clinical histories of *Epidemics* no attempt is made to diagnose the various cases, though of course the common names of various diseases are found to be useful in describing the "constitutions" of the same book. In the Cnidian treatises, on the contrary, diagnosis is carried to extremes.

that prognosis was considered of value for its own sake. We must never forget that the Greek physician was a scientist as well as a practitioner. Like the rest of his race he had a boundless curiosity, and a great eagerness to know "some new thing."

A Greek was always argumentative—even when ill—and a Greek doctor was bound to persuade his patient to undergo the proper treatment. His persuasive powers were particularly necessary when operative surgery was called for, as anaesthetics and anodynes were not available, and the art of nursing was in its infancy. We are therefore not surprised that a doctor wished to impress his patients by stating without being told what had occurred before he was called in. In days when quackery abounded, and when practitioners often wandered from place to place instead of establishing a reputation in one district, such a way of inspiring confidence was doubly needed.

In ancient times the very human desire to know the future was stronger than it is now. Science has to a great extent cleared away the uncertainty that must always, at least partially, obscure the consequences of our acts and experiences, and has above all diminished the risks that attend them. But a Greek must have been tormented by doubts to an extent that can scarcely be appreciated by a modern. To lessen them he had recourse to oracles, divination and augury, and physicians too were expected to relieve fears, or at least to turn them into unpleasant certainties or probabilities.[1]

[1] See *e. g.* Aeschylus, *Prometheus Bound*, 698, 699:

τοῖς νοσοῦσί τοι γλυκύ
τὸ λοιπὸν ἄλγος προὐξεπίστασθαι τορῶς.

The usefulness of prognosis in treatment is easier to understand, and our only surprise is that Hippocrates seems not to make full use of the opportunities it afforded. Meeting dangers by anticipation is not a prominent feature of his regimen.

The most remarkable characteristic of the Hippocratic doctrine of prognosis is the stress laid upon the symptoms common to all acute morbid conditions. This effort to distinguish "disease" from "diseases" may be due in part to the Greek instinct to put the general before the particular, an instinct seen in its extreme form in the Platonic theory of Ideas. But it is not entirely to be accounted for in this way. Hippocrates was comparatively free from the prejudices of his race, and if he thought any view valuable in medical practice it was probably valuable in reality and not a mere fad. It is therefore our duty to inquire whether there was any reason why the study of morbid phenomena in general was of interest in the age in which he lived.[1] I believe the reason lies in the predominance in ancient Greece of two classes of illness.

The most important diseases of the Hippocratic age were the chest complaints, pneumonia and pleurisy (pulmonary tuberculosis was also very general), and the various forms, sub-continuous and remittent, of malaria. Other acute diseases were comparatively rare, as we can see from the enumeration of such given in the fifth chapter of *Prognostic*, and, moreover, in a malarious country most diseases are modified or "coloured" by malarial symptoms. It was therefore natural that Hippocrates should subconsciously regard acute diseases as falling into

[1] Contrast, however, what I say on p. xv.

two main categories, and this point having been reached it was but a step to think that the two might ultimately be resolved into one.

It must also be remembered that the means of treatment available to Hippocrates were few in number. The most he could do was to hinder Nature as little as possible in her efforts to expel a disease, and to assuage pain as far as the limited knowledge of the time permitted. The negative side of medicine was far more prominent than the positive. "To do good, or at least to do no harm," was the true physician's ideal. To make the patient warm and comfortable, to keep up the strength by means of simple food without disturbing the digestion, to prevent auto-intoxication from undigested food—this was about all ancient medicine could accomplish, at least on the material side.[1] The psychological aspect of healing was well recognized in ancient times, as we see *inter alia* from the work *Precepts*,[2] and we must take this into account when we estimate the real value of Hippocratic medicine. But here, too, prognosis came in. By telling the past, and by foretelling the future, an effort was made to arouse and to keep alive the patient's faith in his doctor.

[1] The *vis medicatrix naturae* was the true healer. Whatever the disease, this (so thought Hippocrates) had its chance to operate when hindrances were removed.

[2] See especially Chapter VI (Vol. I. p. 319).

II

THE CNIDIAN SCHOOL OF MEDICINE

When reference is made to the Cnidian physicians there is a great possibility of error, an error which, as a matter of fact, is always liable to occur with designations of this type. Do we mean by a Cnidian a doctor trained at Cnidos or a physician with views of a peculiar kind? The two are by no means the same; a Cnidos-trained man might hold some Coan views, a Cos-trained man might adopt some Cnidian opinions. So we must not suppose either (*a*) that all Cnidians necessarily held the same theories, or (*b*) that treatises containing doctrines which we know to have been popular at Cnidos were written by authors trained in that school. All we can say is that such and such an opinion is in harmony with the teaching known to have been in favour with the Cnidian School of a certain period.

Practically all we know about the Cnidians is the criticism of *Cnidian Sentences* put forward by the author of *Regimen in Acute Diseases*,[1] supplemented by a few remarks in Galen.[2] We are told that the book had been re-edited, and that the second edition

[1] Chapters I–III.

[2] See *e.g.* (Kühn) XV. 363, 419, 427, 428, and V. 760, 761. Littré II. 198–200 gives the chief passages in a translation.

was, in the opinion of the Hippocratic writer, an improvement on the first. The critic alleges that the Cnidians attached too little importance to prognosis, and too much to the discussion of unessential details; that their treatment was faulty,[1] and the number of remedies employed by them in chronic complaints was far too small;[2] that they carried the classification of diseases to extremes,[3] holding that a difference in symptoms constituted a different variety of disease.

The chief Cnidian physician was Euryphon, almost contemporary with Hippocrates, and according to Galen[4] the author of *Cnidian Sentences*. Possibly he wrote one if not two of the works in the *Corpus*, as passages from two of them appear to be attributed to Euryphon by Galen and Soranus respectively.[5]

The question of Cnidian tenets assumes a greater importance from the number of works in the *Corpus* which have been assigned to Cnidian authors by various critics. When a passage found in the Hippocratic collection is assigned to a Cnidian author by ancient authorities it is natural to assume that the whole book in which the passage occurs, and any other books closely related to it, are also

[1] We have a specimen of it in their treatment of pus in the lung; Kühn I. 128: ἐξέλκοντες τὴν γλῶτταν ἐνίεσάν τι εἰς τὴν ἀɔτηρίαν ὑγρὸν τὸ σφοδρὰν βῆχα κινῆσαι δυνάμενον.

[2] They were purges, whey and milk.

[3] See Galen XV. 427 and 363.

[4] XVII., Pt. I. 886.

[5] See W. A. Greenhill's article "Euryphon" in Smith's *Dictionary of Greek and Roman Biography and Mythology*, and also that in Pauly-Wissowa by M. Wellmann. The passage quoted by Galen (XVII., Pt. I. 888) is found in *Diseases II.* Chapter XLVIII (Littré VII. 104).

Cnidian. Ermerins[1] makes a formidable list, amounting in all to about one-third of the *Corpus*, which he assigns to this school. It is easy, however, to pursue this line of argument to extremes. We cannot be sure, if we remember how commonly ancient medical writers copied one another, that the whole book is Cnidian when a passage from it is given a Cnidian origin. Nobody would argue that the second book of *Diseases* is the same as *Cnidian Sentences* just because Galen[2] assigns to the latter a passage to which a parallel is to be found in the former, especially when we remember that *Cnidian Sentences*, at any rate the first edition of it, was probably written in the aphoristic style.

As in other problems connected with the Hippocratic collection, it is important to lay stress upon what we know with tolerable certainty, so as neither to argue in a circle nor to be led astray by will-o'-the-wisps. Now it is clear from the Hippocratic criticisms that the Cnidians had no sympathy with "general pathology" and the doctrine of prognosis founded upon it, and that they did consider the classification of diseases a fundamental principle of medical science. Littré[3] argues at some length that the Hippocratic doctrine was right for the fifth century B.C., and the Cnidian for the nineteenth century A.D. Only with our increased knowledge, he urges, can the Cnidian method

[1] *Hippocrates*, Vol. III. p. viii.

[2] XVII., Pt. I. p. 888. We should also note that Galen (XV. 427, 428) says that the Cnidians recognized (among other varieties of disease) four diseases of the kidneys, three kinds of tetanus and three kinds of consumption. This agrees with *Internal Affections* (Littré VII. 189–207).

[3] Vol. II., pp. 200–205.

bear fruit; with the limited knowledge of the Hippocratic age to cultivate general pathology and prognosis was the correct course. To a certain extent this view is correct; in the Hippocratic age little could be done for patients suffering from acute diseases except to keep them warm and comfortable, and to restrict their diet. Yet we must always remember that "general" pathology really does not exist, and that any prognosis based upon it must be very uncertain indeed. Hippocrates was great because he had the true scientific insight, not because of prognosis but in spite of it. The Cnidians, on the other hand, were truly scientific when they insisted on accurate and even meticulous classification. It is no discredit to them that they classified wrongly, and based on their faulty classification faulty methods of treatment. If diseases are to be classified according to symptoms, variations of symptoms must be held to imply variations of diseases. Modern pathology has proved this classification wrong, and the treatment of symptoms has accordingly fallen into discredit. But it is at least as wise to treat symptoms as it is to build up a fictitious general pathology, and to cultivate the barren prognosis that depends upon it. The Cnidians were comparatively unsuccessful because they had not learned to distinguish the essential from the unessential. Hippocrates was a genius who followed a will-o'-the-wisp; the Cnidians were plodders along the dreary stretch of road that lies before every advance in knowledge. Hippocrates did the wrong thing well; the Cnidians did the right thing badly.

There can be no doubt, although we have no

ancient testimony to this effect, that Cnidian doctrine influenced physicians who did not belong to the school, and in fact medicine generally. A dislike of theory, a careful cataloguing of symptoms and equally careful prescriptions for every sort of illness, are characteristics that appear in several of the works in the *Corpus* generally considered Cnidian. *Diseases II.* and *Internal Affections* are a sort of physician's *vade mecum,* and must have been far more useful to the general practitioner than either *Epidemics* or *Regimen in Acute Diseases.*

If therefore we find in any parts of the Hippocratic collection the characteristics I have mentioned to an unusually marked degree, we may be fairly certain that the writer was influenced by the Cnidian School, though we may not assume that he was Cnidian in training. It is interesting that, if we omit the semi-philosophical treatises, and confine our attention to the severely practical works, the greater part of the *Corpus* shows Cnidian rather than Hippocratic tendencies.[1] In some cases (*Diseases II.* and *Internal Affections*) the influence is very strong, in others it is but slight.

The truth seems to be that the peculiarly Hippocratic doctrines are of greater interest and value to scientists than they are to practising doctors. They are suggestive, they inspire, they win our admiration for their humility in claiming so little for medicine and so much for the recuperative powers of Nature, but they give little help to the doctor on his

[1] In particular the gynaecological treatises seem to have Cnidian characteristics. If gynaecology was a special feature of the Cnidian School it is another instance of the practical nature of its instruction.

rounds. So the practical side of medicine, which demands text-books, produced during the fourth century works with the Cnidian characteristics of diagnosis and prescription, just as it produced the aphoristic books of the fifth century.

III

PROGNOSTIC AND THE APHORISTIC BOOKS

THE mutual relations of three of the works in the Hippocratic collection, *Prorrhetic I.*, *Coan Prenotions* and *Prognostic*, have been of interest to students ever since Ermerins published his dissertation on the subject in 1832.[1] The question is in many respects unique, and is inseparable from the much wider question of the history of the aphoristic style.

The facts are these: *Prorrhetic I.* consists of 170 propositions written in the style characteristic of the work *Aphorisms*. Of these 153 occur almost *verbatim* in *Coan Prenotions* along with 487 others, also expressed aphoristically. *Prognostic* is a finished work, but embodied in it are some 58 propositions from *Coan Prenotions*, but only two or at the most three from *Prorrhetic I.*

The style and language of *Prorrhetic I.* and of *Coan Prenotions* are very similar,[2] but it should be noticed that the former work often inserts particular

[1] *Specimen Historico-medicum inaugurale de Hippocratis Doctrina a Prognostice oriunda.*

[2] A careful examination of the books has not given me any evidence tending to show that the works belong to different periods of medical thought. Both, like *Prognostic*, deal with the question, "What do symptoms portend?" and deal with it in much the same way.

instances of the general propositions, *e.g.* "as happened to Didymarchus of Cos."

The work *Aphorisms* contains 68 propositions found in *Coan Prenotions.*

Ermerins, followed by Littré and Adams, concluded that *Prorrhetic I.* was the earliest work, followed later by *Coan Prenotions,* which was in turn used by Hippocrates when he wrote his treatise *Prognostic.*

Obviously the question is not easy to decide, and certainly cannot be settled in the dogmatic manner adopted by the three scholars I have mentioned.

For the sake of brevity I will call *Prorrhetic I.* A, *Coan Prenotions* B, and *Prognostic* C.

Now let us suppose that there was some common source for all three works. This hypothesis scarcely accounts for the striking likeness of A to B and its equally striking unlikeness to C.

Let us suppose that A and B copied some common source, in itself a most likely hypothesis; but if C copied B (he certainly did not copy A), why did he choose 58 propositions of which only two or at the most three, are to be found in A?

It is most unlikely that B and C copied some common source independent of A, because nearly all A is in B.

Now let us suppose that one or other of the extant works is the primary source of the two others.

If A copied B, why did he choose just those propositions that are not in C?

A certainly did not copy C.

If C copied B, why did he choose just those propositions that are not in A?

C certainly did not copy A.

B may very well have copied both A and C.[1]

Before going any further it will be well to print in parallel columns the passages that are common to all three works. These are certainly two and possibly three in number.

Prorrhetic I.	*Coan Prenotions*	*Prognostic*
ὀδόντων πρίσις ὀλέθριον οἷσι μὴ σύνηθες καὶ ὑγιαίνουσιν. § 48.	ὀδόντας συνερίζειν ἢ πρίειν, ᾧ μὴ σύνηθες ἐκ παιδίου, μανικὸν καὶ θανάσιμον· ἤδη δὲ παραφρονέων ἢν ποιῇ τοῦτο, παντελῶς ὀλέθριον. ὀλέθριον δὲ καὶ ξηραίνεσθαι τοὺς ὄδοντας. § 230.	ὀδόντας δὲ πρίειν ἐν πυρετῷ, ὁκόσοισι μὴ σύνηθές ἐστιν ἀπὸ παίδων, μανικὸν καὶ θανατῶδες· ἢν δὲ καὶ παραφρονέων τοῦτο ποιῇ, ὀλέθριον κάρτα ἤδη γίνεται. Chapter III.
αἱ τρομώδεες, ἀσάφεες, ψηλαφώδεες παρακρούσιες πάνυ φρενιτικαί, ὡς καὶ Διδυμάρχῳ ἐν Κῷ. § 34.	αἱ τρομώδεες, ψηλαφώδεες παρακρούσιες φρενιτικαί. § 76.	See Chapter IV.
καὶ ἔμετοι μετὰ ποικιλίης κακόν, ἄλλως τε καὶ ἐγγὺς ἀλλήλων ἰόντων. § 60.	εἰ δὲ καὶ πάντα τὰ χρώματα ὁ αὐτὸς ἐμέοι, ὀλέθριον. § 545.	εἰ δὲ καὶ πάντα τὰ χρώματα ὁ αὐτὸς ἄνθρωπος ἐμέοι, κάρτα ὀλέθριον ἤδη γίνεται. Chapter XIII.

[1] The problem seems to turn on the dissimilarity of A and C. Whatever hypothesis is taken, other than that B is the latest of the three works, it involves intrinsic improbabilities.

The likeness of *Prorrhetic I.* to *Coan Prenotions* must not be judged by the few cases where there is a third parallel in *Prognostic*. The following selections form a much better test.

Prorrhetic I.	*Coan Prenotions*
Οἱ κωματώδεες ἐν ἀρχῇσι γινόμεναι, μετὰ κεφαλῆς, ὀσφύος, ὑποχονδρίου, τραχήλου ὀδύνης, ἀγρυπνέοντες, ἦρά γε φρενιτικοί εἰσιν; § 1.	οἱ κωματώδεες ἐν ἀρχῇσι γενόμενοι, μετὰ κεφαλῆς, ὀσφύος, ὑποχονδρίου, τραχήλου ὀδύνης, ἀγρυπνέοντες, ἦρά γε φρενιτικοί; § 175.
φάρυγξ ἐπώδυνος, ἰσχνή, μετὰ δυσφορίης, πνιγώδης, ὀλεθρίη ὀξέως. § 86.	φάρυγξ ἐπώδυνος, ἰσχνή μετὰ δυσφορίης, ὀλέθριον ὀξέως. § 260.
ἐν τῇσιν ἀσώδεσιν ἀγρυπνίῃσι τὰ παρ' οὖς μάλιστα. § 157.	ἐν ἀσώδεσιν ἀγρύπνοις, τὰ παρ' οὖς μάλιστα. § 552.
τὰ παρ' ὦτα φλαῦρα τοῖσι παραπληκτικοῖσιν. § 160.	τὰ παρ' οὖς φαῦλα τοῖσι παραπληκτικοῖσιν. § 198.
τὰ σπασμώδεα τρόπον παροξυνόμενα κατόχως τὰ παρ' οὖς ἀνίστησιν. § 161.	τὰ σπασμώδεα τρόπον παροξυνόμενα κατόχως τὰ παρ' οὖς ἐπαίρει. § 346.
ὑποχονδρίου σύντασις μετὰ κώματος ἀσώδεος καὶ κεφαλαλγίης τὰ παρ' οὖς ἐπαίρει. § 169.	ὑποχονδρίων σύντασις μετὰ κώματος ἀσώδεος κεφαλαλγικῷ τὰ παρ' οὖς ἐπαίρει. § 283.

It will be noticed that the textual differences between these two works are no greater, and no more numerous, than those regularly found in the manuscripts of a single treatise.

We have seen that mathematically the most likely supposition is that B is the latest work. If this be true, the writer incorporated A almost in its entirety, and when A was imperfect or deficient had recourse to C or to other documents. One of these was obviously *Aphorisms*—unless, indeed, *Aphorisms* is the borrower. But there remain over 300 propositions in B which are either original or copied from sources either unknown or not yet considered.[1]

The third set of parallel passages seems to indicate how the writer of B went to work. Both A and C point out that the vomiting of matters of different colours is a bad symptom, but C has expressed this much better than A, and in language evidently not borrowed from A. Accordingly B copies C, omitting the unessential words for the sake of brevity.

It is unsafe to draw conclusions from the fuller treatment of the subject matter in B than in A, or in C than in B, as we cannot say whether B is expanding A or A is abbreviating and compressing B. This line of argument leaves us just where we were. Similarly it is uncertain whether A added the names of patients whose cases illustrated a general proposition, or whether B omitted them as unnecessary. Accordingly, although the arguments used by Littré and Ermerins support my hypothesis that B is later than A, I shall make no use of them.

[1] Littré refers to many places in the *Corpus* which are similar to passages in *Coan Prenotions.* Omitting those already considered, I find parallel passages in *Epidemics II.*, *Epidemics IV.*, *Epidemics VI.*, *Epidemics VII.*, *Diseases I.*, *Diseases II.*, *Diseases III.*, and to *Wounds in the Head.* On the whole, it is more probable that all copied some common source.

No very positive opinions on this question are really admissible; we can only incline towards one view or another. I have already stated my belief that *Coan Prenotions* is the latest work, but before attempting to go further the whole question of aphoristic literature must be considered.

It is often said that aphorisms belong to pre-scientific days, that proverbs and similar pithy remarks embody experience, collected and generalized indeed, but not yet reduced to a science. Such a remark is true of moral aphorisms, and of Eastern thought generally; but it needs much modification when we discuss their use in Greek scientific literature.

The aphoristic style was adopted by some early Greek philosophers because it arrests the attention and assists the memory.[1] Partly through the influence of poetry, particularly the style of verse adopted by oracles, and partly because the stirring period of the Persian wars fostered a lofty, inspired type of diction, philosophy tinged the aphorism with sublimity and mysticism.

These features are especially striking in the writings of Heraclitus, but even before him aphoristic sentences occur in the philosophic fragments which still remain. When scientific medicine adopted the style is uncertain, but it became very popular, not only in the Coan School of medicine, but also in that at Cnidos. In *Regimen in Acute Diseases* the writer criticizes a Cnidian work, which had already reached a second edition, called *Cnidian Sentences* (Κνίδιαι γνῶμαι); the mere name shows plainly that it was

[1] See on this question Diels, *Herakleitos von Ephesos*.

written in aphorisms.[1] The Hippocratic collection gives us *Prorrhetic I.*, *Coan Prenotions*, *Aphorisms*, *Dentition* and *Nutriment*.

This popularity can hardly have been fortuitous; the aphoristic style must have been suited to express the work of medical science at this particular epoch. Reasons for its adoption are not far to seek. In the first place prose had not developed by the time of Heraclitus the many various forms which were afterwards available. The aphorism, however, was ready to hand. In the second place it is, as has already been said, a valuable mnemonic aid. But perhaps the chief reason for the adoption of the aphoristic manner is its singular fitness to express scientific thought at certain stages of its development.

There are times when the collection and classification of phenomena are the first interest of scientific minds. Embracing theories and constructive ideas are for the moment in the background. Thought does not soar, but crawls. Such a time came to Greek medical science in the fifth century B.C., when, curiously enough, Greek philosophy, for at least three-quarters of the time, tended towards the opposite extreme. Medicine had received a strong positive bias. Superstition had been vanquished and philosophy was being checked. Medical men clamoured for facts, and yet more facts. Everywhere physicians were busy collecting evidence and classifying it; the absorbing question of the day was for

[1] I am aware that Galen's quotation (XVII., Pt. I. p. 888) does not read like a series of aphorisms; but Galen may be quoting from the later editions. It is hard to believe that a book with the title Κνίδιαι γνῶμαι was not written in aphorisms.

many of them the tracing of sequences in morbid phenomena.

Such men found in the aphorism, purged of its mysticism and obscurity, a most convenient means of expressing their thoughts. It seemed an ideal vehicle of generalized fact.[1]

But towards the end of the fifth century other forms of prose were available, and the scientific treatise became a possibility. Moreover, sophistry and rhetoric threatened to leaven all Greek literature and all Greek thought. From the close of this century for nearly one hundred years philosophic speculation was rapidly destroying the influence that medicine exerted in the direction of positive science. The aphorism became unpopular, even among the Coan physicians, and gave place to the rhetorical prose style characteristic of the early fourth century.[2] An attempt to revive it in its Heraclitean form, with all the obscurity and occult allusiveness of Heraclitus, was made by the author of *Nutriment* about the year 400 B.C., but it was an experiment never repeated, and the aphorism, as a Greek literary form, died out, at least as far as medicine and science generally were concerned.[3]

One is accordingly tempted to believe, as at least a probable working hypothesis, that the aphorism

[1] Aphorisms served as "heads of discourse" for lecturers and as "cram" books for students. A love of fact apart from speculation seems naturally to express itself in aphoristic language.

[2] See, e.g., *The Art* and *Regimen I*. The lecturer's "heads of discourse" also took another form, as we see from *Humours*, which is a work containing matter of this nature.

[3] *Dentition* is a possible exception.

was a common medium of medical thought in the fifth century, but was rarely employed later.

For these reasons I am inclined to place the dates of nearly all the aphoristic writings in the Hippocratic collection between 450 and 400 B.C. The beginning of the period should perhaps be placed a little earlier, but were one of the works written much before 450 we should expect to find it marked by some of the characteristics of the Pindaric period, such as we, in fact, do find in the curious treatise on the *Number Seven*, which Roscher would date about 480 B.C.

If, therefore, one may be allowed to exercise the constructive imagination in this case, we may suppose that *Prorrhetic I.* was the first to be written. It may possibly not be original; it is perhaps a compilation from older material. Then came independently *Prognostic* (not, of course, an aphoristic work) and *Aphorisms*,[1] or at least a great part of that composite book. Finally, the writer of *Coan Prenotions* embodied practically all *Prorrhetic I.* in a work intended to embrace the whole of prognosis in its general outlines. He borrowed extensively from *Prognostic* and *Aphorisms*—possibly from other books in our Hippocratic collection—and perhaps he knew, and made use of, works now no longer extant. The dates—they are purely conjectural—might be :—[2]

[1] Perhaps *Aphorisms* is somewhat older than *Coan Prenotions*; possibly its author used a lost work used also by the writer of *Coan Prenotions*. One cannot be dogmatic or positive.

[2] All that I have said must be taken in conjunction with my remarks (*General Introduction* to Vol. I. p. xxviii.) on publication in ancient times.

Prorrhetic I.	440 B.C.
Prognostic } *Aphorisms* }	415 B.C.
Coan Prenotions	410 B.C.

Nutriment, the latest aphoristic work of importance, was written about 400 B.C.

IV

ANCIENT NURSING

It is typical of the obscurity which veils many problems of ancient medicine that so little is told us of nurses and nursing. The conclusion we are tempted to draw from this silence is that the task of nursing fell to the women, whether slaves or free, of the household. The work of Greek women, important as it was, is rarely described for us, probably because it was not considered sufficiently dignified for literary treatment. This conclusion is not entirely conjectural, as we have some positive evidence from the *Economica* of Xenophon.[1] But it is unsafe to dismiss the question without further inquiry. One piece of evidence is so strong that we are forced to look farther afield for a true explanation of the problem.

The clinical histories in the *Epidemics* contain fairly complete accounts of the symptoms which the patients experienced on the several days of their illness. It is true that all the histories are not equally full, and that gaps of greater or less size occur. But the fact remains that the detail is too great to have been observed by the medical attendant personally. He could not have spared the time from his other practice. We are not left entirely to

[1] VII. 37.

conjecture. Every now and then, by a chance allusion, we can tell that there were attendants waiting on the patient and reporting to the doctor. We may therefore assume that a great deal of the information given in the clinical histories is the result of their observations. It is information which in many instances required a trained eye, one quick to catch the essential and to anticipate the doctor's desire for the necessary information.[1] But there were no trained nurses; therefore there must have been present, or at least in the house, people with some medical qualifications. So far I have been reasoning deductively from the evidence given by *Epidemics* and *Regimen in Acute Diseases.* In another work of the Hippocratic collection, *Decorum,* a hint is dropped which enables us to turn our probable conclusion into something approaching a certainty. In that book the doctor is advised to leave a pupil with a patient.[2] It is plain that such a course would be to the advantage of all concerned. The patient would have a skilled, or partially skilled, attendant who would perform, or at least superintend, the necessary nursing. The doctor had someone upon whom he could rely to carry out his orders and to report to him when necessary, thus saving him many troublesome visits. The pupil had a chance of gaining experience which was very important in a land where it was impossible to "walk the hospitals." The plan, therefore, had many advantages. It had also many

[1] The minute directions given in *Regimen in Acute Diseases* could not have been carried out by unskilled attendants. Only a doctor or partially trained student would have had the necessary skill and knowledge.

[2] Chapter XVII

equally obvious disadvantages. The apprentice might be a mere beginner, and do more harm than good. The master physician could scarcely have had enough pupils to leave one with each patient who was seriously ill. The pupil himself must have gained only a limited experience. Perhaps the last point is not serious, as there were few really important diseases in ancient Greece; but the combination of nurse and medical student is far from ideal, and the development of nursing as an independent profession was a necessary preliminary to the triumph of modern surgery and of modern medicine.

V

ANCIENT MEDICAL ETIQUETTE

Apart from a few chance passages in our ancient authorities, the only sources of information for ancient etiquette are *Oath, Law, Physician, Decorum* and *Precepts*.[1]

Of course in a sense there was no medical etiquette in ancient times. Etiquette implies pains and penalties for the offender, and there was no General Medical Council to act as judge and executioner. It has been thought that *Oath* implies the existence of a medical guild. This is most doubtful, and even if it be true, the guild had no power to prevent a sinning doctor from practising; it could merely exercise care in the selection of its members to be educated.

The Greek physician obeyed the laws of etiquette, not through fear of punishment, but for love of his craft. The better sort of Greek was always an artist first and a man afterwards. The very name for etiquette, εὐσχημοσύνη, shows that it was "good form," rather than a matter of duty, to obey the code of conduct laid down by custom. Etiquette had

[1] It is interesting to note that the "great age" of Greek medicine has left us nothing about etiquette. The gradual decline of medicine and possibly the influx of slaves into the profession made it necessary later to put the rules of etiquette into writing. At first it was an unwritten code, with all the strength, as well as all the weakness, of unwritten codes.

nothing to do with the categorical imperative. Its rules implied "should" or "ought," never "must."

Because its sanction was comparatively weak its scope was comparatively wide. If laws are going to be strictly enforced, they must be narrowed down to a minimum; if their observance is merely a matter of honour and decency, they can be made comprehensive. Ancient medical etiquette, accordingly, was of wider scope than modern; it included many things which would now be regarded as a part of good manners, and some things which come under the cognizance of the law of the land.

Taking the five works mentioned above as our basis, we find that the εὐσχήμων was (*a*) bound to abstain from certain things, and (*b*) bound to perform certain others.

(*a*) The εὐσχήμων ought not—

(1) to give poison, or to be privy to the giving of it;

(2) to cause abortion;

(3) to abuse his position by indulging his sexual appetites;

(4) to tell secrets, whether heard in the course of his practice or in ordinary conversation;[1]

(5) to advertise, at least not in an obtrusive and vulgar manner;[2]

[1] Nowadays only what is learnt professionally must be kept secret.

[2] The lecture or harangue (like that of the cheap-jack at a fair) was the ancient method of advertising. See the discouragement of the ἐπίδειξις in *Precepts*. To act as state-doctor *gratis* was a method of advertising to which no stigma was attached.

(6) to operate—a rule which came into vogue after the "great" period of Greek medicine, though the exact date is very uncertain.[1]

(*b*) The εὐσχήμων was bound—

(1) to call in a consultant when necessary;
(2) to act as consultant when asked to do so;
(3) to take the patient's means into account when charging a fee;
(4) to be clean in person, in particular to abstain from wine when visiting patients;
(5) to cultivate a philosophic frame of mind (dignity, reserve and politeness).[2]

Galen[3] tells us that a surgeon often concealed the person of his patient, not for reasons of modesty, but to prevent other professional doctors present from learning any methods he wished to keep secret. So it was apparently no part of etiquette, though it is nowadays, to make public all new discoveries.

Medical etiquette was and still is intended to protect the patient and to maintain the dignity of the profession. The latter is perhaps the more important consideration nowadays; in Greek times it was rather the welfare of the patient.

The chief difference between ancient etiquette and modern is the absence in ancient times of a strong, external force controlling professional conduct. The moral sanction of ancient etiquette

[1] See the Introduction to *Oath* in Vol. I.

[2] The author of *Epidemics I.* and *III.*, to judge from the style of his work, must have possessed these qualities in a marked degree.

[3] See Kühn XVIII. Pt. II., pp. 685 foll.

accounts for its comprehensiveness ; along with the absence of medical degrees or diplomas it accounts for the prevalence of quackery. Between the scientific physician and the quack there is now a great gulf fixed, but in Greek times quackery and scientific medicine shaded into one another. *Precepts* shows us plainly that in the lower ranks of the profession quackery was common, and, although condemned by the best minds, it did not prevent a man from competing with genuine physicians in ordinary practice.[1]

[1] I may perhaps be allowed to refer to my paper, read before the Royal Society of Medicine in January, 1923, in which the question of ancient medical etiquette is discussed more fully.

VI

"THE ART"

It is not uncommon to hear people say that they "do not believe in medicine," and that "doctors are of no use." But unless they are Christian Scientists or similar faddists they call in a physician when they are really ill, thus proving that their remarks are not the expression of their truest opinions.

But in the time of Hippocrates medicine, in spite of its recent progress, had not yet made good its position, even among educated men. The evidence to show this is overwhelming; I need merely mention the remarks in *Regimen in Acute Diseases,*[1] the treatise called *The Art,* and the well-known hostility of Plato.

What were the reasons for this διαβολή? The writer of *Regimen in Acute Diseases* puts it down to want of medical research; many important points, he says, had not even been mooted by physicians, so that there were wide divergencies of opinion among practitioners. The author of *The Art* brings forward, and answers, two main objections to medicine: (*a*) there are some cases which the physician does not cure, and (*b*) some cases cure themselves without the help of a doctor. Plato's

[1] (Chapter VIII) *καίτοι διαβολήν γε ἔχει ὅλη ἡ τέχνη πρὸς τῶν δημοτέων μεγάλην, ὡς μὴ δοκεῖν ὅλως ἰητρικὴν εἶναι.*

chief criticism is that medicine prolongs useless lives; but we can see, wherever he refers to medical men, that he held them and their craft in no great respect. Plato disparaged arts (τέχναι), and the Greek physician was proud of what he called, with pleasing arrogance, "*the* Art."

We who can view the whole question in better perspective after the lapse of so many centuries see other reasons for the discredit from which medicine suffered. A few of these it will be useful to examine, however briefly, in the hope that we may thus appreciate better the greatness of the Coan and Cnidian schools.

Quackery was common enough in the Hippocratic period, and probably infected all grades of the profession except the highest. As no tests were required before a man could set up in private practice, unqualified doctors caused the whole profession to fall into a certain amount of disrepute. This disrepute would be increased, rather than diminished, by the charlatanism from which the temples of Asclepius were by no means free. Rational medicine suffered along with the art of healing as a whole.

Superstition was rampant in the ancient world, and even doctors were infected by the taint. It is true that there is no superstition in the Hippocratic collection, but it is attacked in two treatises. This attack implies that superstition was still a real danger. The danger was all the greater in days when medicine, while recognizing the psychological factor in treatment, could not yet distinguish legitimate suggestion and auto-suggestion from blind and stupid credulity.

But the worst enemy of rational medicine lay in its connections with philosophy and rhetoric. Greek philosophy in its earlier periods was highly speculative. This is as it should be, but it made philosophy a bad ally.[1] Medical practice must not be based upon speculation, which, though it has its place in the progress of medicine, must be put to the test, not in ordinary practice, but on and by "martyrs to science." Rhetoric, too, allied itself harmfully to rational medicine. Plato in the *Gorgias*[2] tells us under what disadvantage purely medical skill laboured as compared with very inferior qualifications combined with the power of persuasion. Rhetoric enabled a quack to palm himself off as a trained physician. It is only when we remember the disastrous association of rhetoric with the arts and sciences, and its failure to keep strictly to its own province, that we can rightly understand Plato's antipathy to what is, after all, an attractive and useful accomplishment. It is more than doubtful whether the elaborate defence of medicine in *The Art,* with its graceful antithesis and oratorical force, did any good. Even a Greek, with all his love of argument, felt that actions speak louder than words, that cures, and not eloquent writing, really count.

If we may judge from their writings in the *Corpus,* a true Coan and a true Cnidian were wonderfully free from all the faults I have mentioned. There is neither quackery in their works, nor superstition, nor "philosophy," nor yet rhetoric. They were devotees of positive science. They had separated

[1] See on this point the Introduction to *Ancient Medicine* in Vol. I.

[2] 456 B, C.

off medicine from all other branches of learning, so that it could be developed on its own lines, unhampered by extraneous influences and unscientific practices and beliefs. But they suffered not only from the discredit cast upon the art of healing by ignorant or unscrupulous practitioners, but also from the διαβολή which sprang out of their own imperfections. Medicine was yet in its infancy. and the scientific doctor, whether Coan or Cnidian, was a modest man and made no extravagant claims.[1] He fully realized that medicine could do little except remove as many of the hindrances as possible that impede Nature in her efforts to bring about a cure. But the multitude in Greece, like the multitude today, demanded something more spectacular. There is a tendency first to credit the physician with far greater powers than he possesses, and then to blame him because he can really do so little. Disappointment breeds discontent.

In spite of all discouragement the Greek physician persevered. He had a lofty ideal, and he was proud of his art, with a sure confidence in its ultimate victory over disease.

[1] The writer of *Ancient Medicine* claims that medicine is merely a branch of the art of dieting, and grew naturally out of that art.

VII

MEDICAL WRITINGS AND LAYMEN

Greek activities were not so rigidly marked off into classes as are modern activities. Division of labour and specialization were less developed, and the amateur was not so sharply distinguished from the professional.

In certain kinds of arts, indeed, the modern distinction held; a carpenter, a smith or an armourer followed trades which without careful apprenticeship could not be prosecuted with success. But the wide scope of a citizen's public duties often led him to pose as an authority on matters of which he was profoundly ignorant.

Literature in particular tended to be amateurish, a tendency which was encouraged by the ease with which a man could become an author. In the days before printing anybody could "publish" a work without the expenditure of time and money that is now necessary. Much therefore was published which was amateurish, and often second-rate or careless.

Medical literature appears to have suffered as severely in this way as any other. Almost anybody thought he was a fit person to write on medical subjects, even though like Plato he had received no medical training. Some of these efforts—probably the best of them—are to be found in the Hippocratic collection.

The popularity of medicine as a literary subject was of doubtful value to it as a science. Rational medicine was struggling to assert itself, and found that the alliance of enthusiastic amateurs did more harm than good. I have already discussed the disastrous attempt of philosophy to embrace medical theory—disastrous because, instead of adapting philosophy to medicine, it tried to adapt medicine to philosophy. Another enemy in the guise of a friend was rhetoric. It happened that rational medicine was at its best just at the time when sophistry was moulding that beautiful but artificial style which exercised such a potent influence upon Greek prose. As far as we can see, the great physicians were unaffected by sophistry, but sophistry could not refrain from tampering with medicine. The student must not be led by the extravagance of Ermerins, who postulates "sophists" as the authors of a great part of the collection, into the opposite error of minimizing the sophistic [1] character of certain treatises.

Of the treatises that show this characteristic the chief are *The Art*, *Breaths*,[2] and *Nature of Man* (down to the end of Chapter VIII). The first defends the thesis that there *is* an art of medicine; the second tries to prove that πνεῦμα is the cause of diseases;

[1] By "sophistry" here I mean a toying with philosophy and an artificial style of writing which is associated with the school of Gorgias.

[2] In Vol. I. I called this treatise (περὶ φυσῶν) *Airs*, before I realized the difficulty of finding the best English equivalent.

while the third maintains the doctrine of four humours, against those who said that man was composed of a single substance.

It is hard to believe that any one of these was written by a professional physician. It is not that the works contain doctrines which no practitioner could have held, although some of the doctrines put forward are rather strange. The main reason for supposing that they were written by laymen is that the centre of interest is not science but rhetorical argument. *The Art* especially is so full of the tricks of style that we associate with Gorgias and his school that Gomperz is convinced that it was written by Protagoras himself.[1] Professor Taylor[2] points out that there are many Hippocratic works in which the main interest is philosophy. "The persons who play with them (*i.e.* the words *ἰδέα* and *εἶδος*) are the speculative philosophers, the Hegels and Schellings of their day, to whom medicine is not interesting for its own sake, or as a profession by which they have to live, but as a field in which they can give free scope for their love of *Naturphilosophie* and propound undemonstrable theories about the number and nature of the ultimate kinds of body, and support them by biological analogy."

The distinction between lay and professional being ill-defined, it is impossible in all cases to decide confidently whether the writer was a physician or a sophist; a man in fact might very well be both. I think, however, that an unbiassed reader would say that the three works I have mentioned were

[1] See *Die Apologie der Heilkunst s.v.* "Protagoras," p. 179.
[2] *Varia Socratica*, p. 225.

written by laymen, if at least by "laymen" is meant a man who may incidentally know something about medicine, while his main interest lies elsewhere.

That works of no value to medicine should find their way into the Hippocratic collection is not strange. I have given[1] reasons for holding that this collection represents the library of the Coan school. Such a library would not be confined to purely technical treatises, and might well contain books which, while of no medical value, were of great medical interest. Perhaps some were presentation copies from sophistic admirers of the chief physicians of the school.

[1] Vol. I. pp. xxix. and xxx.

VIII

LATER PHILOSOPHY AND MEDICINE

During the fifth century B.C. philosophy made a determined effort to bring medicine within the sphere of its influence, and to impose upon it the method of ὑποθέσεις. Typical of this effort are *Nutriment* (Vol. I.), and *Breaths,* which is included in the present volume.

This effort of philosophy was violently opposed by the chief adherents of the rational school of medicine, and we still have in *Ancient Medicine* a convincing statement of the position held by the empirics.

Why was medicine so determined to throw off the incubus? Simply because an attempt was being made to impose *à priori* opinions upon physical science, which has a method of its own quite incompatible with unverifiable speculation. Medicine was here face to face with a deadly enemy.

A hundred or more years after *Nutriment* was written another wave of philosophy swept over medicine. Its exact date cannot be fixed, but it probably did not begin until the third century B.C. was well advanced.

But this second attempt to influence medicine was not resented, for philosophy had changed its outlook.[1] Ethics, conduct and morality were now its main interest, and in this sphere of thought philosophy had a better chance of success. Aristotle had laid the foundations of moral science, and had pointed out that the facts of this science are the experiences of our emotional life. Our speculations about these experiences are for the most part verifiable, and so the science stands on a sound foundation. Both the Stoics and Epicureans, while differing considerably both from Aristotle and from each other in their views about the *summum bonum*, were at one in that they considered conduct to be the main thing in human life.

So we find that both schools tried to discover what the conduct of the ideal physician should be in the practice of his profession. *Precepts* is distinctly Epicurean, both in its epistemology[2] and general outlook; *Decorum* and *Law* are Stoic, at least they must have been written by authors both well acquainted with Stoic modes of thought and favourably inclined towards them.

In thus insisting upon the moral side of a physician's work these later philosophers—or perhaps it would be more accurate to call them adherents of the later philosophy—made no small contribution to medical etiquette. In Greece at any rate this etiquette did not aim mainly at promoting the

[1] I do not mean to say that the old mistake of the fifth century had disappeared – we have but to read the history of pneumatism to disprove that—but a new aspect of philosophy now became prominent.

[2] The writer of *Precepts* seems eager to point out that the Epicurean theory of knowledge was very similar to the standpoint of empiric medicine.

material interests of the profession, but at raising the morality of its practitioners.

Though we may smile at some of the trivialities in *Precepts* and *Decorum,* there is nevertheless much that is admirable. There are two sentences, one from each of these tracts, which have often seemed to me to sum up admirably the efforts of later philosophy to influence medicine. They are:—

ἢν γὰρ παρῇ φιλανθρωπίη, πάρεστι καὶ φιλοτεχνίη.—*Precepts* VI.

ἰητρὸς γὰρ φιλόσοφος ἰσόθεος.—*Decorum* V.

IX

THE MANUSCRIPT TRADITION OF THE HIPPOCRATIC COLLECTION

When I first began seriously to study the Greek medical writings, some sixteen years ago, I had no idea that the history of the text could be of much importance or interest except to professional palaeographers. Even when I was writing the first volume of my translation for the Loeb Series I was somewhat sceptical of the real value to a translator of Hippocratic textual criticism, and it was only when I saw that the important, but strangely neglected, treatise *Precepts* could not be placed in its proper historical relationship, without a thorough examination of the transmission of the text, that I realized how necessary it is for even a translator to master the problem as far as our imperfect knowledge allows us. A little has been achieved by Gomperz, Wilamowitz and the Teubner editors, but outside their labours there is still an "uncharted region" on to which some light at least must be thrown.

Possibly the most important factor to remember about the transmission of the Hippocratic text is that the treatises composing it are practical textbooks or scientific essays and not literary masterpieces. There were not the same reasons for keeping the text pure that were operative in the case of the great poets, orators, historians and philosophers. The medical school of Cos would not regard its miscellaneous library with the veneration with which the Academy and the Peripatetics regarded the writings of Plato and Aristotle; and the later custodians of the Hippocratic books, the librarians of Alexandria and of other centres of learning, were not as solicitous about them as they were about the text of, say, Homer. On the other hand, there was a succession of medical students and practitioners who needed copies of these books for practical purposes, and were quite content if they could be supplied with handbooks containing the information they required, even though these were textually inaccurate. To what lengths this textual corruption might go is well shown by some of the late Latin translations, for instance that in the library of Corpus Christi College, Cambridge. I have examined this manuscript personally, and the text is almost unrecognizable.

Of course many manuscripts continued to give a comparatively pure text. But at some time or other, probably before Galen,[1] the manuscripts resolved themselves into at least two classes, one of which differs from the other in the order of words, and in slight alterations (*e. g.* κάμνων or ἀσθενέων for νοσέων) which make no essential difference to the general sense.

Both classes of manuscripts contain a large number of glosses. It is obvious that few authors were so likely to collect a crop of glosses as were the medical

[1] Galen mentions readings belonging to both classes of MSS. See *e. g.* my notes on *Regimen in Acute Diseases LVI.* and *LVIII.*

writers. Every reader would be tempted to annotate his copy, and any annotation might find its way into the text. A careful comparison of A with the other manuscripts shows that the latter contain scores of glosses, and we may be sure that A, our purest authority, must contain many others that we cannot eliminate by reference to other documents, but only by careful subjective criticism.

So at the present day there are roughly two classes. One class, in spite of many atrocities of spelling, gives a text which, both in dialect and in sense, is in some 70 per cent. of the cases where differences arise greatly superior to that of the other class, which seems to have aimed more at smoothness and regularity, and to have adopted many Ionic forms, whether genuine or sham, from which the first class is comparatively free.

To the first class belong, in particular, A, θ, C′ and B.

To the second class belong M, V, and the later Paris manuscripts.

The classes, as one might have expected, are not rigidly divided. A sometimes agrees with M against V or with V against M. Moreover, sometimes the second class presents readings which are obviously more likely to be correct than those of the first. In not a few cases all the manuscripts agree in giving a reading which is most unlikely to be right. Nevertheless, the broad distinction between the two classes remains.

In the older editions (Zwinger, Foes, Mack, etc.) there are recorded many variants from manuscripts now lost. As far as we can see these manuscripts belonged mostly to the second, or inferior, class.

This inferior class is divided into two main subclasses, represented respectively by V and M. In the first the works appear roughly in the order in which they are given in the V index, namely *ὅρκος, νόμος, ἀφορισμοί, προγνωστικόν, περὶ διαίτης ὀξέων, κατ' ἰητρεῖον, περὶ ἀγμῶν κ.τ.λ.* See Kühlewein I. xv. and Littré I. 529. The other class tends to reproduce the order given by M, namely, *ὅρκος, νόμος, περὶ τέχνης, περὶ ἀρχαίης ἰητρικῆς, παραγγελίαι, περὶ εὐσχημοσύνης, κ.τ.λ.* See Kühlewein I. xix.

How good a test of the tradition the order of the works may be is well shown by my experience of the manuscript Holkhamensis 282. The librarian of the Earl of Leicester, to whom the manuscript belongs, sent me a list of the works which it contains, and it was obvious that the order was for the first half of the manuscript that of V. When the manuscript was afterwards sent to Cambridge for my inspection, it occurred to me that, the order of the treatises being the same, the manuscript was probably allied to V. So I chose some thirty test passages from V where that manuscript differs from A or M. In every instance Holk. 282 had the same reading, even misspellings and the lacuna in *Airs Waters Places III.* (after *ψυχρά*). We are not perhaps justified in saying that Holk. 282 was copied from V, but the two must be very nearly allied.[1]

[1] I did not know of the existence of this manuscript when I wrote Vol. I., so perhaps some description of it may not be out of place, as Baroccian 204 and Holkhamensis 282 are the only important manuscripts of Hippocrates in Great Britain. It is written on European paper in a careful but rather difficult hand. Octavo and unfoliated. The date is approximately 1500 A.D., and it was probably written in Italy. Dr. F. C. Unger (*Mnemosyne* LI., Part I., 1923) does not think it

Besides these authorities, which may be considered primary, we have also the surviving works of Erotian and Galen, which may be considered of secondary

---

actually copied from V, and this is probably true of περὶ καρδίης, the only part Dr. Unger has yet seen. In many places however the likeness between the two is almost startling. The manuscript contains (after a glossary) ἀφορισμοί. προγνωστικόν. πρὸς Κνιδίας γνώμας. κατ' ἰητρεῖον. περὶ ἀγμῶν. περὶ ἄρθρων. περὶ τῶν ἐν κεφαλῇ τρωμάτων. περὶ ἀέρων ὑδάτων τόπων. περὶ ἐπιδημιῶν. περὶ φύσιος ἀνθρώπου. περὶ φύσιος παιδίου· περὶ γονῆς. περὶ ἑπταμήνου. περὶ ὀκταμήνου. περὶ τόπων τῶν κατὰ ἄνθρωπον. περὶ ἰητροῦ. περὶ κρισίων. περὶ καρδίης. περὶ σαρκῶν. περὶ ἀδένων οὐλομελίης. περὶ ἀνατομῆς. περὶ παρθενίων. περὶ ὀδοντοφυΐας. γυναικείων α and β. περὶ ἀφόρων. περὶ ἐγκατατομῆς παιδίου.

Baroccian 204 is a very legible fifteenth century manuscript. The order of the treatises it contains is that of the M class. Holkhamensis 282 is closely allied to Vaticanus Gr. 276; Baroccian 204 is similar to Marcianus Venetus 269.

Baroccian 204 is 30·5 cm. by 22·5 cm.; the scribe wrote forty lines to the page, leaving a wide margin. Although the writing is not very large, and there is but a small space between the lines, it is, next to θ, the easiest manuscript to read of those that I have seen. The writer of the part I examined (there are many hands) has a habit of placing two dots over iota, and sometimes over upsilon.

The value of Baroccian 204 to the textual critic may perhaps be estimated from the following statistics.

I have compared the readings of A, V and M with those of Baroccian 204 from the end of *Prognostic* (beginning at τὴν χολήν) to the beginning of *Regimen in Acute Diseases* (ending at καὶ τοῖσιν ὑγιαίνουσιν in Chapter IX).

The title of *Regimen in Acute Diseases* agrees with that of M, except for the position of the author's name, which is not first, as in M, but after ὀξέων. It runs: περὶ διαίτης ὀξέων Ἱπποκράτους· οἱ δὲ περὶ πτισάνης· οἱ δὲ πρὸς τὰς κνιδίας γνώμας. A's title is Ἱπποκράτους περὶ πτισάνης: V's Ἱπποκράτους πρὸς τὰς κνιδίας γνώμας ἢ περὶ πτισάνης.

value. With regard to these I have little to add to what has been put so well by I. Ilberg in the second chapter of the introduction to the Teubner edition (*de memoria secundaria*). I would remark however that:—

(1) Galen's comments sometimes seem to imply that the differences between the A group and the MV group existed in his day;
(2) Galen's explanations sometimes seem to apply to readings now lost. See for instance my note on *Regimen in Acute Diseases XXXII.*

The remarks I have just made are the result of independent study of (*a*) recorded readings and (*b*) manuscripts or photographs of manuscripts. As far as it is possible I have kept my mind uninfluenced by the labours of Gomperz, Nelson and Ilberg. If my results confirm theirs they are probably right; in so far as I may disagree I am probably wrong.

In the eighty lines thus compared:

(1) Baroccian 204 agrees with M as against AV in 24 places;
(2) It agrees with MV as against A in 17 places;
(3) It agrees with AM as against V in 2 places;
(4) It agrees with AV as against M in 1 place;
(5) It agrees with A as against MV in 2 places.

In one place only is it peculiar. At the end of *Prognostic* it has *τῶ μὴ οὐ* where M has *τὸ /// μὴ οὐ*.

There seems to be a great similarity between Baroccian 204 and Paris 2254 in the passage indicated above.

My heartiest thanks are due to the Earl of Leicester and to Mr. C. W. James, his librarian; to the officials of the Cambridge University Library; and to Dr. Minns for his kind help in dating Holkhamensis 282 for me.

The results to which I have come are:—

(1) The readings of the A, θ, C′ class are *ceteris paribus* to be preferred to those of the MV class.

(2) We cannot hope to restore the text beyond reaching the best textual tradition current in the time of Galen. Occasionally even this aim cannot be reached.

(3) It is futile to attempt to restore the exact dialect actually written by the authors. They probably did not all write exactly the same kind of Ionic, as it was a literary and not a spoken dialect as far as medicine and science generally are concerned. It is more than futile to think that we know whether the author wrote *e. g.* τοῖς, τοῖσι or τοῖσιν.

When I translated *Precepts* in Vol. I. I was forced to rely upon the collations which Cobet and (I believe) Daremberg made for Ermerins and Littré. I have now an excellent photograph of M, the only first-class manuscript containing *Precepts.* The strange words and constructions I have noticed on pages 308 and 309 of Vol. I. are in general confirmed. I must, however, note one or two points.

(1) In Chapter XII M reads clearly ἱστορεομένην for the monstrosity ἱστοριευμένην of the vulgate.

(2) In Chapter IX M reads μέγα ἂν τεκμήριον φανείη μέγα ξὺν τῇ οὐσίῃ τῆς τέχνης. This seems to suggest that the clause was copied from Chapter V of the treatise περὶ τέχνης in the form μέγα τεκμήριον τῇ οὐσίῃ, which in some MS. or MSS. appeared as τεκμήριον μέγα.

Then a later scribe, combining, wrote *μέγα τεκμήριον μέγα τῇ οὐσίῃ*. A later scribe thought that the second *μέγα* was *μετά*, and he or a still later scribe changed *μετὰ* to *ξὺν* because of the dative following.

(3) In Chapter I, where the vulgate has *ἢν τὰ ἐπίχειρα ἐκομίζοντο*, the scribe of M began to write an *η*, and then changed this to his contraction for *ει*. Dr. Minns confirms my view after inspection of the photograph. Apparently, then, the scribe of M had *ἢν* before him, and changed it to *εἰ* when he saw the indicative following.

(4) In the other places, so far as I can see, M agrees with the vulgate, having *ἢν δὲ καιρὸς εἴη* in Chapter VI, and *μὴ εἴη ἐπαύρασθαι* in Chapter II. It is suspicious however that in both cases the optative is that of the verb *εἰμί*. I suggest that the author wrote in both cases *ᾖ*, that a later scribe "ionized" to *ἔῃ*, and a later one still read this as *εἴη*. But in two other places (see section 7) M has *ἢν* with optative.

(5) On the whole, however, my view is confirmed that the work is very late, and was probably written by an imperfect Greek scholar. The negative *μὴ* is ousting *οὐ*, and the strange readings *ὃ ἂν ἐρέω* (VIII) and *ὅποι ἂν καὶ ἐπιστατήσαιμι* (XIII) occur in M, except that *ὅπῃι* is written (correctly) instead of *ὅποι*.

(6) In Chapter II (end) M has *μετ' ἀπρηξίης*, thus confirming my conjecture.

(7) I give here the chief variants M shows in *Precepts* other than those already noted.

Vol. I. p. 312, l. 10. ταῦτα after ὁκοίως.
l. 15. ἦν omitted.
l. 16. γὰρ after εὑρίσκεται.
l. 18. τὴν after φύσιν.
p. 314, l. 10. ἤν τι δοκοίη.
p. 316, l. 6. ὑποθήσηι, not ὑποθήσεις.
l. 9. ἦν ὀξὺ for ἐν ὀξεῖ.

At the beginning of Chapter V the manuscript shows many smudges and signs of disturbance. προκρίνοντες occurs again after κολάσιος. The reading a little later (p. 318 top) is ἠδελφισμένως (corrected to ἠδελφισμένος) ἰητρεύοι πίστει ἢ ἀτεραμνίηι.

p. 322, Chapter VIII. παρὰ (not περὶ) σημασίης. A little later on, οὐδ' ἤν τις . . . κελεύοι.

p. 326 (top). παντί τε πάντηι τε καὶ πάσηι . . . δεδημιουργημένηι . . . ἔλιμμα . . . ἠδ' ἂν γένηται τι.

Chapter X. κέκτησαι and (lower) πάσηι for πᾶσι.

p. 328, Chapter XIII (l. 4). λυμίης corrected to λοιμίης.

l. 12. ξυλλόγου· αἰτήσαιμι δ' ἄν. βούσθην is at the end of a line, and smudged.

l. 18. μήτε (second hand apparently μετὰ) χειροτριβίης ἀτρεμιότητι.

Chapter XIV. ξυνεσταμένης.

p. 330 (top). Second hand has a correction κχω (?) over ἐγχειρεῖν and then νοσέοντος occurs for τοῦ κάμνοντος.

p 330, l. 14. παμπουλὺς corrected to παμπουλὴ by second hand.

l. 17. The first hand had ξυμπάθησιν corrected by the second hand,[1] who has apparently tried to change the ν to ς.

p. 332. τρόπος is read, not τόπος.

The Order of the Books in the Manuscripts

The order of the treatises in our manuscripts is a good clue to the "family" to which any particular manuscript belongs. I have already noticed the help this truth afforded in the study of Holkhamensis 282. It may therefore be useful to give the order in which the works are arranged in our most valuable manuscripts.

One truth at least cannot escape our notice. The "V" type and the "M" type are very clearly marked, and most of the less important manuscripts conform more or less exactly to one or the other of these types. A combination of these two classes of manuscripts gave us our "vulgate" text.

It is also probable that each separate order (M, V, and so on) represents a different "collection" of Hippocratic works. Possibly some of these orders go back to the days of the great libraries at Alexandria and other places, and represent the order of the rolls in the bookcases.

A (Paris 2253)

1. *Coan Prenotions.*
2. *Ptisan.*
3. *Humours.*
4. *Use of Liquids.*
5. *Address at the Altar.*
6. *The Art.*
7. *Nature of Man.*
8. *Breaths.*
9. *Places in Man.*
10. *Ancient Medicine.*
11. *Epidemics I.*

[1] I am not sure whether the correcting hand is the same as that of the original scribe, but I think it is not.

Vindobonensis med. IV (θ)

1. *Internal Affections.*
2. *Affections.*
3. *Sacred Disease.*
4. *Diseases I.*
5. *Diseases III.*
6. *Diseases II.*
7. *Regimen I.*
8. *Regimen II.*
9. *Regimen III.*
10. *Dreams.*
11. *Diseases of Women I.*
12. *Diseases of Women II.*
13. *Nature of Women.*

Laurentianus 74, 7 (B)

1. *Surgery.*
2. *Fractures.*
3. *Articulations.*
4. *Wounds in the Head.*

Marcianus Venetus 269 (M)

1. *Oath.*
2. *Law.*
3. *The Art.*
4. *Ancient Medicine.*
5. *Precepts.*
6. *Decorum.*
7. *Nature of Man.*
8. *Generation.*
9. *Nature of the Child.*
10. *Articulations.*
11. *Humours.*
12. *Nutriment.*
13. *Sores.*
14. *Sacred Disease.*
15. *Diseases I.*
16. *Diseases II.*
17. *Diseases III.*
18. *Diseases IV.*
19. *Affections.*
20. *Internal Affections.*
21. *Regimen I.*
22. *Regimen II.*
23. *Regimen III.*
24. *Dreams.*
25. *Sight.*
26. *Critical Days.*
27. *Aphorisms.*
28. *Prognostic.*
29. *Regimen in Acute Diseases*
30. *Breaths.*
31. *Instruments of Reduction.*
32. *Nature of Bones.*
33. *Fractures.*
34. *Surgery.*
35. *Excision of the Foetus.*
36. *Diseases of Women I.*
37. *Diseases of Women II.*
38. *Barrenness.*
39. *Superfoetation.*
40. *Seven Months' Child.*
41. *Eight Months' Child.*
42. *Diseases of Girls.*
43. *Nature of Women.*
44. *Epidemics VI.*
45. *Epidemics VII.*
46. *Letters.*
47. *Discourse on Madness.*
48. *Decree of the Athenians.*
49. *Speech at the Altar.*
50. *Speech of the Envoy.*

The Index in Vaticanus Graecus 276 (V)

This index appears in V before the works themselves.

1. *Oath.*
2. *Law.*
3. *Aphorisms.*
4. *Prognostic.*
5. *Surgery.*
6. *Fractures.*
7. *Articulations.*
8. *Wounds in the Head.*
9. *Airs Waters Places.*
10. *Epidemics.*
11. *Nature of Man.*
12. *Nature of the Child.*
13. *Nature of Generation.*
14. *Superfoetation.*
15. *Seven Months' Child.*
16. *Eight Months' Child.*
17. *Diseases of Girls.*
18. *Nature of Women.*
19. *Dentition.*
20. *Places in Man.*
21. *Diseases of Women.*
21. *Barrenness.*
23. *Excision of the Foetus.*
24. *Use of Liquids.*
25. *Nutriment.*
26. *Regimen.*
27. *Regimen in Health.*
28. *Diseases.*
29. *Affections.*
30. *Internal Affections.*
31. *Sacred Disease.*
32. *Sevens.*
33. *Critical Days.*
34. *Sores.*
35. *Deadly Wounds.*
36. *Withdrawal of Missiles.*
37. *Hemorrhoids.*
38. *Fistulae.*
39. *Purges.*
40. *Hellebore.*
41. *Clysters.*
42. *Glands.*
43. *Instruments of Reduction.*
44. *Nature of Bones.*
45. *Sight.*
46. *Heart.*
47. *Coition.*
48. *Fleshes.*
49. *Crisis.*
50. *Prorrhetic I* and *II.*
51. *Coan Prenotions.*
52. *Humours.*
53. *Natures.*
54. *Ancient Medicine.*
55. *The Art.*
56. *The Physician.*
57. *Precepts.*
58. *Decorum.*
59. *Anatomy.*
60. *Letters.*
61. *Speech at the Altar.*
62. *Speech of the Envoy.*

Vaticanus Graecus 276 (V)

1. *Oath.*
2. *Law.*
3. *Aphorisms.*
4. *Prognostic.*
5. *Regimen in Acute Diseases.*
6. *Surgery.*
7. *Fractures.*
8. *Articulations.*
9. *Wounds in the Head.*
10. *Airs Waters Places.*

11. *Epidemics.*
12. *Nature of Man.*
13. *Nature of the Child.*
14. *Generation.*
15. *Superfoetation.*
16. *Seven Months' Child.*
17. *Eight Months' Child.*
18. *Girls.*
19. *Nature of Women.*
20. *Dentition.*
21. *Places in Man.*
22. *Diseases of Women.*
23. *Barrenness.*
24. *Superfoetation* (repeated see above).
25. *Excision of the Foetus.*
26. *Physician.*
27. *Crises.*
28. *Heart.*
29. *Fleshes.*
30. *Glands.*
31. *Anatomy.*
32. *Letters.*
33. *Decree of the Athenians.*
34. *Speech at the Altar.*
35. *Speech of the Envoy.*

Paris 2255 and 2254 (E and D)

These two MSS. are complementary, 2255 being the first.

2255

1. *Oath.*
2. *Law.*
3. *Art.*
4. *Ancient Medicine.*
5. *Precepts.*
6. *Decorum.*
7. *Nature of Man* and *Regimen in Health.*
8. *Generation.*
9. *Nature of the Child.*
10. *Articulations.*
11. *Humours.*
12. *Nutriment.*
13. *Sores.*
14. *Sacred Disease.*
15. *Diseases* (four books).
16. *Affections.*
17. *Internal Affections.*
18. *Regimen* (three books).
19. *Dreams.*
20. *Sight.*
21. *Critical Days.*
22. *Physician.*
23. *Fleshes.*
24. *Dentition.*
25. *Anatomy.*
26. *Heart.*
27. *Glands.*
28. *Places in Man.*
29. *Airs Waters Places.*
30. *Use of Liquids.*
31. *Crisis.*
32. *Aphorisms.*
33. *Prognostic.*
34. *Wounds in the Head.*
35. *Prognosis of Years.*[1]

[1] Littré remarks (I. p. 520): "Ceci est un fragment, mis hors de sa place, du traité des *Airs, des Eaux et des Lieux*, et un indice de la manière dont il arrivait aux copistes de déranger l'ordre d'un livre et de faire de nouveaux traités."

2254

1. *Regimen in Acute Diseases.*
2. *Breaths.*
3. *Instruments of Reduction.*
4. *Nature of Bones.*
5. *Fractures.*
6. *Surgery.*
7. *Excision of the Embryo.*
8. *Diseases of Women.*
9. *Barrenness.*
10. *Superfoetation.*
11. *Seven Months' Child.*
12. *Eight Months' Child.*
13. *Diseases of Girls.*
14. *Nature of Women.*
15. *Excision of the Foetus.*
16. *Prorrhetic* (two books).
17. *Fistulae.*
18. *Hemorrhoids.*
19. *Coan Prenotions.*
20. *Epidemics* (seven books).
21. *Letters.*

Paris 2146 (Index)

1. *Oath.*
2. *Law.*
3. *Aphorisms.*
4. *Prognostic.*
5. *Surgery.*
6. *Fractures.*
7. *Articulations.*
8. *Wounds in the Head.*
9. *Airs Waters Places.*
10. *Epidemics.*
11. *Nature of Man.*
12. *Nature of the Child.*
13. *Nature of Generation.*
14. *Superfoetation.*
15. *Seven Months' Child.*
16. *Eight Months' Child.*
17. *Girls.*
18. *Nature of Women.*
19. *Dentition.*
20. *Places in Man.*
21. *Diseases of Women I.* and *II.*
22. *Barrenness.*
23. *Excision of the Foetus.*
24. *Use of Liquids.*
25. *Nutriment.*
26. *Regimen I., II., III.* and *in Health.*
27. *Diseases I., II., III.*
28. *Affections.*
29. *Internal Affections.*
30. *Sacred Disease.*
31. *Sevens.*
32. *Critical Days.*
33. *Sores.*
34. *Deadly Wounds.*
35. *Withdrawal of Missiles.*
36. *Hemorrhoids.*
37. *Purges.*
38. *Hellebore.*
39. *Clysters.*
40. *Glands.*
41. *Instruments of Reduction.*
42. *Nature of Bones.*
43. *Sight.*
44. *Heart.*
45. *Coition.*
46. *Fleshes.*
47. *Crisis.*
48. *Prorrhetic I., II.*
49. *Coan Prenotions.*
50. *Humours.*
51. *Nature.*
52. *Ancient Medicine.*
53. *The Art.*

54. *Physician.*
55. *Precepts.*
56. *Decorum.*
57. *Mind* (περὶ γνώμης *i.e.* ἀνατομῆς).
58. *Letters.*
59. *Address at the Altar.*
60. *Speech of the Envoy.*

This list is practically the same as that of the index i[n] Vaticanus 276.

Paris 2142 (H)

1. *Oath.*
2. *Law.*
3. *The Art.*
4. *Ancient Medicine.*
5. *Precepts.*
6. *Decorum.*
7. *Nature of Man.*
8. *Generation.*
9. *Nature of the Child.*
10. *Articulations.*
11. *Humours.*
12. *Nutriment.*
13. *Sores.*
14. *Sacred Disease.*
15. *Diseases.*
16. *Affections.*
17. *Internal Affections.*
18. *Regimen.*
19. *Dreams.*
20. *Sight.*
21. *Critical Days.*
22. *Aphorisms.*
23. *Prognostic.*
24. *Regimen in Acute Diseases.*
25. *Breaths.*
26. *Instruments of Reduction.*
27. *Nature of Bones.*
28. *Fractures.*
29. *Surgery.*
30. *Excision of Embryo.*
31. *Diseases of Women.*
32. *Barrenness.*
33. *Superfoetation.*
34. *Seven Months' Child.*
35. *Eight Months' Child.*
36. *Diseases of Girls.*
37. *Nature of Women.*
38. *Excision of Foetus.*
39. *Prorrhetic I., II.*
40. *Fistulae.*
41. *Hemorrhoids.*
42. *Coan Prenotions.*
43. *Epidemics.*
44. *Letters.*
45. *Address at the Altar.*
46. *Speech of the Envoy.*
47. *Decree.*
48. *Letters of Democritus.*

This list conforms to the "M" type.

The manuscripts Paris 2140, 2143 and 2145 (I, J and K) are very similar. I give here the list in 2145. It is of the "M" type.

1. *Oath.*
2. *Law.*
3. *The Art.*
4. *Ancient Medicine.*
5. *Precepts.*
6. *Decorum.*

7. *Nature of Man* with *Regimen in Health.*
8. *Generation* and *Nature of the Child.*
9. *Nature of the Child.*
10. *Articulations.*
11. *Humours.*
12. *Nutriment.*
13. *Sores.*
14. *Sacred Disease.*
15. *Diseases.*
16. *Affections.*
17. *Internal Affections.*
18. *Regimen.*
19. *Dreams.*
20. *Sight.*
21. *Critical Days.*
22. *Aphorisms.*
23. *Prognostic.*
24. *Regimen in Acute Diseases.*
25. *Breaths.*
26. *Instruments of Reduction.*
27. *Nature of Bones.*
28. *Fractures.*
29. *Surgery.*
30. *Excision of the Embryo.*
31. *Diseases of Women.*
32. *Barrenness.*
33. *Superfoetation.*
34. *Seven Months' Child.*
35. *Eight Months' Child.*
36. *Diseases of Girls.*
37. *Nature of Women.*
38. *Excision of the Foetus.*
39. *Prorrhetic* (two books).
40. *Fistulae.*
41. *Hemorrhoids.*
42. *Coan Prenotions.*
43. *Epidemics.*
44. *Letters.*
45. *Address at the Altar.*
46. *Speech of the Envoy.*

Holkhamensis 282

1. *Aphorisms.*
2. *Prognostic.*
3. *Regimen in Acute Diseases.*
4. *Surgery.*
5. *Fractures.*
6. *Articulations.*
7. *Wounds in the Head.*
8. *Airs Waters Places.*
9. *Epidemics.*
10. *Nature of Man.*
11. *Nature of the Child.*
12. *Generation.*
13. *Seven Months' Child.*
14. *Eight Months' Child.*
15. *Places in Man.*
16. *Physician.*
17. *Crisis.*
18. *Heart.*
19. *Fleshes.*
20. *Glands.*
21. *Anatomy.*
22. *Girls.*
23. *Dentition.*
24. *Diseases of Women I.* and *II.*
25. *Barrenness.*
26. *Excision of the Foetus.*

This list down to *Eight Months' Child* agrees with V. After this point it does not.

The Aldine Index

1. *Iusiurandum Hippocratis.*
2. *De arte.*
3. *De prisca medicina.*
4. *De medico.*
5. *De probitate.*
6. *Hippocratis praecepta.*
7. *Lex Hippocratis.*
8. *De natura hominis.*
9. *De ratione victus salubris. Polybi discipuli Hippocratis.*
10. *De semine. Polybi.*
11. *De natura foetus.*
12. *De carne.*
13. *De septimestri partu.*
14. *De octomestri partu.*
15. *De superfoetatione.*
16. *De extractione foetus.*
17. *De dentitione.*
18. *De dissectione.*
19. *De corde.*
20. *De glandibus.*
21. *De natura ossium.*
22. *De locis in homine.*
23. *De aere, aqua, locis.*
24. *De victus ratione.*
25. *De insomniis.*
26. *De alimento quem esse Hippocratis negat Galenus.*
27. *De usu humidorum.*
28. *De humoribus.*
29. *De flatibus.*
30. *De sacro morbo, docti cuiusdam.*
31. *De morbis.*
32. *De affectibus. Polybi.*
33. *De internarum partium affectibus.*
34. *De morbis virginum.*
35. *De natura muliebri.*
36. *De morbis mulierum.*
37. *De sterilibus.*
38. *Supposititia quaedam calci primi de morbis mulierum adscripta.*
39. *De morbis passim grassantibus.*
40. *De ratione victus acutorum.*
41. *De iuditiis.*
42. *De diebus iudicialibus.*
43. *Hippocratis definitae sententiae.*
44. *Hippocratis praenotiones.*
45. *Hippocratis praedictiones.*
46. *Coacae praecognitiones.*
47. *De vulneribus capitis.*
48. *De fracturis.*
49. *De articulis.*
50. *Hippocratis de medici munere.*
51. *Hippocratis de curandis luxatis.*
52. *De ulceribus.*
53. *De fistulis.*
54. *De haemorrhoidibus.*
55. *De visu.*
56. *Hippocratis epistolae.*
57. *Decretum Atheniensium.*
58. *Epibomios.*
59. *Oratio Thessali Hippocratis filii legati ad Athenienses.*

The Index in the Edition of Foes

1. *Hippocratis iusiurandum.*
2. *Hippocratis lex.*
3. *De arte lib. I.*
4. *De prisca Medicina, libr. I.*

5. *De Medico, lib. I.*
6. *De decente habitu, aut decoro libr. I.*
7. *Praeceptiones.*

---

8. *Praenotionum, libr. I.*
9. *De humoribus, libr. I.*
10. *De iudicationibus, libr. I.*
11. *De diebus iudicatoriis, libr. I.*
12. *Praedictorum, libr. II.*
13. *Coacae Praenotiones in breves sententias distinctae.*

---

14. *De natura hominis.*
15. *De genitura.*
16. *De natura pueri.*
17. *De carnibus.*
18. *De septimestri partu.*
19. *De octimestri partu.*
20. *De superfoetatione.*
21. *De dentitione.*
22. *De corde.*
23. *De glandulis.*
24. *De ossium natura.*
25. *De aëre, locis & aquis.*
26. *De flatibus.*
27. *De morbo sacro.*

---

28. *De salubri victus ratione.*
29. *De victus ratione, libr. III.*
30. *De insomniis.*
31. *De alimento.*
32. *De victus ratione in morbis acutis.*
33. *De locis in homine.*
34. *De liquidorum usu.*

---

35. *De morbis, libr. IV.*
36. *De affectionibus, libr. I.*
37. *De internis affectionibus, libr. I.*
38. *De his quae ad virgines spectant, libr. I.*
39. *De natura muliebri, libr. I.*
40. *De mulierum morbis, libr. II.*
41. *De his quae uterum non gerunt, libr. I.*
42. *De videndi acie, lib. I.*

---

43. *Medicina officina, aut de officio Medici, lib. I.*
44. *De fracturis, libr. I.*
45. *De articulis, libr. I.*
46. *Vectiarium, hoc est, de ossium per molitionem impellendorum ratione, libr. I.*
47. *De ulceribus, libr. I.*
48. *De fistulis, libr. I.*
49. *De haemorrhoidibus, hoc est, de venis in ano sanguinem fundere solitis, libr. I.*
50. *De capitis vulneribus, libr. I.*
51. *De foetus in utero mortui exectione, lib. I.*
52. *De corporum resectione, libr. I.*

---

53. *hoc est, De morbis populariter grassantibus, libr. VII. Quorum Primus, Tertius & Sextus, post Galeni Commentarios, Annotationibus sunt illustrati. Secundus verò ante annos triginta cum Commentariis editus, denuo ab authore est recognitus. Reliqui iustis Annotationibus donati.*

54. *hoc est, Aphorismorum, lib. I. cum brevibus notis.*

---

55. *Epistolae aliquot.*
56. *Atheniensium Senatusconsultum.*
57. *Oratio ad aram.*
58. *Thessali Legati Oratio.*
59. *Genus & vita Hippocratis, secundum Soranum.*
60. *De purgatoriis remediis.*
61. *De structura Hominis.*

HIPPOCRATES

PROGNOSTIC

INTRODUCTION

This work has never been attributed to any author except Hippocrates, but we must remember that some modern scholars use the term "Hippocrates" in a somewhat peculiar sense.

Its subject is the prognosis of acute diseases in general, which Hippocrates made his special province. I have dealt with prognosis already, and it only remains to say a few words about the manuscripts and editions.

The chief authorities for the construction of the text are M, V, and a tenth-century manuscript[1] called "446 supplément" by Littré and C′ by Kühlewein. Holkhamensis 282, which I have examined, is here practically identical with V, and has not helped towards the construction of the text. There is an invaluable commentary by Galen.

C′ is carelessly written, being full of misspellings which often appear due to writing from dictation.[2] On the other hand, there are omissions which prove conclusively that a scribe's eye passed from one word to another, omitting all the intervening syllables.[3] The obvious conclusion to draw is that both tran-

[1] It contains *Prognostic*, part of *Aphorisms*, *Epistle to Ptolemy*, and several works of Galen. See Littré II. 103.

[2] *E. g.* ηισῶν for ἦσσον, εἴη for ᾖ, ᾗ for εἴη, αἰμείσθω for ἐμείσθω, εὔκριτοι for εὔκρητοι.

[3] See *e. g.* pp. 23, 26, 45, 50.

scription and dictation played their part in the early transmission of the text.

The text of C′ differs considerably from that of M and V. These very often agree when C′ presents either a completely different version or else a different order of words. The remarkable point about the variations is that they rarely affect the sense to any appreciable degree. For instance, in Chapter I C′ has *τῶν τοιουτέων νοσιμάτων* (*sic*), while M V have *τῶν παθέων τῶν τοιουτέων*. Similar variations are very common, and point to a time when the text was copied with close attention to the sense and with little care for verbal fidelity. One would be tempted to postulate two editions of the work were the variations of greater intrinsic importance. They are, however, in no sense corrections, and it is hard to imagine that the author would have taken the trouble to make such trivial alterations intentionally. It is more probable that between the writer's date and that of Galen there was a period when copies of Hippocrates were made without attention to verbal accuracy. From one of these are descended M and V, from another is descended C′. This lack of respect for the actual words of Hippocrates provided that the general sense is unaffected may perhaps be connected with assimilation of the dialect of all the Hippocratic collection to an Ionic model. An age which did not scruple to alter words would probably not scruple to alter their form.

It is not easy to decide whether C′ or M V represents the more ancient tradition. A few variations, however, are distinctly in favour of C′, and I have adopted this manuscript as my primary authority in constructing the text.

INTRODUCTION

There are, besides C′, twenty-one Paris manuscripts containing *Prognostic*.

The early editions and translations, the first two translations being into Latin from the Arabic, are very numerous.[1] The dates show that from 1500 to about 1650 this work was used by doctors throughout Europe as a practical text-book.[2] The first English translation was written by Peter Low (London, 1597), and was followed by that of Francis Clifton (London, 1734), of John Moffat (London, 1788), and of Francis Adams (London, 1849). Littré's edition and translation in the second volume are among his best work, and the text of Kühlewein is a great improvement on all his predecessors'. I have adopted his principles of spelling while constructing an independent text.

[1] Galen's commentary is often added, as are also notes by more modern editors.

[2] See Littré II, 103–109.

ΠΡΟΓΝΩΣΤΙΚΟΝ

I. Τὸν ἰητρὸν δοκεῖ μοι ἄριστον εἶναι πρόνοιαν ἐπιτηδεύειν· προγινώσκων γὰρ καὶ προλέγων παρὰ τοῖσι νοσέουσι τά τε παρεόντα καὶ τὰ προγεγονότα καὶ τὰ μέλλοντα ἔσεσθαι, ὁκόσα τε παραλείπουσιν οἱ ἀσθενέοντες ἐκδιηγεύμενος πιστεύοιτο ἂν μᾶλλον γινώσκειν τὰ τῶν νοσεύντων πρήγματα, ὥστε τολμᾶν ἐπιτρέπειν τοὺς ἀνθρώπους σφᾶς αὐτοὺς τῷ ἰητρῷ. τὴν δὲ[1] θεραπείην ἄριστα ἂν ποιέοιτο προειδὼς τὰ ἐσόμενα 10 ἐκ τῶν παρεόντων παθημάτων. ὑγιέας μὲν γὰρ ποιεῖν ἅπαντας τοὺς νοσέοντας[2] ἀδύνατον· τοῦτο γὰρ καὶ τοῦ προγινώσκειν τὰ μέλλοντα ἀποβήσεσθαι κρέσσον ἂν ἦν· ἐπειδὴ δὲ οἱ ἄνθρωποι ἀποθνήσκουσιν, οἱ μὲν πρὶν ἢ καλέσαι τὸν ἰητρὸν ὑπὸ τῆς ἰσχύος τῆς νούσου, οἱ δὲ καὶ ἐσκαλεσάμενοι παραχρῆμα ἐτελεύτησαν, οἱ μὲν ἡμέρην μίαν ζήσαντες, οἱ δὲ ὀλίγῳ πλείονα χρόνον, πρὶν ἢ τὸν ἰητρὸν τῇ τέχνῃ πρὸς ἕκαστον νόσημα ἀνταγωνίσασθαι. γνῶναι[3] οὖν χρὴ τῶν τοιούτων

[1] For δὲ Wilamowitz reads τε.

[2] νοσέοντας C′: ἀσθενέοντας MV.

[3] γνῶναι Littré from Paris 2269: γνῶντα C′: γνόντα MV.

PROGNOSTIC

I. I HOLD that it is an excellent thing for a physician to practise forecasting. For if he discover and declare unaided[1] by the side of his patients the present, the past and the future, and fill in the gaps in the account given by the sick, he will be the more believed to understand the cases, so that men will confidently entrust themselves to him for treatment. Furthermore, he will carry out the treatment best if he know beforehand from the present symptoms what will take place later. Now to restore every patient to health is impossible. To do so indeed would have been better even than forecasting the future. But as a matter of fact men do die, some owing to the severity of the disease before they summon the physician, others expiring immediately after calling him in—living one day or a little longer—before the physician by his art can combat each disease. It is necessary, therefore, to learn the natures of such diseases, how much they

[1] I try by this word to represent the preposition προ- in the compound verbs, which means "before being told" in reference to τὰ παρέοντα and τὰ προγεγονότα, and "before the events occur" in reference to τὰ μέλλοντα ἔσεσθαι. πρόνοια is equivalent to πρόγνωσις.

20 νοσημάτων[1] τὰς φύσιας, ὁκόσον ὑπὲρ τὴν δύναμίν εἰσιν τῶν σωμάτων[2] καὶ τούτων τὴν πρόνοιαν ἐκμανθάνειν. οὕτω γὰρ ἄν τις θαυμάζοιτο δικαίως καὶ ἰητρὸς ἀγαθὸς ἂν εἴη· καὶ γὰρ οὓς οἷόν τε περιγίνεσθαι ἔτι μᾶλλον ἂν δύναιτο διαφυλάσσειν ἐκ πλείονος χρόνου προβουλευόμενος πρὸς ἕκαστα, καὶ τοὺς ἀποθανευμένους τε καὶ σωθησομένους
27 προγινώσκων τε καὶ προλέγων ἀναίτιος ἂν εἴη.

II. Σκέπτεσθαι δὲ χρὴ ὧδε ἐν τοῖσιν ὀξέσι νοσήμασιν· πρῶτον μὲν τὸ πρόσωπον τοῦ νοσέοντος, εἰ ὅμοιόν ἐστι τοῖσι τῶν ὑγιαινόντων, μάλιστα δέ, εἰ αὐτὸ ἑωυτῷ· οὕτω γὰρ ἂν εἴη ἄριστον, τὸ δὲ ἐναντιώτατον τοῦ ὁμοίου δεινότατον. εἴη δ' ἂν τὸ τοιόνδε· ῥὶς ὀξεῖα, ὀφθαλμοὶ κοῖλοι, κρόταφοι συμπεπτωκότες, ὦτα ψυχρὰ καὶ συνεσταλμένα καὶ οἱ λοβοὶ τῶν ὤτων ἀπεστραμμένοι καὶ τὸ δέρμα τὸ περὶ τὸ πρόσωπον σκληρὸν καὶ περιτε-
10 ταμένον καὶ καρφαλέον ἐόν· καὶ τὸ χρῶμα τοῦ σύμπαντος προσώπου χλωρὸν ἢ μέλαν ἐόν.[3] ἢν μὲν[4] ἐν ἀρχῇ τῆς νούσου τὸ πρόσωπον τοιοῦτον ᾖ καὶ μήπω οἷόν τε ᾖ τοῖσιν ἄλλοισι σημείοισι συντεκμαίρεσθαι, ἐπανερέσθαι χρή, μὴ ἠγρύπνηκεν ὁ ἄνθρωπος ἢ τὰ τῆς κοιλίης ἐξυγρασμένα ἦν[5] ἰσχυρῶς, ἤ λιμῶδές τι ἔχει αὐτόν. καὶ ἢν μέν τι τούτων ὁμολογῇ, ἧσσον νομίζειν δεινὸν εἶναι· κρίνεται δὲ ταῦτα ἐν ἡμέρῃ τε καὶ νυκτί, ἢν διὰ

[1] τῶν τοιούτων νοσημάτων C′ (with misspelling): τῶν παθέων τῶν τοιουτέων MV.

[2] After σωμάτων all the MSS. have ἅμα δὲ καὶ εἴ τι θεῖον ἔνεστιν ἐν τῇσι νούσοισι. It is regarded as an interpolation by Kühlewein.

[3] ἦν MV: εἰ C′.

[4] μὲν C′: μὲν οὖν MV.

[5] ἦν Kühlewein: εἴη C′: ᾖ MV.

exceed the strength of men's bodies,[1] and to learn how to forecast them. For in this way you will justly win respect and be an able physician. For the longer time you plan to meet each emergency the greater your power to save those who have a chance of recovery, while you will be blameless if you learn and declare beforehand those who will die and those who will get better.

II. In acute diseases the physician must conduct his inquiries in the following way. First he must examine the face of the patient, and see whether it is like the faces of healthy people, and especially whether it is like its usual self. Such likeness will be the best sign, and the greatest unlikeness will be the most dangerous sign. The latter will be as follows. Nose sharp, eyes hollow, temples sunken, ears cold and contracted with their lobes turned outwards, the skin about the face hard and tense and parched, the colour of the face as a whole being yellow or black.[2] If at the beginning of the disease the face be like this, and if it be not yet possible with the other symptoms to make a complete prognosis, you must go on to inquire whether the patient has been sleepless, whether his bowels have been very loose, and whether he suffers at all from hunger. And if anything of the kind be confessed, you must consider the danger to be less. The crisis comes

[1] The clause omitted by Kühlewein, "and at the same time whether there is anything divine in the diseases," is found in all MSS. It is contrary to Hippocratic doctrine, and to suppose that τὸ θεῖον means λοιμός has no Hippocratic authority, nor would a reference to plague be in place here.

[2] *I. e.* very dark. Similarly μέλανα οὖρα is dark urine, of the colour of port wine, as I ought to have remarked in Vol. I. when translating *Epidemics*. So frequently.

ταύτας τὰς προφάσιας τὸ πρόσωπον τοιοῦτον ᾖ·
20 ἢν δὲ μηδὲν τούτων φῇ μηδὲ ἐν τῷ χρόνῳ τῷ
προειρημένῳ καταστῇ, εἰδέναι τοῦτο τὸ σημεῖον
θανατῶδες ἐόν.[1] ἢν δὲ καὶ παλαιοτέρου ἐόντος
τοῦ νοσήματος ἢ τριταίου[2] τὸ πρόσωπον τοιοῦτον
ᾖ, περί τε τούτων ἐπανερέσθαι, περὶ ὧν καὶ πρότερον ἐκέλευσα καὶ τὰ ἄλλα σημεῖα σκέπτεσθαι,
τά τε ἐν τῷ σύμπαντι[3] σώματι καὶ τὰ ἐν τοῖσι
ὀφθαλμοῖσιν. ἢν γὰρ τὴν αὐγὴν φεύγωσιν ἢ δα-
κρύωσιν ἀπροαιρέτως ἢ διαστρέφωνται ἢ ὁ ἕτερος
τοῦ ἑτέρου ἐλάσσων γίνηται ἢ τὰ λευκὰ ἐρυθρὰ
30 ἴσχωσιν ἢ πελιδνὰ ἢ φλέβια μέλανα ἐν αὐτοῖσιν[4]
ἢ λῆμαι φαίνωνται περὶ τὰς ὄψιας ἢ καὶ ἐναιωρεύ-
μενοι ἢ ἐξίσχοντες ἢ ἔγκοιλοι ἰσχυρῶς γινόμενοι[5]
ἢ τὸ χρῶμα τοῦ σύμπαντος προσώπου ἠλλοιω-
μένον, ταῦτα πάντα κακὰ νομίζειν εἶναι καὶ ὀλέθρια. σκοπεῖν δὲ χρὴ καὶ τὰς ὑποφάσιας τῶν
ὀφθαλμῶν ἐν τοῖσιν ὕπνοισιν· ἢν γάρ τι ὑποφαί-
νηται συμβαλλομένων τῶν βλεφάρων τοῦ λευκοῦ,
μὴ ἐκ διαρροίης ἢ φαρμακοποσίης ἐόντι ἢ μὴ εἰθισμένῳ οὕτω καθεύδειν, φαῦλον τὸ σημεῖον καὶ
40 θανατῶδες σφόδρα. ἢν δὲ καμπύλον γένηται ἢ
πελιδνὸν[6] βλέφαρον ἢ χεῖλος ἢ ῥὶς μετά τινος
τῶν ἄλλων σημείων, εἰδέναι χρὴ ἐγγὺς ἐόντα τοῦ
θανάτου· θανατῶδες δὲ καὶ χείλεα ἀπολυόμενα
44 καὶ κρεμάμενα καὶ ψυχρὰ καὶ ἔκλευκα γινόμενα.

III. Κεκλιμένον δὲ χρὴ καταλαμβάνεσθαι τὸν

[1] εἰδέναι τοῦτο τὸ σημεῖον θανατῶδες ἐόν. For this M has εἰδέναι χρὴ ἐγγὺς ἐόντα τοῦ θανάτου.

[2] After τριταίου M adds ἢ τεταρταίου.

[3] After σύμπαντι MV add προσώπῳ καὶ τὰ ἐν τῷ.

[4] After αὐτοῖσιν MV add ἔχωσιν.

after a day and a night if through these causes the face has such an appearance. But should no such confession be made, and should a recovery not take place within this period, know that it is a sign of death. If the disease be of longer standing than three days[1] when the face has these characteristics, go on to make the same inquiries as I ordered in the previous case, and also examine the other symptoms, both of the body generally and those of the eyes. For if they shun the light, or weep involuntarily, or are distorted, or if one becomes less than the other, if the whites be red or livid or have black veins in them, should rheum appear around the eyeballs, should they be restless or protruding or very sunken, or if the complexion of the whole face be changed—all these symptoms must be considered bad, in fact fatal.[2] You must also examine the partial appearance of the eyes in sleep. For if a part of the white appear when the lids are closed, should the cause not be diarrhoea or purging, or should the patient not be in the habit of so sleeping, it is an unfavourable, in fact a very deadly symptom.[3] But if, along with one of the other symptoms, eyelid, lip or nose be bent or livid, you must know that death is close at hand. It is also a deadly sign when the lips are loose, hanging, cold and very white.

III. The patient ought to be found by the

[1] *I. e.* if more than two complete days have elapsed.
[2] Or, "if not fatal."
[3] Or, "if not a very deadly symptom."

---

[5] After γινόμενοι M adds ἢ αἱ ὄψιες αὐχμῶσαι καὶ ἀλαμπεῖς.
[6] After πελιδνὸν M adds ἢ ὠχρόν.

νοσέοντα ὑπὸ τοῦ ἰητροῦ ἐπὶ τὸ πλευρὸν τὸ δεξιὸν ἢ τὸ ἀριστερὸν καὶ τὰς χεῖρας καὶ τὸν τράχηλον καὶ τὰ σκέλεα ὀλίγον ἐπικεκαμμένα ἔχοντα καὶ τὸ σύμπαν σῶμα ὑγρὸν κείμενον· οὕτω γὰρ καὶ οἱ πλεῖστοι τῶν ὑγιαινόντων κατακλίνονται· ἄρισται δὲ τῶν κατακλισίων αἱ ὁμοιόταται τῇσι τῶν ὑγιαινόντων. ὕπτιον δὲ κεῖσθαι καὶ τὰς χεῖρας[1] καὶ τὰ σκέλεα ἐκτεταμένα ἔχοντα ἧσσον
10 ἀγαθόν. εἰ δὲ καὶ προπετὴς γένοιτο καὶ καταρρέοι ἀπὸ τῆς κλίνης ἐπὶ τοὺς πόδας, δεινότερόν ἐστι τοῦτο ἐκείνου.[2] εἰ δὲ καὶ γυμνοὺς τοὺς πόδας εὑρίσκοιτο ἔχων μὴ θερμοὺς κάρτα ἐόντας καὶ τὰς χεῖρας[3] καὶ τὰ σκέλεα ἀνωμάλως διερριμμένα καὶ γυμνά, κακόν· ἀλυσμὸν γὰρ σημαίνει. θανατῶδες δὲ καὶ τὸ κεχηνότα καθεύδειν αἰεὶ καὶ τὰ σκέλεα ὑπτίου κειμένου συγκεκαμμένα εἶναι ἰσχυρῶς καὶ διαπεπλεγμένα. ἐπὶ γαστέρα δὲ κεῖσθαι, ᾧ μὴ σύνηθές ἐστι καὶ ὑγιαίνοντι
20 κοιμᾶσθαι οὕτω, κακόν·[4] παραφροσύνην γὰρ[5] σημαίνει ἢ ὀδύνην τινὰ τῶν περὶ τὴν γαστέρα τόπων. ἀνακαθίζειν δὲ βούλεσθαι τὸν νοσέοντα τῆς νούσου ἀκμαζούσης πονηρὸν μὲν ἐν πᾶσι τοῖσιν ὀξέσι νοσήμασιν, κάκιστον δὲ ἐν τοῖσι περιπνευμονικοῖσιν. ὀδόντας δὲ πρίειν ἐν πυρετῷ, ὁκόσοισι μὴ σύνηθές ἐστιν ἀπὸ παίδων, μανικὸν καὶ θανατῶδες·[6] ἢν δὲ καὶ παραφρονέων τοῦτο ποιῇ, ὀλέθριον κάρτα ἤδη γίνεται.

[1] After χεῖρας M adds καὶ τὸν τράχηλον.
[2] τοῦτο ἐκείνου omitted by MV.
[3] After χεῖρας MV add καὶ τὸν τράχηλον.
[4] κακόν is omitted by MV.
[5] For γὰρ MV have τινὰ followed by σημαίνει ἢ ὀδύνην τῶν ἀμφὶ τὴν κοιλίην τύπων.

physician reclining on his right or left side, with his arms, neck and legs slightly bent, and the whole body lying relaxed; for so also recline the majority of men when in health, and the best postures to recline in are most similar to those of men in health. But to lie on the back, with the arms and the legs stretched out, is less good. And if the patient should actually bend forward, and sink foot-wards away from the bed,[1] the posture should arouse more fear than the last. And if the patient should be found with his feet bare without their being very hot, and with arms and legs flung about anyhow and bare, it is a bad sign, for it signifies distress. It is a deadly symptom also to sleep always with the mouth open, and to lie on the back with the legs very much bent and folded together. To lie on the belly, when the patient is not accustomed so to sleep when in health, is bad, for it signifies delirium, or pain in the region of the belly. But for the patient to wish to sit up when the disease is at its height is a bad sign in all acute diseases, but it is worst in cases of pneumonia. To grind the teeth in fevers, when this has not been a habit from childhood, signifies madness and death; and if the grinding be also accompanied by delirium it is a very deadly sign indeed.

[1] This means apparently that the patient cannot lie back, and so slips towards the foot of the bed. It perhaps corresponds to our "sinking down in the bed" in a state of collapse or great weakness.

---

[6] After *θανατῶδες·* the MSS. have, with slight variations, *ἀλλὰ χρὴ προλέγειν κίνδυνον ἐπ' ἀμφοτέρων ἐσόμενον.* The sentence is deleted by Ermerins and transposed by Gomperz to after *τόπων* (l. 22.).

Ἕλκος δέ, ἤν τε προγεγονὸς τύχῃ ἔχων, ἤν τε
30 καὶ ἐν τῇ νούσῳ γίνηται, καταμανθάνειν· ἢν γὰρ
μέλλῃ ἀπολεῖσθαι ὁ ἀσθενῶν, πρὸ τοῦ θανάτου
32 ἢ πελιδνὸν καὶ ξηρὸν ἔσται ἢ ὠχρὸν καὶ σκληρόν.

IV. Περὶ δὲ χειρῶν φορῆς τάδε γινώσκω·[1] ἐν
πυρετοῖσιν ὀξέσιν ἢ ἐν περιπνευμονίῃσι καὶ ἐν
φρενίτισι καὶ ἐν κεφαλαλγίῃσι πρὸ τοῦ προσώπου
φερομένας καὶ θηρευούσας διὰ κενῆς καὶ κροκύδας
ἀπὸ τῶν ἱματιων ἀποτιλλούσας καὶ καρφολογεούσας[2] καὶ ἀπὸ τῶν τοίχων ἄχυρα ἀποσπώσας,
7 πάσας εἶναι κακὰς καὶ θανατώδεας.

V. Πνεῦμα δὲ πυκνὸν μὲν ἐὸν πόνον σημαίνει
ἢ φλεγμονὴν ἐν τοῖσιν ὑπὲρ τῶν φρενῶν χωρίοι-
σιν· μέγα δὲ ἀναπνεόμενον καὶ διὰ πολλοῦ χρόνου
παραφροσύνην σημαίνει·[3] ψυχρὸν δὲ ἐκπνεό-
μενον ἐκ τῶν ῥινῶν καὶ τοῦ στόματος ὀλέθριον
κάρτα ἤδη γίνεται. εὔπνοιαν δὲ χρὴ νομίζειν
κάρτα μεγάλην δύναμιν ἔχειν ἐς σωτηρίην ἐν
πᾶσι τοῖσιν ὀξέσι νοσήμασιν, ὁκόσα σὺν πυρετοῖς
9 ἐστιν καὶ ἐν τεσσαράκοντα ἡμέρῃσι κρίνεται.

VI. Οἱ δὲ ἱδρῶτες ἄριστοι μέν εἰσιν ἐν πᾶσι
τοῖσιν ὀξέσι νοσήμασιν, ὁκόσοι ἂν ἐν ἡμέρῃσι
κρισίμῃσι γίνωνται καὶ τελέως τοῦ πυρετοῦ
ἀπαλλάσσωσιν. ἀγαθοὶ δὲ καὶ ὁκόσοι διὰ παντὸς τοῦ σώματος γινόμενοι ἀπέδειξαν τὸν ἄνθρωπον εὐπετέστερον φέροντα τὸ νόσημα. οἳ δ'
ἂν μὴ τούτων τι ἀπεργάζωνται,[4] οὐ λυσιτελέες.
κάκιστοι δὲ οἱ ψυχροὶ καὶ μοῦνον περὶ τὴν
κεφαλὴν[5] γινόμενοι καὶ τὸν αὐχένα· οὗτοι γὰρ
10 σὺν μὲν ὀξεῖ πυρετῷ θάνατον σημαίνουσιν, σὺν
11 πρηϋτέρῳ δέ, μῆκος νούσου.

[1] Before ἐν πυρετοῖσιν the MSS. have ὅσοισιν or ὁκόσοισιν. Wilamowitz deletes.

If the patient had a sore before the illness, or if a sore arises during it, pay great attention; for if the sick man is going to die, before death it will be either livid and dry or pale and hard.

IV. As to the motions of the arms, I observe the following facts. In acute fevers, pneumonia, phrenitis and headache,[1] if they move before the face, hunt in the empty air, pluck nap from the bedclothes, pick up bits, and snatch chaff from the walls—all these signs are bad, in fact deadly.[2]

V. Rapid respiration indicates pain or inflammation in the parts above the diaphragm. Deep and slow respiration indicates delirium. Cold breath from the nostrils and mouth is a very fatal sign indeed. Good respiration must be considered to have a very great influence on recovery in all the acute diseases that are accompanied by fever and reach a crisis in forty days.

VI. In all the acute diseases those sweats are best that occur on critical days and completely get rid of the fever. Those too are good that occur all over the body, showing that the patient is bearing the disease better. Sweats without one of these characteristics are not beneficial. Worst are the cold sweats that break out only around the head and neck; for these with acute fever indicate death, with a milder fever a long illness.

[1] Obviously not ordinary headaches, but such as accompany high fever.

[2] Or, "if not deadly."

---

[2] MV omit *καὶ καρφολογεούσας* but insert (before *καὶ κροκύδας*) the words *καὶ ἀποκαρφολογούσας*.

[3] *σημαίνει* C′: *δηλοῖ* MV.

[4] *ἀπεργάζωνται* C′ (with *ο* for *ω*): *ἐξεργάσωνται* MV.

[5] After *κεφαλὴν* MV add *καὶ τὸ πρόσωπον*.

VII. Ὑποχόνδριον δὲ ἄριστον μὲν ἀνώδυνόν τε ἐὸν καὶ μαλθακὸν καὶ ὁμαλὸν καὶ ἐπὶ δεξιὰ καὶ ἐπ᾽ ἀριστερά· φλεγμαῖνον δὲ καὶ ὀδύνην παρέχον ἢ ἐντεταμένον ἢ ἀνωμάλως διακείμενον τὰ δεξιὰ πρὸς τὰ ἀριστερά, ταῦτα πάντα φυλάσσεσθαι χρή. εἰ δὲ καὶ σφυγμὸς ἐνείη ἐν τῷ ὑποχονδρίῳ, θόρυβον σημαίνει ἢ παραφροσύνην· ἀλλὰ τοὺς ὀφθαλμοὺς τῶν τοιούτων ἐπικατιδεῖν χρή· ἢν γὰρ αἱ ὄψιες πυκνὰ κινέωνται, μανῆναι τὸν 10 κάμνοντα[1] ἐλπίς.

Οἴδημα δὲ ἐν τῷ ὑποχονδρίῳ σκληρόν τε ἐὸν καὶ ἐπώδυνον κάκιστον μέν, εἰ παρ᾽ ἅπαν εἴη τὸ ὑποχόνδριον. εἰ δὲ εἴη ἐν τῷ ἑτέρῳ πλευρῷ, ἀκινδυνότερόν ἐστιν ἐν τῷ ἐπ᾽ ἀριστερὰ ἐόν.[2] σημαίνει δὲ τὰ τοιαῦτα οἰδήματα ἐν ἀρχῇ μὲν κίνδυνον θανάτου ὀλιγοχρόνιον·[3] εἰ δὲ ὑπερβάλλοι εἴκοσιν ἡμέρας ὅ τε πυρετὸς ἔχων καὶ τὸ οἴδημα μὴ καθιστάμενον, ἐς διαπύησιν τρέπεται. γίνεται δὲ τούτοισιν ἐν τῇ πρώτῃ περιόδῳ καὶ 20 αἵματος ῥῆξις διὰ ῥινῶν καὶ κάρτα ὠφελεῖ· ἀλλ᾽ ἐπανέρεσθαι χρή, εἰ κεφαλὴν ἀλγέουσιν ἢ ἀμβλυώσσουσιν· εἰ γὰρ εἴη τι τούτων,[4] ἐνταῦθα ἂν ῥέποι. μᾶλλον δὲ τοῖσι νεωτέροισι πέντε καὶ τριήκοντα ἐτέων τοῦ αἵματος τὴν ῥῆξιν προσδέχεσθαι.

Τὰ δὲ μαλθακὰ τῶν οἰδημάτων καὶ ἀνώδυνα καὶ τῷ δακτύλῳ ὑπείκοντα χρονιωτέρας τὰς κρίσιας ποιεῖται καὶ ἧσσον ἐκείνων δεινότερά

[1] For τὸν κάμνοντα MV read τοῦτον.

[2] ἐόν Wilamowitz from ἐόντι of C′. Omitted by MV.

[3] After ὀλιγοχρόνιον the MSS. read ἔσεσθαι, which Wilamowitz deletes.

VII. It is best for the hypochondrium to be free from pain, soft, and with the right and left sides even; but should it be inflamed, painful, distended, or should it have the right side uneven with the left —all these signs are warnings. If there should be throbbing as well in the hypochondrium, it indicates a disturbance or delirium. The eyes of such patients ought to be examined, for if the eyeballs move rapidly you may expect the patient to go mad.

A swelling in the hypochondrium that is hard and painful is the worst, if it extend all over the hypochondrium; should it be on one side only it is less dangerous on the left.[1] Such swellings at the commencement indicate that soon there will be a danger of death, but should the fever continue for more than twenty days without the swelling subsiding, it turns to suppuration. Such patients in the first period experience epistaxis also, which is very beneficial to them. But one should ask them further if they have a headache or dimness of vision, for if one of these symptoms occur the disease will be determined in that direction. The epistaxis is more likely to happen when the patients are younger than thirty-five years.

Swellings that are soft and painless, yielding to the finger, cause the crises to be later,[2] and are less dangerous than those just described. But if the

[1] The sentence implies that the swelling is more dangerous on the right; probably the first reference to appendicitis in Greek literature.

[2] Or, "to be more protracted."

---

[4] For εἰ . . . τούτων MV read ἢν γάρ τι τοιοῦτον εἴη.

ἐστιν· εἰ δὲ ὑπερβάλλοι ἑξήκοντα ἡμέρας ὅ τε
30 πυρετὸς ἔχων καὶ τὸ οἴδημα μὴ καθιστάμενον,
ἔμπυον ἔσεσθαι σημαίνει· καὶ τοῦτο καὶ τὸ ἐν
τῇ ἄλλῃ κοιλίῃ κατὰ τὸ αὐτό. ὁκόσα μὲν οὖν
ἐπώδυνά τέ ἐστιν καὶ σκληρὰ καὶ μεγάλα,
σημαίνει κίνδυνον θανάτου ὀλιγοχρονίου, ὁκόσα
δὲ μαλθακά τε καὶ ἀνώδυνα καὶ τῷ δακτύλῳ
πιεζόμενα ὑπείκει, χρονιώτερα.

Τὰς δὲ ἀποστάσιας ἧσσον τὰ ἐν τῇ γαστρὶ
οἰδήματα ποιεῖται τῶν ἐν τοῖσιν ὑποχονδρίοισιν,
ἥκιστα δὲ τὰ ὑποκάτω τοῦ ὀμφαλοῦ ἐς διαπύησιν
40 τρέπεται· αἵματος δὲ ῥῆξιν μάλιστα ἐκ τῶν ἄνω
τόπων προσδέχεσθαι. ἁπάντων δὲ χρὴ τῶν
οἰδημάτων χρονιζόντων περὶ ταῦτα τὰ χωρία
ὑποσκέπτεσθαι τὰς ἐμπυήσιας. τὰ δὲ διαπυή-
ματα ὧδε χρὴ σκέπτεσθαι τὰ ἐντεῦθεν· ὁκόσα μὲν
αὐτῶν ἔξω τρέπεται, ἄριστά ἐστι σμικρά τε ἐόντα
καὶ ὡς μάλιστα ἔξω ἐκκλίνοντα καὶ ἐς ὀξὺ ἀποκο-
ρυφούμενα·[1] τὰ δὲ μεγάλα τε ἐόντα καὶ πλατέα
καὶ ἥκιστα ἐς ὀξὺ ἀποκορυφούμενα κάκιστα· ὁκόσα
δὲ ἔσω ῥήγνυται, ἄριστά ἐστιν, ἃ τῷ ἔξω χωρίῳ
50 μηδὲν ἐπικοινωνεῖ, ἀλλὰ ἔστιν προσεσταλμένα
τε καὶ ἀνώδυνα καὶ ὁμόχροον ἅπαν τὸ ἔξω χωρίον
φαίνεται. τὸ δὲ πῦον ἄριστόν ἐστιν λευκόν τε
καὶ λεῖον καὶ ὁμαλὸν καὶ ὡς ἥκιστα δυσῶδες· τὸ
54 δὲ ἐναντίον τῷ τοιούτῳ κάκιστον.

VIII. Οἱ δὲ ὕδρωπες οἱ ἐκ τῶν ὀξέων νοσημάτων πάντες κακοί· οὔτε γὰρ τοῦ πυρὸς ἀπαλλάσ-
σουσιν ἐπώδυνοί τέ εἰσιν κάρτα καὶ θανατώδεες.
ἄρχονται δὲ οἱ πλεῖστοι ἀπὸ τῶν κενεώνων τε καὶ
τῆς ὀσφύος, οἱ δὲ καὶ ἀπὸ τοῦ ἥπατος. ὁκόσοισι
μὲν οὖν ἐκ τῶν κενεώνων αἱ ἀρχαὶ καὶ τῆς ὀσφύος

fever continue longer than sixty days, and the swelling does not subside, it is a sign that there will be suppuration, and a swelling in any other part of the cavity will have the same history. Now swellings that are painful, hard, and big, indicate a danger of death in the near future; such as are soft and painless, yielding to the pressure of the finger, are of a more chronic character.

Abscessions are less frequently the result of swellings in the belly than of swellings in the hypochondria; least likely to turn to suppuration are swellings below the navel, but expect hemorrhage, most probably from the upper parts. But whenever the swellings in these regions are protracted one must suspect suppurations. Collections of pus there ought to be judged of thus. Such of them as turn outwards are most favourable when they are small, and bend as far as possible outwards, and come to a point; the worst are those which are large and broad, sloping least to a point. Such as break inwards are most favourable when they are not communicated at all to the outside, but do not project and are painless, while all the outside appears of one uniform colour. The pus is most favourable that is white and smooth, uniform and least evil-smelling. Pus of the opposite character is the worst.

VIII. Dropsies that result from acute diseases are all unfavourable, for they do not get rid of the fever and they are very painful and fatal. Most of them begin at the flanks and loins, though some begin also at the liver. Now whenever they begin in the flanks and loins the feet swell, and chronic diar-

[1] ἀποκορυφούμενα C′: ἀποκυρτούμενα MV.

γίνονται, οἵ τε πόδες οἰδέουσιν καὶ διάρροιαι πολυχρόνιοι ἴσχουσιν οὔτε τὰς ὀδύνας λύουσαι τὰς ἐκ τῶν κενεώνων τε καὶ τῆς ὀσφύος οὔτε τὴν
10 γαστέρα λαπάσσουσαι· ὁκόσοισι δὲ ἀπὸ τοῦ ἥπατος γίνονται, βῆξαί τε θυμὸς αὐτοῖς ἐγγίνεται καὶ ἀποπτύουσιν οὐδὲν ἄξιον λόγου καὶ οἱ πόδες οἰδέουσιν καὶ ἡ γαστὴρ οὐ διαχωρεῖ, εἰ μὴ σκληρά τε καὶ ἐπώδυνα καὶ πρὸς ἀνάγκην, καὶ περὶ τὴν κοιλίην γίνεται οἰδήματα, τὰ μὲν ἐπὶ δεξιά, τὰ δὲ ἐπ' ἀριστερά, ἱστάμενά τε
17 καὶ καταπαυόμενα.

IX. Κεφαλὴ δὲ καὶ χεῖρες καὶ πόδες ψυχρὰ ἐόντα κακὸν τῆς τε κοιλίης καὶ τῶν πλευρῶν θερμῶν ἐόντων. ἄριστον δὲ ἅπαν τὸ σῶμα θερμόν τε εἶναι καὶ μαλθακὸν ὁμαλῶς.

Στρέφεσθαι δὲ δεῖ τὸν νοσέοντα ῥηϊδίως καὶ ἐν τοῖσι μετεωρισμοῖσιν ἐλαφρὸν εἶναι· εἰ δὲ βαρὺς[1] ἐὼν φαίνοιτο καὶ τὸ ἄλλο σῶμα καὶ τὰς χεῖρας καὶ τοὺς πόδας, ἐπικινδυνότερόν ἐστιν. εἰ δὲ πρὸς τῷ βάρει καὶ οἱ ὄνυχες καὶ οἱ δάκτυλοι
10 πελιδνοὶ γίνονται,[2] προσδόκιμος ὁ θάνατος αὐτίκα· μελαινόμενοι δὲ παντελῶς οἱ δάκτυλοι καὶ[3] οἱ πόδες ἧσσον ὀλέθριοι τῶν πελιδνῶν· ἀλλὰ καὶ[4] τὰ ἄλλα σημεῖα σκέπτεσθαι χρή· ἢν γὰρ εὐπετέως φαίνηται φέρων τὸ κακὸν[5] ἤ καὶ ἄλλο τι τῶν περιεστικῶν σημείων πρὸς τούτοισιν ἐπιδεικνύῃ, τὸ νόσημα ἐς ἀπόστασιν τραπῆναι ἐλπίς, ὥστε τὸν μὲν ἄνθρωπον περιγενέσθαι, τὰ δὲ μελανθέντα τοῦ σώματος ἀποπεσεῖν.

[1] βαρὺς MV: βαρύτερος C′. [2] γίνονται: γίνοιντο C′.
[3] Some MSS. read ἢ for καί. καὶ must often be translated "or."

rhoeas afflict the patient, which neither relieve the pains in the flanks and loins nor soften the belly. But whenever the dropsies begin in the liver, the patient experiences a desire to cough without bringing up any sputum worth speaking of, while the feet swell and the bowels pass no excreta except such as are hard, painful and forced,[1] and swellings rise around the belly, some to the right and some to the left, growing and subsiding.

IX. For the head, hands, and feet to be cold is a bad sign if the belly and sides be warm; but it is a very good sign when the whole body is evenly warm and soft.

The patient ought to turn easily and to be light when lifted up. But if he should prove to be heavy in the body generally, especially in the hands and feet, it is a rather dangerous sign. And if in addition to the heaviness both the nails and fingers turn livid, death may be expected forthwith; but when fingers or feet become quite black it is a less fatal sign than their becoming livid. But the other symptoms also must be attended to. For if the patient should show himself bearing up against the illness, or manifest, in addition to the signs mentioned before, some other symptom indicating recovery, the illness may be expected to turn to an abscession, with the result that the patient loses the blackened members but recovers.

[1] Either by purging or (more probably) through constipation.

---

[4] καὶ is omitted by C′. [5] κακὸν MV: νόσημα C′.

Ὄρχιες δὲ καὶ αἰδοῖον ἀνεσπασμένα σημαίνει
20 πόνον ἢ θάνατον.[1]

X. Περὶ δε ὕπνου ὥσπερ καὶ κατὰ φύσιν ἡμῖν σύνηθές ἐστιν, τὴν μὲν ἡμέρην ἐγρηγορέναι χρή, τὴν δὲ νύκτα καθεύδειν· ἢν δὲ τοῦτο μεταβεβλημένον ᾖ,[2] κάκιον γίνεται· ἥκιστα δὲ ἂν λυπέοι, εἰ κοιμῷτο πρωὶ ἐς τὸ τρίτον μέρος τῆς ἡμέρης· οἱ δὲ ἀπὸ τούτου τοῦ χρόνου ὕπνοι πονηρότεροί εἰσιν· κάκιστον δὲ μὴ κοιμᾶσθαι μήτε τῆς ἡμέρης μήτε τῆς νυκτός· ἢ γὰρ ὑπὸ ὀδύνης τε καὶ πόνου ἀγρυπνοίη ἂν ἢ παραφροσύνη ἔσται ἀπὸ τούτου
10 τοῦ σημείου.

XI. Διαχώρημα δὲ ἄριστόν ἐστιν μαλθακόν τε καὶ συνεστηκὸς καὶ τὴν ὥρην, ἥνπερ καὶ ὑγιαίνοντι διεχώρει, πλῆθος δὲ πρὸς λόγον τῶν ἐσιόντων· τοιαύτης γὰρ ἐούσης τῆς διεξόδου ἡ κάτω κοιλίη ὑγιαίνοι ἄν. εἰ δὲ εἴη ὑγρὸν τὸ διαχώρημα, συμφέρει μήτε τρύζειν μήτε πυκνόν τε καὶ κατ᾽ ὀλίγον διαχωρεῖν· κοπιῶν γὰρ ὁ ἄνθρωπος ὑπὸ τῆς συνεχέος ἐξαναστάσιος καὶ ἀγρυπνοίη ἄν· εἰ δὲ ἀθρόον πολλάκις διαχωρέοι,
10 κίνδυνος λειποθυμῆσαι. ἀλλὰ χρὴ κατὰ τὸ πλῆθος τῶν ἐσιόντων ὑποχωρεῖν δὶς ἢ τρὶς τῆς ἡμέρης καὶ τῆς νυκτὸς ἅπαξ, τὸ δὲ πλεῖστον ὑπίτω πρωί, ὥσπερ καὶ σύνηθες ἦν τῷ ἀνθρώπῳ. παχύνεσθαι δὲ χρὴ τὸ διαχώρημα πρὸς τὴν κρίσιν ἰούσης τῆς νούσου. ὑπόπυρρον δὲ ἔστω καὶ μὴ λίην δυσῶδες· ἐπιτήδειον δὲ καὶ ἕλμινθας στρογγύλας διεξιέναι μετὰ τοῦ διαχωρήματος πρὸς τὴν κρίσιν ἰούσης τῆς νούσου.[3] δεῖ δὲ ἐν

[1] σημαίνει πόνον ἢ θάνατον C′ : πόνους ἰσχυροὺς σημαίνει καὶ κίνδυνον θανατώδεα MV.

Testicles or member being drawn up is a sign of pain or death.

X. As for sleep, the patient ought to follow the natural custom of being awake during the day and asleep during the night. Should this be changed it is rather a bad sign. Least harm will result if the patient sleep from early morning for a third part of the day. Sleep after this time is rather bad. The worst thing is not to sleep either during the day or during the night. For either it will be pain and distress that cause the sleeplessness or delirium will follow this symptom.

XI. Stools are best when soft and consistent, passed at the time usual in health, and in quantity proportional to the food taken; for when the discharges have this character the lower belly is healthy. If the bowels be loose, it is a favourable sign that there should be no noise, and that the stools should not be frequent and scanty. For if the patient be continually getting up he will be fatigued and suffer from lack of sleep, while if he often pass copious stools there is a danger of fainting. But he should go to stool twice or three times during the day, according to the quantity of food taken, and once during the night; most copiously, however, early in the morning, as his custom also was. The stool ought to grow thicker as the disease nears the crisis. It should be reddish-yellow, and not over-fetid. It is a favourable sign when round worms pass with the discharge as the disease nears the crisis. In every illness the bowels

---

[2] For ᾗ MV read εἴη.

[3] C′ omits this and the preceding sentence, the eye of the scribe passing from one *νούσου* to the other.

παντὶ νοσήματι λαπαρήν τε εἶναι τὴν κοιλίην
20 καὶ εὔογκον. ὑδαρὲς δὲ κάρτα ἢ λευκὸν ἢ
χλωρὸν[1] ἰσχυρῶς ἢ ἀφρῶδες διαχωρεῖν, πονηρὰ
ταῦτα πάντα. πονηρὸν δὲ καὶ σμικρόν τε ἐὸν
καὶ γλίσχρον καὶ λευκὸν καὶ ὑπόχλωρον καὶ
λεῖον. τούτων δὲ θανατωδέστερα ἂν εἴη τὰ
μέλανα ἢ πελιδνὰ ἢ λιπαρὰ ἢ ἰώδεα καὶ κάκο-
δμα. τὰ δὲ ποικίλα χρονιώτερα μὲν τούτων,
ὀλέθρια δὲ οὐδὲν ἧσσον· ἔστιν δὲ ταῦτα ξυσματώδεα[2] καὶ χολώδεα καὶ πρασοειδέα καὶ μέλανα,
ποτὲ μὲν ὁμοῦ διεξερχόμενα, ποτὲ δὲ καὶ κατὰ
30 μέρος.

Φῦσαν δὲ ἄνευ ψόφου καὶ πραδήσιος διεξιέναι
ἄριστον· κρέσσον δὲ καὶ σὺν ψόφῳ διεξελθεῖν ἢ
αὐτοῦ ἐναπειλῆφθαι καὶ συνειλεῖσθαι· καίτοι καὶ
οὕτω διεξελθοῦσα σημαίνει πονεῖν τι τὸν ἄνθρωπον ἢ παραφρονεῖν, ἢν μὴ ἑκὼν οὕτω ποιῆται
ὁ ἄνθρωπος τὴν ἄφεσιν τῆς φύσης. τοὺς δὲ ἐκ
τῶν ὑποχονδρίων πόνους τε καὶ τὰ κυρτώματα,
ἢν ᾖ νεαρά τε καὶ μὴ σὺν φλεγμονῇ, λύει βορβο-
ρυγμὸς ἐγγενόμενος ἐν τῷ ὑποχονδρίῳ καὶ μά-
40 λιστα μὲν διεξιὼν[3] σὺν κόπρῳ τε καὶ οὔρῳ· εἰ
δὲ μή, καὶ αὐτὸς διαπεραιωθείς· ὠφελεῖ δὲ καὶ
42 ὑποκαταβὰς ἐς τὰ κάτω χωρία.

XII. Οὖρον δὲ ἄριστόν ἐστιν, ὅταν ᾖ[4] λευκὴ ἡ
ὑπόστασις καὶ λείη καὶ ὁμαλὴ παρὰ πάντα τὸν
χρόνον, ἔστ᾽ ἂν κριθῇ ἡ νοῦσος· σημαίνει γὰρ
ἀσφάλειαν καὶ νόσημα ὀλιγοχρόνιον ἔσεσθαι. εἰ
δὲ διαλείποι καὶ ποτὲ μὲν καθαρὸν οὐρέοι, ποτὲ

[1] After χλωρὸν MV add ἢ ἐρυθρόν.

[2] After ξυσματώδεα Kühlewein reads (from Galen) τε καὶ αἱματώδεα.

should be soft and distended. But for stools to be very fluid, or white, or exceedingly green,[1] or frothy, are all bad signs. It is a bad sign too when they are scanty and viscid, white, greenish and smooth. But more deadly than these will be stools that are black, or livid, or oily, or verdigris-coloured[2] and fetid. Varied stools indicate an illness which, while longer than those just referred to, will be no less dangerous; such are like scrapings, bilious, leek-green, and black, exhibiting these characteristics sometimes all at once and sometimes by turns.

It is best for flatulence to pass without noise and breaking, though it is better for it to pass even with noise than to be intercepted and accumulated internally; yet even if passed thus it indicates that the patient is suffering or delirious, unless he emits the flatulence wittingly. But pains and swellings in the hypochondria, if they be recent and without inflammation, are cured by a rumbling occurring in the hypochondrium, which is most favourable when it passes along with stools and urine, though it is beneficial even if it merely passes by itself. It is also beneficial when it descends into the lower parts.

XII. Urine is best when the sediment is white, smooth and even for the whole period of the illness until the crisis, for it indicates a short sickness and a sure recovery. But should the sediment intermit, and the urine sometimes be clear and sometimes show the white, smooth, even deposit, the illness will

[1] That is, "yellowish green." [2] Or, "rust-coloured."

---

[3] διεξιὼν M: διεξελθὼν C′. [4] C′ reads εἴη for ᾖ.

δὲ ὑφίσταιτο τὸ λευκόν τε καὶ λεῖον καὶ ὁμαλόν, χρονιωτέρη γίνεται ἡ νοῦσος καὶ ἧσσον ἀσφαλής. εἰ δὲ εἴη τό τε οὖρον ὑπέρυθρον καὶ ἡ ὑπόστασις ὑπέρυθρός τε καὶ λείη, πολυχρονιώτερον μὲν
10 τοῦτο τοῦ προτέρου γίνεται, σωτήριον δὲ κάρτα. κριμνώδεες δὲ ἐν τοῖσιν οὔροισιν ὑποστάσιες πονηραί· τούτων δὲ ἔτι κακίους αἱ πεταλώδεες· λεπταὶ δὲ καὶ λευκαὶ κάρτα φλαῦραι· τούτων δὲ ἔτι κακίους αἱ[1] πιτυρώδεες. νεφέλαι δὲ ἐναιωρεύμεναι τοῖσιν οὔροισι λευκαὶ μὲν ἀγαθαί, μέλαιναι δὲ φλαῦραι. ἔστ' ἂν δὲ λεπτὸν ᾖ τὸ οὖρον καὶ πυρρόν, ἄπεπτον σημαίνει τὸ νόσημα εἶναι· εἰ δὲ καὶ πολυχρόνιον εἴη τὸ νόσημα, τὸ δὲ οὖρον τοιοῦτον ἐόν, κίνδυνος μὴ οὐ δυνήσεται ὁ ἄν-
20 θρωπος διαρκέσαι, ἔστ' ἂν πεπανθῇ ἡ νοῦσος. θανατωδέστερα δὲ τῶν οὔρων τά τε δυσώδεα καὶ ὑδατώδεα καὶ μέλανα καὶ παχέα· ἔστι δὲ τῇσι μὲν γυναιξὶ καὶ τοῖσιν ἀνδράσι τὰ μέλανα τῶν οὔρων κάκιστα, τοῖσι δὲ παιδίοισι τὰ ὑδατώδεα. ὁκόσοι δὲ οὖρα λεπτὰ καὶ ὠμὰ οὐρέουσι πολὺν χρόνον, ἢν καὶ τὰ ἄλλα σημεῖα ὡς περιεσομένοις ᾖ, τούτοισιν ἀπόστασιν δεῖ προσδέχεσθαι ἐς τὰ κάτω τῶν φρενῶν χωρία. καὶ τὰς λιπαρότητας δὲ τὰς ἄνω ἐφισταμένας ἀραχνοει-
30 δέας μέμφεσθαι· συντήξιος γὰρ σημεῖα. σκοπεῖν δὲ τῶν οὔρων, ἐν οἷς εἰσιν αἱ νεφέλαι,[2] ἤν τε κάτω ἔωσιν ἤν τε ἄνω, καὶ τὰ χρώματα ὁκοῖα ἴσχουσιν καὶ τὰς μὲν κάτω φερομένας σὺν τοῖσι χρώμασιν, οἷα εἴρηται ἀγαθὰ εἶναι, ἐπαινεῖν, τὰς δὲ ἄνω

[1] C′ omits πεταλώδεες . . . . κακίους αἱ, the scribe passing from the first κακίους αἱ to the second, omitting the intervening words.

be longer and recovery less likely. Should the urine be reddish and the sediment reddish and smooth, recovery will be sure, although the illness will be longer than in the former case. Sediments in urine which are like coarse meal are bad, and even worse than these are flaky sediments. Thin, white sediments are very bad, and even worse than these are those like bran. Clouds suspended in the urine are good when white but bad when black.[1] So long as the urine is thin and of a yellowish-red colour, it is a sign that the disease is unconcocted; and if the disease should also be protracted, while the urine is of this nature, there is a danger lest the patient will not be able to hold out until the disease is concocted. The more fatal kinds of urine are the fetid, watery, black[1] and thick; for men and women black urine is the worst, for children watery urine. Whenever the urine is for a long time thin and crude, should the other symptoms too be those of recovery, an abscession is to be expected to the parts below the diaphragm. Fatty substances like spiders' webs settling on the surface are alarming, as they are signs of wasting. The urine in which the clouds are, whether these be on the bottom or at the top, must be examined, as well as the colours of these clouds, and those that float at the bottom with the colours I have stated to be good, should be welcomed, while clouds on the top,

[1] *I. e.* like port wine. See p. 9.

---

[2] After νεφέλαι C′ has σὺν τοῖς χρώμασιν ὡς εἴρηται, and omits the phrase σὺν . . . εἴρηται lower down. The text in this part is very uncertain, the variants being numerous but unimportant. I follow Kühlewein, but with no confidence. Fortunately the sense is quite clear.

σὺν τοῖσι χρώμασιν, οἷα εἴρηται κακὰ εἶναι, μέμφεσθαι. μὴ ἐξαπατάτω δέ σε, ἤν τι αὐτὴ ἡ κύστις νόσημα ἔχουσα τῶν οὔρων τὰ τοιαῦτα ἀποδιδῷ· οὐ γὰρ τοῦ ὅλου σώματος σημεῖον,
39 ἀλλ' αὐτῆς καθ' ἑωυτήν.

XIII. Ἔμετος δὲ ὠφελιμώτατος φλέγματός τε καὶ χολῆς συμμεμειγμένων ὡς μάλιστα καὶ μὴ παχὺς μηδὲ πολὺς κάρτα ἐμείσθω· οἱ δὲ ἀκρητέστεροι κακίους. εἰ δὲ εἴη τὸ ἐμεύμενον πρασοειδὲς ἢ πελιδνὸν ἢ μέλαν, ὅ τι ἂν ᾖ τούτων τῶν χρωμάτων, νομίζειν χρὴ πονηρὸν εἶναι· εἰ δὲ καὶ πάντα τὰ χρώματα ὁ αὐτὸς ἄνθρωπος ἐμέοι, κάρτα ὀλέθριον ἤδη γίνεται· τάχιστον δὲ θάνατον σημαίνει τὸ πελιδνὸν τῶν ἐμεσμάτων, εἰ ὄζοι δυσ-
10 ῶδες· πᾶσαι δὲ αἱ ὑπόσαπροι καὶ δυσώδεες ὀδμαὶ
11 κακαὶ ἐπὶ πᾶσι τοῖσιν ἐμεομένοισι.

XIV. Πτύελον χρὴ ἐπὶ πᾶσι τοῖσιν ἀλγήμασι τοῖσι περὶ τὸν πνεύμονά τε καὶ τὰς πλευρὰς ταχέως τε ἀναπτύεσθαι καὶ εὐπετέως, συμμεμειγμένον τε φαίνεσθαι τὸ ξανθὸν ἰσχυρῶς τῷ πτυέλῳ· εἰ γὰρ πολλῷ ὕστερον μετὰ τὴν ἀρχὴν τῆς ὀδύνης ἀναπτύοιτο ξανθὸν ἐὸν ἢ πυρρὸν ἢ πολλὴν βῆχα παρέχον ἢ μὴ ἰσχυρῶς συμμεμειγμένον, κάκιον γίνεται· τό τε γὰρ ξανθὸν ἄκρητον ἐὸν κινδυνῶδες, τὸ δὲ λευκὸν καὶ γλίσ-
10 χρον καὶ στρογγύλον ἀλυσιτελές· κακὸν δὲ καὶ χλωρόν τε ἐὸν κάρτα καὶ ἀφρῶδες· εἰ δὲ εἴη οὕτως ἄκρητον, ὥστε καὶ μέλαν φαίνεσθαι, δεινότερόν ἐστιν τοῦτο ἐκείνων·[1] κακὸν δὲ καὶ ἢν μηδὲν ἀνακαθαίρηται μηδὲ προΐῃ ὁ πνεύμων, ἀλλὰ

[1] ἐκείνων MV : ἐκείνου C'.

with the colours I have stated to be bad, should be considered unfavourable. But be not deceived if the urine have these bad characters because the bladder itself is diseased; for they will not be a symptom of the general health,[1] but only of the bladder by itself.

XIII. That vomit is most useful which is most thoroughly compounded of phlegm and bile, and it must not be thick nor brought up in too great quantity. Less compounded vomits are worse. And if that which is brought up be of the colour of leeks, or livid, or black,[2] in all cases vomit of these colours must be considered bad. If the same patient brings up vomit of all these colours, he is quite at death's door. Of the vomits, the livid indicates the earliest death, should the odour be foul; but all odours which are rather putrid and foul are bad in the case of all vomits.

XIV. Sputum, in all pains of the lungs and ribs, should be quickly and easily brought up, and the yellow should appear thoroughly compounded with the sputum; for if long after the beginning of the pain yellow sputum should be coughed up, or reddish-yellow, or causing much coughing, or not thoroughly compounded, it is a rather bad sign. For yellow sputum, uncompounded, is dangerous, and the white, viscous and round bodes no good. Pale green, if pronounced, and frothy sputum is also bad. If it should be so uncompounded as to appear actually black,[2] this is a more alarming sign than the others. It is bad too if nothing be brought up, and the lungs eject nothing, but are full, and bubble in the throat. In

[1] Hippocratic prognosis is concerned only with "general" pathology. [2] See p. 9.

πλήρης ἐὼν ζέῃ ἐν τῇ φάρυγγι. κορύζας δὲ καὶ πταρμοὺς ἐπὶ πᾶσι τοῖσι περὶ τὸν πνεύμονα νοσήμασιν κακὸν καὶ προγεγονέναι καὶ ἐπιγενέσθαι· ἀλλ' ἐν τοῖσιν ἄλλοισι τοῖσι θανατώδεσι νοσήμασιν οἱ πταρμοὶ λυσιτελέες.[1] αἵματι δὲ
20 συμμεμειγμένον μὴ πολλῷ πτύελον ξανθὸν ἐν τοῖσι περιπνευμονικοῖσιν ἐν ἀρχῇ μὲν τῆς νούσου ἀναπτυόμενον περιεστικὸν κάρτα· ἑβδομαίῳ δὲ ἐόντι ἢ παλαιοτέρῳ ἧσσον ἀσφαλές. πάντα δὲ τὰ πτύελα πονηρά ἐστιν, ὁκόσα ἂν τὴν ὀδύνην μὴ παύῃ· κάκιστα δὲ τὰ μέλανα, ὡς διαγέγραπται· παύοντα δὲ τὴν ὀδύνην πάντα ἀμείνω[2]
27 πτυόμενα.

XV. Ὁκόσα δὲ τῶν ἀλγημάτων ἐκ τούτων τῶν χωρίων μὴ παύεται μήτε πρὸς τὰς τῶν πτυέλων καθάρσιας μήτε πρὸς τὴν τῆς κοιλίης ἐκκόπρωσιν μήτε πρὸς τὰς φλεβοτομίας τε καὶ φαρμακείας καὶ διαίτας, εἰδέναι δεῖ ἐκπυήσοντα. τῶν δὲ ἐκπυημάτων ὁκόσα μὲν ἔτι χολώδεος ἐόντος τοῦ πτυέλου ἐκπυίσκεται, ὀλέθρια κάρτα, εἴτε ἐν μέρει τὸ χολῶδες τῷ πύῳ ἀναπτύοιτο εἴτε ὁμοῦ. μάλιστα δέ, ἢν ἄρξηται χωρεῖν τὸ ἐκπύημα ἀπὸ
10 τούτου τοῦ πτυέλου, ἑβδομαίου ἐόντος τοῦ νοσήματος, ἐλπὶς τὸν ἀλγέοντα[3] ἀποθανεῖσθαι τεσσαρεσκαιδεκαταῖον, ἢν μή τι αὐτῷ ἐπιγένηται ἀγαθόν. ἔστιν δὲ τὰ μὲν ἀγαθὰ τάδε· εὐπετέως φέρειν τὸ νόσημα, εὔπνοον εἶναι, τῆς ὀδύνης ἀπηλλάχθαι, τὸ πτύελον ῥηϊδίως ἀναβήσσειν, τὸ σῶμα πᾶν ὁμαλῶς θερμόν τε εἶναι καὶ μαλθακὸν καὶ δίψαν μὴ ἔχειν, οὖρα δὲ καὶ διαχωρήματα καὶ ὕπνους καὶ ἱδρῶτας, ὡς διαγέγραπται ἕκαστα[4] ἀγαθὰ ἐόντα, ταῦτα ἐπιγενέσθαι· οὕτω μὲν γὰρ

all lung diseases it is bad for catarrhs and sneezing either to precede or to follow, but all other dangerous diseases are benefited by sneezing. For a little blood mixed with yellow sputum to be brought up in cases of pneumonia at the beginning of the disease is a very favourable sign of recovery, but less favourable on the seventh day or later. All sputum is bad if it does not remove the pain, but the worst, as I have said, is the black, while in all cases the removal of the pain by expectoration is a better sign.

XV. Such pains in these parts as do not give way before either purging of sputum, or evacuation of the bowels, or venesection, purges and regimen, must be regarded as about to turn to empyema. Such empyemas as form while the sputum is still bilious are very fatal, whether the bile and pus be brought up by turns or together. Especially should the empyema begin from sputum of this character when the disease has reached the seventh day, the patient may be expected to die on the fourteenth day unless some good symptom happen to him. The good symptoms are these: to bear up easily against the disease; to have good respiration; to be free from the pain; to cough up the sputum readily; the whole body to be evenly warm and soft; to have no thirst; urine, stools, sleep and sweat to get the characters that have been severally described as good.

---

[1] Ermerins transposes the whole passage *κορύζας δὲ* . . . . *λυσιτελέες* to the end of the chapter.

[2] After *ἀμείνω* Kühlewein adds *τὰ* (perhaps rightly).

[3] *ἀλγέοντα* C′: *τὰ τοιαῦτα πτύοντα* MV and other MSS.

[4] After *ἕκαστα* the MSS. have *εἰδέναι*. Deleted by Ermerins and Reinhold.

20 τούτων πάντων τῶν σημείων ἐπιγενομένων οὐκ ἂν ἀποθάνοι ὁ ἄνθρωπος· ἢν δὲ τὰ μὲν τούτων ἐπιγένηται, τὰ δὲ μή, πλείω χρόνον ζήσας ἢ τεσσαρεσκαίδεκα ἡμέρας ἀπόλοιτ' ἄν. κακὰ δὲ τἀναντία τούτων· δυσπετέως φέρειν τὴν νοῦσον, πνεῦμα μέγα καὶ πυκνὸν εἶναι, τὴν ὀδύνην μὴ πεπαῦσθαι, τὸ πτύελον μόλις ἀναβήσσειν, διψῆν κάρτα, τὸ σῶμα ὑπὸ τοῦ πυρὸς ἀνωμάλως ἔχεσθαι καὶ τὴν μὲν γαστέρα[1] καὶ τὰς πλευρὰς θερμὰς εἶναι ἰσχυρῶς, τὸ δὲ μέτωπον καὶ τὰς χεῖρας καὶ 30 τοὺς πόδας ψυχρά, οὖρα δὲ καὶ διαχωρήματα καὶ ὕπνους καὶ ἱδρῶτας, ὡς διαγέγραπται ἕκαστα κακὰ ἐόντα, τούτων εἴ τι ἐπιγίνοιτο τῷ πτυέλῳ τούτῳ, ἀπόλοιτ' ἂν ὁ ἄνθρωπος, πρὶν ἢ ἐς τὰς τεσσαρεσκαίδεκα ἡμέρας ἀφικέσθαι, ἢ ἐναταῖος ἢ ἑνδεκαταῖος. οὕτως οὖν συμβάλλεσθαι χρή, ὡς τοῦ πτυέλου τούτου θανατώδεος ἐόντος μάλα καὶ οὐ περιάγοντος ἐς τὰς τεσσαρεσκαίδεκα ἡμέρας ἀφικνεῖσθαι. τὰ δὲ ἐπιγινόμενα ἀγαθά τε καὶ κακὰ συλλογιζόμενον ἐκ τούτων χρὴ τὰς 40 προρρήσιας ποιεῖσθαι· οὕτω γὰρ ἂν μάλιστα ἀληθεύοις. αἱ δὲ ἄλλαι ἐκπυήσιες αἱ πλεῖσται ῥήγνυνται, αἱ μὲν εἰκοσταῖαι, αἱ δὲ τριηκοσταῖαι, αἱ δὲ τεσσαρακονθήμεροι, αἱ δὲ πρὸς τὰς ἑξήκοντα 44 ἡμέρας ἀφικνέονται.

XVI. Ἐπισκέπτεσθαι δὲ χρὴ τὴν ἀρχὴν τοῦ ἐμπυήματος[2] λογιζόμενον ἀπὸ τῆς ἡμέρης, ᾗ τὰ πρῶτα ὁ ἄνθρωπος ἐπύρεξεν ἢ ᾗ ποτε αὐτὸν ῥῖγος ἔλαβεν καὶ ᾗ φαίη[3] ἀντὶ τῆς ὀδύνης αὐτῷ βάρος ἐγγενέσθαι ἐν τῷ τόπῳ, ᾧ ἤλγει·

[1] γαστέρα C′: κοιλίην MV.

If all these symptoms supervene, the patient will not die; if some, but not all, supervene, the patient will die after living for longer than fourteen days. Bad symptoms are the opposite of those I have just given: to bear up against the disease with difficulty; respiration to be deep and rapid; the pain not to have ceased; to cough up the sputum with difficulty; to be very thirsty; the body to be unevenly affected by the fever, the belly and the sides being exceedingly warm, and the forehead, hands and feet cold; urine, stools, sleep and sweat to have the characters already described severally as bad—should sputum of the kind mentioned above be followed by any of these symptoms the patient will die before completing the fourteen days, on the ninth or eleventh day. So that must be the conclusion drawn, as this sputum is very deadly, and does not allow the patient to survive fourteen days. You must take into account both the good signs and the bad that occur and from them make your predictions; for in this way you will prophesy aright. Most other empyemas break, some on the twentieth day, some on the thirtieth, some on the fortieth, while others last sixty days.

XVI. Consider that the beginning of the empyema dates from the day on which the patient was first attacked by fever or by rigor, or on which he said that a heaviness took the place of the pain in that

---

[2] After ἐμπυήματος the MSS. have ἔσεσθαι. It is deleted by Wilamowitz. Perhaps γενέσθαι should be read.

[3] Possibly ἂν has here fallen out before ἀντί. In the Hippocratic collection, however, the optative is not seldom found with the sense of optative with ἄν. ᾖ is an emendation of Wilamowitz; C′ has ἐὰν and MV have εἰ.

ταῦτα γὰρ ἐν ἀρχῇσι γίνεται τῶν ἐμπυημάτων. ἐξ οὖν τούτου τοῦ χρόνου[1] χρὴ προσδέχεσθαι τοῦ πύου ἔσεσθαι τὰς ῥήξιας ἐς τοὺς χρόνους τοὺς προειρημένους. εἰ δὲ εἴη τὸ ἐμπύημα ἐπὶ
10 θάτερα μοῦνον, στρέφειν τε καὶ καταμανθάνειν χρὴ ἐπὶ τούτοισι, μή τι ἔχει ἄλγημα ἐν τῷ πλευρῷ· καὶ ἤν τι θερμότερον ᾖ τὸ ἕτερον τοῦ ἑτέρου, κατακλινομένου ἐπὶ τὸ ὑγιαῖνον πλευρὸν ἐρωτᾶν, εἴ τι δοκεῖ βάρος αὐτῷ ἐκκρέμασθαι ἐκ τοῦ ἄνωθεν. εἰ γὰρ εἴη τοῦτο, ἐκ τοῦ ἐπὶ θάτερόν ἐστι τὸ ἐμπύημα, ἐφ᾽ ὁκοτέρῳ ἂν πλευρῷ τὸ
17 βάρος ἐγγίνηται.[2]

XVII. Τοὺς δὲ σύμπαντας ἐμπύους γινώσκειν χρὴ τοῖσδε τοῖς σημείοισι· πρῶτον μὲν ὁ πυρετὸς οὐκ ἀφίησιν, ἀλλὰ τὴν μὲν ἡμέρην λεπτὸς ἴσχει, ἐς νύκτα δὲ πλείων, καὶ ἱδρῶτες πολλοὶ γίνονται, βῆξαί τε θυμὸς αὐτοῖσιν ἐγγίνεται καὶ ἀποπτύουσιν οὐδὲν ἄξιον λόγου, καὶ οἱ μὲν ὀφθαλμοὶ ἔγκοιλοι γίνονται, αἱ δὲ γνάθοι ἐρυθήματα ἴσχουσιν, καὶ οἱ ὄνυχες τῶν χειρῶν γρυποῦνται καὶ οἱ δάκτυλοι θερμαίνονται καὶ μάλιστα τὰ
10 ἄκρα, καὶ ἐν τοῖσι ποσὶν οἰδήματα γίνεται[3] καὶ φλύκταιναι γίνονται ἀνὰ τὸ σῶμα καὶ σιτίων οὐκ ἐπιθυμέουσιν.

Ὁκόσα μὲν οὖν ἐγχρονίζει τῶν ἐμπυημάτων, ἴσχει τὰ σημεῖα ταῦτα καὶ πιστεύειν αὐτοῖσι χρὴ κάρτα· ὁκόσα δὲ ὀλιγοχρόνιά ἐστι τούτοισιν

[1] MV have τουτέων τῶν χρόνων.

[2] I have followed C′ here, but I feel sure that the text must remain uncertain, since it is probably mutilated, with gaps from εἰ δὲ εἴη to the end of the chapter.

[3] After γίνεται C′ has ἱστάμενα καὶ καταπαυόμενα.

part in which he had been aching. These symptoms occur at the beginning of empyema. Expect then that the gathering will break after the intervals mentioned above from the date of the beginning. Should the empyema be one-sided only, turn the patient in this case, and inquire whether he has a pain in the side. And if one side be somewhat hotter than the other, ask the patient, while he is lying on the sound side, if he feels a weight hanging from the upper part. Should this be so, the empyema is one-sided, on whichever side the weight occurs.[1]

XVII. All sufferers from empyema may be distinguished by the following symptoms. In the first place the fever never stops, being slight during the day but more severe at night; copious sweats occur; the patient has a desire to cough, without bringing up any sputum worth speaking of; the eyes become sunken; the cheeks are flushed; the finger-nails are bent and the fingers grow hot, especially at the tips; the feet swell up; blisters rise about the body, and the appetite fails.

Prolonged empyema has these symptoms, which may be implicitly relied on; when recent it is indicated by the same signs, should there appear those

[1] I have done my best to make sense out of this very obscure passage. Why should the physician make these experiments, if he know sthat the empyema is on one side, and knows also which is "the sound side"? Was it to confirm his suspicions? Was it to persuade the patient that he had empyema, and so get his consent to an operation, should one prove necessary? I have long suspected that the text is very mutilated, and that several sentences have dropped out. If the text could be restored, we should probably see that the writer considered not one case only, but two or three.

ἐπισημαίνεται, τοιούτων ἤν τι ἐπιφαίνηται οἷα καὶ τοῖσιν ἐξ ἀρχῆς γινομένοισιν, ἅμα δὲ καὶ ἤν τι δυσπνούστερος ᾖ ὁ ἄνθρωπος. τὰ δὲ ταχύτερόν τε καὶ βραδύτερον ῥηγνύμενα γινώσκειν
20 χρὴ τοῖσδε τοῖς σημείοισι· ἢν μὲν ὁ πόνος ἐν ἀρχῇσι γίνηται καὶ ἡ δύσπνοια καὶ ἡ βὴξ καὶ ὁ πτυελισμὸς διατελῇ[1] ἔχων, ἐς τὰς εἴκοσι ἡμέρας προσδέχεσθαι τὴν ῥῆξιν ἢ καὶ ἔτι πρόσθεν· ἢν δὲ ἡσυχέστερος ὁ πόνος ᾖ καὶ τὰ ἄλλα πάντα κατὰ λόγον, τούτοισι προσδέχεσθαι τὴν ῥῆξιν ὕστερον· προγενέσθαι δὲ ἀνάγκη καὶ πόνον καὶ δύσπνοιαν καὶ πτυελισμὸν πρὸ τῆς τοῦ πύου ῥήξιος.

Περιγίνονται δὲ τούτων μάλιστα οὓς ἂν ἀφῇ ὁ
30 πυρετὸς αὐθημερὸν μετὰ τὴν ῥῆξιν καὶ σιτίων ταχέως ἐπιθυμέωσιν καὶ δίψης ἀπηλλαγμένοι ἔωσιν καὶ ἡ γαστὴρ σμικρά τε καὶ συνεστηκότα διαχωρῇ καὶ τὸ πῦον λευκόν τε καὶ λεῖον καὶ ὁμόχροον ἐκχωρῇ καὶ φλέγματος ἀπηλλαγμένον καὶ ἄνευ πόνου τε καὶ βηχὸς ἀνακαθαίρηται.[2] ἄριστα μὲν οὕτω καὶ τάχιστα ἀπαλλάσσουσιν· εἰ δὲ μή, οἷσιν ἂν ἐγγυτάτω τούτων γίνηται. ἀπόλλυνται δὲ οὓς ἂν ὁ πυρετὸς αὐθημερὸν μὴ ἀφῇ, ἀλλὰ δοκέων ἀφιέναι αὖθις φαίνηται ἀνα-
40 θερμαινόμενος, καὶ δίψαν μὲν ἔχωσι, σιτίων δὲ μὴ ἐπιθυμέωσιν καὶ ἡ κοιλίη ὑγρὴ ᾖ καὶ τὸ πῦον χλωρὸν καὶ πελιδνὸν ἢ φλεγματῶδες καὶ ἀφρῶδες· οἷσι ταῦτα πάντα γίνεται, ἀπόλλυνται· ὁκόσοισι δὲ τούτων τὰ μὲν γίνεται, τὰ δὲ μή, οἱ μὲν αὐτῶν ἀπόλλυνται, οἱ δὲ ἐν πολλῷ χρόνῳ περιγίνονται. ἀλλ' ἐκ πάντων τῶν τεκμηρίων τῶν ἐόντων ἐν τού-
47 τοισι τεκμαίρεσθαι[3] καὶ τοῖσιν ἄλλοισιν ἅπασιν.

symptoms which occur at the beginning, if at the same time there be some difficulty of breathing. Whether the gathering will break earlier or later may be determined by the following signs. If the pain take place at the beginning, and if the difficulty of breathing, the coughing and the expectoration be continued,[1] expect the breaking by the twentieth day or even earlier. If, however, the pain be milder, and all the signs be proportionately mild, expect the breaking later. Before the gathering breaks there must occur pain, difficulty of breathing and expectoration.

Those chiefly recover who lose the fever on the same day after the gathering breaks, quickly recover their appetite, and are rid of thirst; when the bowels pass small, solid motions, and the pus evacuated is white, smooth, uniform in colour, rid of phlegm and brought up without pain and coughing. These make the best and quickest recovery; the nearer the approximation to their symptoms the better. Those die who are not left on the same day by the fever, which seems to leave them and then appears again with renewal of heat; who are thirsty but have no appetite; whose bowels are loose, and who evacuate pus that is yellow and livid or full of phlegm and froth. Those who show all these symptoms die; those who show some only either die or recover after a long illness. In these cases, as in all others, it is from the sum-total of the symptoms that an appreciation of the illness should be made.

[1] Or, reading διατείνῃ, "severe."

---

[1] διατελῆ ἔχων C′ (with the spelling διατελέει): διατείνῃ MV.
[2] ἀνακαθαίρηται omitted by MV.
[3] τεκμαίρεσθαι C′; σημαίνεσθαι MV.

XVIII. Ὁκόσοισι δὲ ἀποστάσιες γίνονται ἐκ τῶν περιπνευμονικῶν νοσημάτων παρὰ τὰ ὦτα καὶ ἐκπυέουσιν ἐς τὰ κάτω χωρία καὶ συριγγοῦνται, οὗτοι δὲ περιγίνονται. ὑποσκεπτεσθαι δὲ χρὴ τὰ τοιαῦτα ὧδε· ἢν ὅ τε πυρετὸς ἔχῃ καὶ ἡ ὀδύνη μὴ πεπαυμένη ᾖ καὶ τὸ πτύελον μὴ ἐκχωρῇ κατὰ λόγον, μηδὲ χολώδεες αἱ διαχωρήσιες τῆς κοιλίης ἔωσι μηδὲ εὔλυτοι καὶ εὔκρητοι γίνωνται, μηδὲ τὸ οὖρον παχύ τε κάρτα καὶ
10 πολλὴν ὑπόστασιν ἔχον, ὑπηρετῆται δὲ περιεστικῶς ὑπὸ τῶν λοιπῶν πάντων τῶν περιεστικῶν σημείων, τούτοισι χρὴ τὰς τοιαύτας ἀποστάσιας ἐλπίζειν ἔσεσθαι. γίνονται δὲ αἱ μὲν ἐς τὰ κάτω χωρία, οἷσιν ἄν τι περὶ τὸ ὑποχόνδριον τοῦ φλέγματος ἐγγίνηται, αἱ δὲ ἄνω, οἷσιν ἂν τὸ μὲν ὑποχόνδριον λαπαρόν τε καὶ ἀνώδυνον διατελῇ ἐόν, δύσπνοος δέ τινα χρόνον γενόμενος παύσηται ἄτερ φανερῆς προφάσιος ἄλλης.

Αἱ δὲ ἀποστάσιες αἱ ἐς τὰ σκέλεα ἐν τῇσι
20 περιπνευμονίῃσι τῇσιν ἰσχυρῇσι καὶ ἐπικινδύνοισι λυσιτελέες μὲν πᾶσαι, ἄρισται δὲ αἱ τοῦ πτυέλου ἐν μεταβολῇ ἐόντος ἤδη γινόμεναι· εἰ γὰρ τὸ οἴδημα καὶ ἡ ὀδύνη γίνοιτο, τοῦ πτυέλου ἀντὶ τοῦ ξανθοῦ πυώδεος γινομένου καὶ ἐκχωρέοντος ἔξω, οὕτως ἂν ἀσφαλέστατα ὅ τε ἄνθρωπος περιγίνοιτο, καὶ ἡ ἀπόστασις τάχιστα ἀνώδυνος ἂν παύσαιτο· εἰ δὲ τὸ πτύελον μὴ ἐκχωρέοι καλῶς, μηδὲ τὸ οὖρον ὑπόστασιν ἀγαθὴν ἔχον φαίνοιτο, κίνδυνος γενέσθαι χωλὸν τὸ ἄρθρον ἢ πολλὰ
30 πρήγματα παρασχεῖν. εἰ δὲ ἀφανίζοιντο αἱ ἀποστάσιες τοῦ πτυέλου μὴ ἐκχωρέοντος τοῦ τε πυρετοῦ ἔχοντος, δεινόν· κίνδυνος γὰρ μὴ παρα-

XVIII. Whenever from pneumonia an abscession takes place to the ears, while gatherings occur in the lower parts and fistula forms, the patient recovers. Judge of such cases in the following way. Expect abscessions of this kind when the fever holds, if the pain have not ceased and the expectoration be not normal, if the stools be not bilious, nor become loose and concocted, if the urine have not a very thick, copious deposit, but be assisted favourably by all the other favourable symptoms. The abscessions occur, some to the lower parts, whenever some of the phlegm appears in the region of the hypochondrium, others to the upper parts, whenever the hypochondrium continues to be soft and painless, and the patient suffers from a temporary shortness of breath which ceases without any manifest cause.

Abscessions to the legs in severe and critical pneumonia are all beneficial, but the best are those that occur when the sputum is already changing. For if the swelling and the pain take place at the same time as the sputum is turning from yellow to purulent and is being evacuated, the patient is quite certain to recover, and the abscession will very quickly come to an end without pain. Should, however, the sputum be not well evacuated, and the urine do not show a good deposit, there is a danger that the limb will be lamed or else cause much trouble. Should, however, the abscessions disappear without the evacuation of sputum and while the fever lasts, the prognosis is bad, as there is a danger lest the patient become delirious and die. When empyema occurs as the result of pneu-

φρονήσῃ καὶ ἀποθάνῃ ὁ ἄνθρωπος. τῶν δὲ ἐμπύων τῶν ἐκ τῶν περιπνευμονικῶν οἱ γεραίτεροι μᾶλλον ἀπόλλυνται· ἐκ δὲ τῶν ἄλλων ἐμπυημά-
36 των οἱ νεώτεροι μᾶλλον ἀποθνῄσκουσιν.[1]

XIX. Αἱ δὲ σὺν πυρετῷ ὀδύναι γινόμεναι περὶ τὴν ὀσφύν τε καὶ τὰ κάτω χωρία, ἢν τῶν φρενῶν ἅπτωνται, ἐκλείπουσαι τὰ κάτω χωρία, ὀλέθριαι κάρτα. προσέχειν οὖν δεῖ τὸν νόον καὶ τοῖσιν ἄλλοισι σημείοισιν, ὡς ἤν τι καὶ τῶν ἄλλων σημείων πονηρὸν ἐπιφαίνηται, ἀνέλπιστος ὁ ἄνθρωπος· εἰ δὲ ἀναίσσοντος τοῦ νοσήματος[2] πρὸς τὰς φρένας τὰ ἄλλα σημεῖα μὴ πονηρὰ ἐπιγίνοιτο, ἔμπυον ἔσεσθαι πολλαὶ ἐλπίδες
10 τοῦτον.

Κύστιες δὲ σκληραί τε καὶ ἐπώδυνοι δειναὶ μὲν πᾶσαι· ὀλεθριώταται δὲ ὁκόσαι σὺν πυρετῷ συνεχεῖ γίνονται· καὶ γὰρ οἱ ἀπ' αὐτέων τῶν κυστίων πόνοι ἱκανοὶ ἀποκτεῖναι, καὶ αἱ κοιλίαι οὐ διαχωρέουσιν ἐπὶ τῶν τοιούτων, εἰ μὴ σκληρά τε καὶ πρὸς ἀνάγκην. λύει δὲ οὖρον πυῶδες οὐρηθέν, λευκὴν καὶ λείην ἔχον ὑπόστασιν· ἢν δὲ μήτε τὸ οὖρον μηδὲν ἐνδῷ μήτε ἡ κύστις μαλθαχθῇ ὅ τε πυρετὸς συνεχὴς ᾖ, ἐν τῇσι πρώτῃσι
20 περιόδοισι τοῦ νοσήματος ἐλπὶς τὸν ἀλγέοντα ἀποθανεῖσθαι· ὁ δὲ τρόπος οὗτος μάλιστα τῶν παιδίων ἅπτεται τῶν ἀπὸ ἑπτὰ ἐτέων, ἔστ' ἂν
23 πεντεκαιδεκαετέες γένωνται.

[1] After ἀποθνῄσκουσιν many of the MSS. have (with slight variations) ὁκόσοι δὲ τῶν ἐμπύων καίονται ἢ τέμνονται, οἷσιν ἂν καθαρὸν μὲν τὸ πῦον ᾖ καὶ λευκὸν καὶ μὴ δυσῶδες, σῴζονται· οἷσι δὲ ὕφαιμόν τε καὶ βορβορῶδες ἀπόλλυνται. Neither the scholiast nor Galen comments upon the words,

monia, older patients are the more likely to die; with other kinds of empyema younger people more easily succumb.

XIX. Pains occurring with fever in the region of the loins and lower parts, if they leave the lower parts and attack the diaphragm, are very mortal. So pay attention to the other symptoms also, since, if another bad symptom supervene, the case is hopeless; but if, when the disorder jumps to[1] the diaphragm, the other symptoms that supervene are not bad, confidently expect that empyema will occur in this case.

Hardness and pain in the bladder are always serious, and whenever attended with continuous fever, very fatal. In fact, the pains from the bladder alone are enough to cause death, and in such cases the bowels are not moved, except with hard and forced[2] stools. The disease is resolved by the passing of purulent urine, with a white, smooth sediment. If, however, neither the urine becomes favourable nor the bladder be softened, while the fever is continuous, expect the patient to die in the first periods of the illness. This form attacks especially children between the ages of seven and fifteen years.

[1] *ὡς πρὸς τὰς φρένας* would suggest that the determination of the pain to the diaphragm was only apparent—which is contrary to the first sentence of the chapter.

[2] Either through constipation, or by the use of purgatives.

---

and they are omitted in the Paris MS. 2269. They are deleted by Ermerins, Reinhold and Kühlewein. See also Littré's long note on the passage.

[2] After *νοσήματος* the MSS. have *ὡς*, which I delete as a repetition of the last syllable of *νοσήματος*.

XX. Οἱ δὲ πυρετοὶ κρίνονται ἐν τῇσιν αὐτῇσιν ἡμέρῃσι τὸν ἀριθμόν, ἐξ ὧν τε περιγίνονται οἱ ἄνθρωποι καὶ ἐξ ὧν ἀπόλλυνται. οἵ τε γὰρ εὐηθέστατοι τῶν πυρετῶν καὶ ἐπὶ σημείων ἀσφαλεστάτων βεβῶτες τεταρταῖοι παύονται ἢ πρόσθεν. οἱ δὲ κακοηθέστατοι τῶν πυρετῶν[1] καὶ ἐπὶ σημείων δεινοτάτων γινόμενοι τεταρταῖοι κτείνουσιν ἢ πρόσθεν. ἡ μὲν οὖν πρώτη ἔφοδος αὐτῶν οὕτω τελευτᾷ· ἡ δὲ δευτέρη ἐς τὴν ἑβδόμην 10 περιάγει, ἡ δὲ τρίτη ἐς τὴν ἑνδεκάτην, ἡ δὲ τετάρτη ἐς τὴν τεσσαρεσκαιδεκάτην, ἡ δὲ πέμπτη ἐς τὴν ἑπτακαιδεκάτην, ἡ δὲ ἕκτη ἐς τὴν εἰκοστήν. αὗται μὲν ἐπὶ τῶν ὀξυτάτων νοσημάτων διὰ τεσσάρων ἐς τὰς εἴκοσιν ἐκ προσθέσιος τελευτῶσιν. οὐ δύναται δὲ ὅλῃσιν ἡμέρῃσιν ἀριθμεῖσθαι οὐδὲν τούτων ἀτρεκέως· οὐδὲ γὰρ ὁ ἐνιαυτός τε καὶ οἱ μῆνες ὅλῃσιν ἡμέρῃσιν πεφύκασιν ἀριθμεῖσθαι.

Μετὰ δὲ ταῦτα ἐν τῷ αὐτῷ τρόπῳ κατὰ τὴν 20 αὐτὴν πρόσθεσιν ἡ μὲν πρώτη περίοδος τεσσάρων καὶ τριήκοντα ἡμερέων, ἡ δὲ δευτέρη τεσσαράκοντα ἡμερέων, ἡ δὲ τρίτη ἑξήκοντα ἡμερέων. τούτων δὲ ἐν ἀρχῇσίν ἐστι χαλεπώτατα προγινώσκειν τὰ μέλλοντα ἐν πλείονι χρόνῳ κρίνεσθαι· ὁμοιόταται γὰρ αἱ ἀρχαὶ αὐτῶν εἰσιν· ἀλλὰ χρὴ ἀπὸ τῆς πρώτης ἡμέρης ἐνθυμεῖσθαι καὶ καθ' ἑκάστην τετράδα προστιθεμένην σκέπτεσθαι καὶ οὐ λήσει, ὅπῃ τρέψεται. γίνεται δὲ καὶ τῶν τεταρταίων ἡ κατάστασις ἐκ τούτου τοῦ κόσμου. 30 τὰ δὲ ἐν ἐλαχίστῳ χρόνῳ μέλλοντα κρίνεσθαι

[1] τῶν πυρετῶν, C'.

XX. Fevers come to a crisis on the same days, both those from which patients recover and those from which they die. The mildest fevers, with the most favourable symptoms, cease on the fourth day or earlier. The most malignant fevers, with the most dangerous symptoms, end fatally on the fourth day or earlier. The first assault of fevers ends at this time; the second lasts until the seventh day, the third until the eleventh, the fourth until the fourteenth, the fifth until the seventeenth, and the sixth until the twentieth day. So in the most acute diseases keep on adding periods of four[1] days, up to twenty, to find the time when the attacks end. None of them, however, can be exactly calculated in whole days; neither can whole days be used to measure the solar year and the lunar month.

Afterwards, in the same manner and by the same increment, the first period is one of thirty-four days, the second of forty days and the third of sixty days.[2] At the commencement of these it is very difficult to forecast those which will come to a crisis after a protracted interval, for at the beginning they are very much alike. From the first day, however, you must pay attention, and consider the question at the end of every four days, and then the issue will not escape you. The constitution[3] of quartans too

[1] In the modern way of counting, three.

[2] The series apparently are these:—

1, 4, 7, 11, 14, 17, 20
[24, 27, 31,] 34
[37] 40
[44, 47, 51, 54, 57,] 60.

The whole question, however, is involved in uncertainty, as critical days are not discussed elsewhere, except incidentally in *Epidemics*. See Vol. I., General Introduction, p. liv.

[3] Κατάστασις is here practically equivalent to φύσις. See Vol. I. p. 141 (note).

εὐπετέστερα γινώσκεσθαι· μέγιστα γὰρ τὰ διαφέροντα αὐτῶν ἐστιν ἀπ' ἀρχῆς· οἱ μὲν γὰρ περιεσόμενοι εὔπνοοί τε καὶ ἀνώδυνοί εἰσιν καὶ κοιμῶνται τὰς νύκτας τά τε ἄλλα σημεῖα ἔχουσιν ἀσφαλέστατα· οἱ δὲ ἀπολλύμενοι[1] δύσπνοοι γίνονται, ἀγρυπνέοντες, ἀλλοφάσσοντες τά τε ἄλλα σημεῖα ἔχοντες κάκιστα. ὡς οὖν τούτων προγινωσκομένων συμβάλλεσθαι χρὴ κατά τε τὸν χρόνον καὶ κατὰ τὴν πρόσθεσιν ἑκάστην ἐπὶ τὴν κρίσιν
40 ἰόντων τῶν νοσημάτων. κατὰ δὲ τὸν αὐτὸν τρόπον καὶ τῇσι γυναιξὶν αἱ κρίσιες ἐκ τῶν τόκων
42 γίνονται.

XXI. Κεφαλῆς δὲ ὀδύναι ἰσχυραί τε καὶ συνεχέες σὺν πυρετῷ, εἰ μέν τι τῶν θανατωδέων σημείων προσγίνοιτο, ὀλέθριον κάρτα· εἰ δὲ ἄτερ σημείων τοιούτων ἡ ὀδύνη ὑπερβάλλοι εἴκοσιν ἡμέρας ὅ τε πυρετὸς ἔχοι, ὑποσκέπτεσθαι χρὴ αἵματος ῥῆξιν διὰ ῥινῶν ἢ ἄλλην ἀπόστασιν ἐς τὰ κάτω χωρία. ἔστ' ἂν δὲ ἡ ὀδύνη ᾖ νεαρά, προσδέχεσθαι χρὴ αἵματος ῥῆξιν διὰ ῥινῶν[2] ἢ ἐκπύησιν, ἄλλως τε καὶ ἢν ἡ ὀδύνη περὶ τοὺς
10 κροτάφους ᾖ καὶ τὸ μέτωπον. μᾶλλον δὲ χρὴ τοῦ μὲν αἵματος τὴν ῥῆξιν προσδέχεσθαι τοῖσι νεωτέροισι πέντε καὶ τριήκοντα ἐτέων, τοῖσι δὲ
13 γεραιτέροισι τὴν ἐκπύησιν.

XXII. Ὠτὸς δὲ ὀδύνη ὀξεῖα σὺν πυρετῷ συνεχεῖ τε καὶ ἰσχυρῷ δεινόν· παραφρονῆσαι γὰρ κίνδυνος τὸν ἄνθρωπον καὶ ἀπολέσθαι. ὡς οὖν τούτου τοῦ τρόπου σφαλεροῦ ἐόντος ὀξέως[3] δεῖ προσέχειν τὸν νόον καὶ τοῖσιν ἄλλοισι

[1] ἀπολλύμενοι MC′ : ἀπολούμενοι Littré and Kühlewein from

is of this order. Those that will reach a crisis after the shortest interval are easier to determine, for their differences are very great from the commencement. Those who will recover breathe easily, are free from pain, sleep during the night, and show generally the most favourable symptoms; those who will die have difficulty in breathing, are sleepless and delirious, and show generally the worst symptoms. Learning these things beforehand you must make your conjectures at the end of each increment as the illness advances to the crisis. In the case of women too after delivery, the crises occur according to the same rules.

XXI. Violent and continuous headaches, should there be in addition one of the deadly signs, is a very fatal symptom. But if without such signs the pain continue more than twenty days and the fever last, hemorrhage through the nose is to be expected, or some abscession to the lower parts. And while the pain is recent, one must look for hemorrhage through the nose, or a suppuration, especially if the pain be in the temples and forehead; hemorrhage is rather to be expected in patients under thirty-five years, suppuration in older patients.

XXII. Acute pain of the ear with continuous high fever is dangerous, for the patient is likely to become delirious and die. Since then this type of illness is treacherous, the doctor must pay sharp attention to all the other symptoms also from the

---

Galen's commentary. Kühlewein would spell it ἀπολεύμενοι. I take ἀπολλύμενοι to be a present with future sense.

[2] From ἢ to ῥινῶν is omitted by C′, the eye of the scribe passing from the first διὰ ῥινῶν to the second.

[3] ὀξέως C′, ταχέως MV (apparently a gloss on ὀξέως).

σημείοισι ἅπασιν ἀπὸ τῆς πρώτης ἡμέρης. ἀπόλλυνται δὲ οἱ μὲν νεώτεροι τῶν ἀνθρώπων ἑβδομαῖοι καὶ ἔτι θᾶσσον ὑπὸ τούτου τοῦ νοσήματος, οἱ δὲ γέροντες πολλῷ βραδύτερον· οἱ
10 γὰρ πυρετοὶ καὶ αἱ παραφροσύναι ἧσσον αὐτοῖσιν ἐπιγίνονται, καὶ τὰ ὦτα αὐτοῖσι διὰ τοῦτο φθάνει ἐκπυεύμενα· ἀλλὰ ταύτῃσι μὲν τῇσιν ἡλικίῃσιν ὑποστροφαὶ τοῦ νοσήματος ἐπιγενόμεναι ἀποκτείνουσιν· τοὺς πλείστους· οἱ δὲ νεώτεροι, πρὶν ἐκπυῆσαι τὸ οὖς, ἀπόλλυνται. ἐπὴν δὲ ῥυῇ πῦον λευκὸν ἐκ τοῦ ὠτός, ἐλπὶς περιγενέσθαι τῷ νέῳ, ἤν
17 τι καὶ ἄλλο χρηστὸν αὐτῷ ἐπιγένηται σημεῖον.

XXIII. Φάρυγξ δὲ ἑλκουμένη σὺν πυρετῷ δεινόν· ἀλλ᾽ ἤν τι καὶ ἄλλο σημεῖον ἐπιγένηται τῶν προκεκριμένων πονηρῶν εἶναι, προλέγειν ὡς ἐν κινδύνῳ ἐόντος τοῦ ἀνθρώπου. αἱ δὲ κυνάγχαι δεινόταται μέν εἰσι καὶ τάχιστα ἀναιρέουσιν, ὁκόσαι μήτε ἐν τῇ φάρυγγι μηδὲν ἔκδηλον ποιέουσι μήτε ἐν τῷ αὐχένι, πλεῖστον δὲ πόνον παρέχουσι καὶ ὀρθόπνοιαν· αὗται γὰρ καὶ αὐθημερὸν ἀποπνίγουσι καὶ δευτεραῖαι καὶ τριταῖαι καὶ
10 τεταρταῖαι. ὁκόσαι δὲ τὰ μὲν ἄλλα παραπλησίως ἔχουσι πόνον τε παρέχουσιν, ἐπαίρονται δὲ καὶ ἐρύθημα ἐν τῇ φάρυγγι ἐμποιέουσιν, αὗται ὀλέθριαι μὲν κάρτα, χρονιώτεραι δὲ μᾶλλον τῶν πρόσθεν.[1] ὁκόσοισι δὲ συνεξερεύθει ἡ φάρυγξ καὶ ὁ αὐχήν, αὗται μὲν χρονιώτεραι, καὶ μάλιστα ἐξ αὐτέων περιγίνονται, ἤν ὅ τε αὐχὴν καὶ τὸ στῆθος ἐρύθημα ἴσχωσιν καὶ μὴ παλινδρομῇ τὸ ἐρυσίπελας ἔσω. ἢν δὲ μήτε ἐν ἡμέρῃσι κρισίμῃσι

[1] After πρόσθεν M adds ἢν τὸ ἐρύθημα μέγα γίγνεται.

very first day. Younger patients die from this disease on the seventh day or even earlier; old men die much later, for the fever and the delirium attack them less, and for this reason their ears quickly suppurate. At this time of life, however, relapses occur and prove fatal to most, while younger men die before the ear suppurates. When white pus flows from the ear, you may hope that a young man may recover, if besides he show some other favourable symptom.

XXIII. An ulcerated throat with fever is serious; but if some other sympton also supervene that has been already classed as bad, forecast that the patient is in danger. Angina is very serious and rapidly fatal, when no lesion is to be seen in either throat or neck, and, moreover, it causes very great pain and orthopnoea;[1] it may suffocate the patient even on the first day, or on the second, third or fourth. Such cases as show swelling and redness in the throat, while they are generally similar, and cause pain, are very deadly, though they tend to be more protracted than the former. When throat and neck are both red, the illness is more protracted, and recovery is most likely should neck and chest be red and the erysipelas[2] does not turn back[3] inwards. Should, however, the erysipelas disappear neither on the critical days nor with the formation

[1] Difficulty of respiration, when the patient can breathe only in an upright condition.

[2] See Vol. I., General Introduction, p. lviii.

[3] The word so translated is used to describe the action of peccant humours when, instead of "working off" in an abscess or eruption, etc., they return into the system and cause a relapse or another form of illness.

τὸ ἐρυσίπελας ἀφανίζηται μήτε φύματος συστρα-
20 φέντος ἐν τῷ ἔξω χωρίῳ, μήτε πῦον ἀποβήσσῃ
ῥηϊδίως τε καὶ ἀπόνως,[1] θάνατον σημαίνει ἢ
ὑποστροφὴν τοῦ ἐρυθήματος. ἀσφαλέστατον δὲ
τὸ ἐρύθημα ὡς μάλιστα ἔξω τρέπεσθαι· ἢν δὲ
ἐς τὸν πνεύμονα τρέπηται, παράνοιάν τε ποιεῖ
καὶ ἔμπυοι ἐξ αὐτῶν[2] γίνονται ὡς τὰ πολλά.

Οἱ δὲ γαργαρεῶνες ἐπικίνδυνοι καὶ ἀποτά-
μνεσθαι καὶ ἀποσχάζεσθαι, ἔστ᾿ ἂν ἐρυθροί τ᾿
ἔωσι καὶ μεγάλοι· καὶ γὰρ φλεγμοναὶ ἐπιγίνονται
τούτοισι καὶ αἱμορραγίαι· ἀλλὰ χρὴ τὰ τοιαῦτα
30 τοῖσιν ἄλλοισι μηχανήμασι πειρῆσθαι κατισχναί-
νειν ἐν τούτῳ τῷ χρόνῳ. ὁκόταν δὲ ἀποκριθῇ
ἤδη, ὃ δὴ σταφυλὴν καλέουσι, καὶ γένηται τὸ
μὲν ἄκρον τοῦ γαργαρεῶνος μέζον καὶ πελιδνόν,
τὸ δὲ ἀνωτέρω λεπτότερον, ἐν τούτῳ τῷ καιρῷ
ἀσφαλὲς διαχειρίζειν. ἄμεινον δὲ καὶ ὑπο-
κενώσαντα τὴν κοιλίην τῇ χειρουργίῃ χρῆσθαι,
ἢν ὅ τε χρόνος συγχωρῇ καὶ μὴ ἀποπνίγηται ὁ
38 ἄνθρωπος.[3]

XXIV. Ὁκόσοισι δ᾿ ἂν οἱ πυρετοὶ παύωνται
μήτε σημείων γενομένων λυτηρίων μήτε ἐν ἡμέρῃσι
κρισίμῃσιν, ὑποστροφὴν προσδέχεσθαι τούτοισιν.
ὅστις δ᾿ ἂν τῶν πυρετῶν μηκύνῃ περιεστικῶς
διακειμένου τοῦ ἀνθρώπου, μήτε ὀδύνης ἐχούσης
διὰ φλεγμονήν τινα μήτε διὰ πρόφασιν ἄλλην
μηδεμίαν ἐμφανέα, τούτῳ προσδέχεσθαι ἀπόστασιν μετ᾿ οἰδήματός τε καὶ ὀδύνης ἔς τι τῶν
ἄρθρων καὶ οὐχ ἧσσον τῶν κάτω. μᾶλλον δὲ
10 γίνονται καὶ ἐν ἐλάσσονι χρόνῳ αἱ τοιαῦται

[1] For τε καὶ ἀπόνως C′ reads ὅ τε ἄνθρωπος ἀπόνως ἔχειν δοκέει.

of an abscess on the exterior, and if the patient should not cough up pus easily and without pain, it is a sign of death or of a relapse of the redness. The most hopeful sign is for the redness to be determined as much as possible outwards; but if it be determined to the lungs it produces delirium, and such cases usually result in empyema.

It is dangerous to cut away or lance the uvula while it is red and enlarged, for inflammation and hemorrhage supervene after such treatment; but at this time try to reduce such swellings by the other means. When, however, the gathering is now complete, forming what is called "the grape," that is, when the point of the uvula is enlarged and livid, while the upper part is thinner, it is then safe to operate. It is better, too, to move the bowels gently before the operation, if time permit and the patient be not suffocating.[1]

XXIV. In all cases where the fevers cease neither with signs of recovery nor on critical days a relapse may be expected. If a fever be protracted, although the patient is in a state indicating recovery, and pain do not persist through inflammation or any other obvious cause, you may expect an abscession, with swelling and pain, to one of the joints, especially to the lower ones. Such abscessions come more often, and earlier, when patients are under thirty. You must suspect

[1] See note 3 below.

[2] ἐξ αὐτῶν is bracketed by Kühlewein.

[3] The whole of this section is bracketed by Kühlewein and deleted by Ermerins. The reason for so doing is that it deals with treatment rather than prognosis.

ἀποστάσιες τοῖσι νεωτέροισι τριήκοντα ἐτέων. ὑποσκέπτεσθαι δὲ χρὴ εὐθέως τὰ περὶ τῆς ἀποστάσιος, ἢν εἴκοσιν ἡμέρας ὁ πυρετὸς ἔχων ὑπερβάλλῃ. τοῖσι δὲ γεραιτέροισιν ἧσσον γίνεται πολυχρονιωτέρου ἐόντος τοῦ πυρετοῦ. χρὴ δὲ τὴν μὲν τοιαύτην ἀπόστασιν προσδέχεσθαι συνεχέος ἐόντος τοῦ[1] πυρετοῦ, ἐς δὲ τεταρταῖον καταστήσεσθαι, ἢν διαλείπῃ τε καὶ καταλαμβάνῃ πεπλανημένον τρόπον καὶ ταῦτα ποιέων τῷ 20 φθινοπώρῳ πελάζῃ. ὥσπερ δὲ τοῖσι νεωτέροισι τριήκοντα ἐτέων αἱ ἀποστάσιες γίνονται, οὕτως οἱ τεταρταῖοι μᾶλλον τοῖσι τριηκονταέτεσι καὶ γεραιτέροισιν. τὰς δὲ ἀποστάσιας εἰδέναι χρὴ τοῦ χειμῶνος μᾶλλον γινομένας χρονιώτερόν τε παυομένας, ἧσσον δὲ παλινδρομεούσας.

Ὅστις δ' ἂν ἐν πυρετῷ μὴ θανατώδει φῇ κεφαλὴν ἀλγεῖν καὶ ὀρφνῶδές τι πρὸ τῶν ὀφθαλμῶν γίνεσθαι, ἢν[2] καὶ καρδιωγμὸς τούτῳ προσγένηται, χολώδης ἔμετος παρέσται· ἢν δὲ καὶ 30 ῥῖγος προσγένηται καὶ τὰ κάτω τοῦ ὑποχονδρίου ψυχρὰ ἔχῃ, καὶ θᾶσσον ἔτι ὁ ἔμετος παρέσται· ἢν δέ τι πίῃ ἢ φάγῃ ὑπὸ τοῦτον τὸν χρόνον, κάρτα ταχέως ἐμεῖται. τούτων δὲ οἷσιν ἂν ἄρξηται ὁ πόνος τῇ πρώτῃ ἡμέρῃ γίνεσθαι, τεταρταῖοι πιεζεῦνται μάλιστα καὶ πεμπταῖοι· ἐς δὲ τὴν ἑβδόμην ἀπαλλάσσονται· οἱ μέντοι πλεῖστοι αὐτῶν ἄρχονται μὲν πονεῖσθαι τριταῖοι, χειμάζονται δὲ μάλιστα πεμπταῖοι· ἀπαλλάσσονται δὲ ἐναταῖοι ἢ ἑνδεκαταῖοι· οἳ δ' ἂν 40 ἄρξωνται πεμπταῖοι πονεῖσθαι καὶ τὰ ἄλλα κατὰ

[1] V omits from the preceding τοῦ to this. The scribe passed over the intervening words.

at once the occurrence of an abscession if the fever last longer than twenty days; but in older patients it is less likely, even if the fever be more protracted. If the fever be continuous you must expect the abscession to be of this type, but the disease will resolve into a quartan if it intermit and attack in an irregular fashion, and if autumn approach while it acts in this way. Just as the abscessions occur when the patients are under thirty, so the quartans supervene more often when they are thirty or over. You must know that in winter the abscessions are more likely to occur and are longer in coming to an end, though there is less risk of a relapse.

If a patient in a fever that is not mortal says that his head aches, and that a darkness appears before his eyes, should he also feel heart-burn, a bilious vomiting will soon occur. If a rigor also supervene, and the parts below the hypochondrium be cold, the vomiting will occur sooner still; while if the patient eat or drink something at this time he will vomit very soon indeed. When in such cases the pain begins on the first day, the patients are most distressed on the fourth and fifth, recovering on the seventh. Most of them, however, begin to feel pain on the third day, are at their worst on the fifth, recovering on the ninth or eleventh. When they begin to feel pain on the fifth day, and the

[2] ἦν is my emendation. The MSS. have ἤ, but the scholiast, I find, has ἦν δὲ καί.

λόγον αὐτοῖσι τῶν πρόσθεν γίνηται, ἐς τὴν τεσσαρεσκαιδεκάτην κρίνεται ἡ νοῦσος. γίνεται δὲ ταῦτα τοῖσι μὲν ἀνδράσι καὶ τῇσι γυναιξὶν ἐν τοῖσι τριταίοισι μάλιστα· τοῖσι δὲ νεωτέροισι γίνεται μὲν καὶ ἐν τούτοισι, μᾶλλον δὲ ἐν τοῖσι συνεχεστέροισι πυρετοῖσι καὶ ἐν τοῖσι γνησίοισι τριταίοισιν.

Οἷσι δ' ἂν ἐν τοιουτοτρόπῳ πυρετῷ κεφαλὴν ἀλγέουσιν ἀντὶ μὲν τοῦ ὀρφνῶδές τι πρὸ τῶν
50 ὀφθαλμῶν φαίνεσθαι ἀμβλυωγμὸς γίνηται ἢ μαρμαρυγαὶ προφαίνωνται, ἀντὶ δὲ τοῦ καρδιώσσειν ἐν τῷ ὑποχονδρίῳ ἐπὶ δεξιὰ ἢ ἐπ' ἀριστερὰ συντείνηταί τι μήτε σὺν ὀδύνῃ μήτε σὺν φλεγμονῇ, αἷμα διὰ ῥινῶν τούτοισι ῥαγῆναι προσδόκιμον ἀντὶ τοῦ ἐμέτου. μᾶλλον δὲ καὶ ἐνταῦθα τοῖσι νέοισι τοῦ αἵματος τὴν ῥῆξιν προσδέχεσθαι· τοῖσι δὲ τριηκονταέτεσι καὶ γεραιτέροισιν ἧσσον, ἀλλὰ τοὺς ἐμέτους τούτοισι προσδέχεσθαι.

60 Τοῖσι δὲ παιδίοισι σπασμοὶ γίνονται, ἢν ὅ τε πυρετὸς ὀξὺς ᾖ καὶ ἡ γαστὴρ μὴ διαχωρῇ καὶ ἀγρυπνέωσί τε καὶ ἐκπλαγέωσι καὶ κλαυθμυρίζωσι καὶ τὸ χρῶμα μεταβάλλωσι καὶ χλωρὸν ἢ πελιδνὸν ἢ ἐρυθρὸν ἴσχωσιν. γίνεται δὲ ταῦτα ἐξ ἑτοιμοτάτου μὲν τοῖσι παιδίοισι τοῖσι νεωτάτοισι ἐς τὰ ἑπτὰ ἔτεα· τὰ δὲ πρεσβύτερα τῶν παιδίων καὶ οἱ ἄνδρες οὐκ ἔτι ἐν τοῖσι πυρετοῖσιν ὑπὸ τῶν σπασμῶν ἁλίσκονται, ἢν μή τι τῶν σημείων προσγένηται τῶν ἰσχυροτάτων τε
70 καὶ κακίστων, οἷά περ ἐπὶ τῇσι φρενίτισι γίνεται. τοὺς δὲ περιεσομένους τε καὶ ἀπολλυμένους [1] τῶν παιδίων τε καὶ τῶν ἄλλων τεκμαίρεσθαι τοῖσι

symptoms proceed after the manner I have described, the disease reaches a crisis on the fourteenth day. Men and women experience these symptoms mostly in tertian fevers; younger people too experience them in tertians, but more often in the more continuous fevers and in genuine[1] tertians.

All those who with headache in a fever of this character experience not a darkness before the eyes but a dimness of vision, or see flashes of light, while instead of heart-burn there is a tension of the right or left hypochondrium without pain or inflammation, these you may expect will not vomit but bleed from the nose. In this case too expect the hemorrhage more especially in young people. It occurs less frequently if the patient be of thirty years or more; in these cases expect the vomiting.

Children suffer from convulsions if the fever be acute and the alvine discharges cease; if they cannot sleep but are terrified and moan; if they change their colour and become yellow, livid or red. Convulsions are most likely to attack very young children before they are seven years old; older children and adults are not attacked by convulsions in fevers unless some of the worst and most violent symptoms supervene, as happens in cases of phrenitis. Whether children and whether adults will survive or die you must infer from a combination of all the symptoms,

[1] *I. e.* tertians that *intermit*, the fever ceasing entirely every other day. Many tertians *remit* only, the fever growing less instead of ceasing altogether.

---

[1] ἀπολλυμένους C′M: ἀπολουμένους many MSS. I take ἀπολλυμένους to be a present with future sense.

σύμπασι σημείοισιν, ὡς ἐφ᾽ ἑκάστοις ἕκαστα διαγέγραπται. ταῦτα δὲ λέγω περὶ τῶν ὀξέων 75 *νοσημάτων καὶ ὅσα ἐκ τούτων γίνεται.*

XXV. *Χρὴ δὲ τὸν μέλλοντα ὀρθῶς προγινώσκειν τούς τε περιεσομένους καὶ τοὺς ἀποθανευμένους οἷσί τε ἂν μέλλῃ τὸ νόσημα πλείονας ἡμέρας παραμένειν καὶ οἷσιν ἂν ἐλάσσους, τὰ σημεῖα ἐκμανθάνοντα πάντα δύνασθαι κρίνειν ἐκλογιζόμενον τὰς δυνάμιας αὐτῶν πρὸς ἀλλήλας, ὥσπερ διαγέγραπται περί τε τῶν ἄλλων καὶ τῶν οὔρων καὶ τῶν πτυέλων.*[1] *χρὴ δὲ καὶ τὰς φορὰς τῶν νοσημάτων τῶν αἰεὶ ἐπιδημεόντων* 10 *ταχέως ἐνθυμεῖσθαι καὶ μὴ λανθάνειν τὴν τῆς ὥρης κατάστασιν. εὖ μέντοι χρὴ εἰδέναι περὶ τῶν τεκμηρίων καὶ τῶν ἄλλων σημείων,*[2] *ὅτι ἐν παντὶ ἔτει καὶ πάσῃ χώρῃ*[3] *τά τε κακὰ κακόν τι σημαίνει καὶ τὰ χρηστὰ ἀγαθόν, ἐπεὶ καὶ ἐν Λιβύῃ καὶ ἐν Δήλῳ καὶ ἐν Σκυθίῃ φαίνεται τὰ προγεγραμμένα σημεῖα ἀληθεύοντα. εὖ οὖν χρὴ εἰδέναι, ὅτι ἐν τοῖς αὐτοῖσι χωρίοισιν οὐδὲν δεινὸν τὸ μὴ οὐχὶ τὰ πολλαπλάσια ἐπιτυγχάνειν, ἢν ἐκμαθών τις αὐτὰ κρίνειν τε καὶ ἐκλογίζεσθαι* 20 *ὀρθῶς ἐπίστηται. ποθεῖν δὲ χρὴ οὐδενὸς νοσήματος ὄνομα, ὅ τι μὴ τυγχάνει ἐνθάδε γεγραμμένον· πάντα γάρ, ὁκόσα ἐν τοῖσι χρόνοισι τοῖσι προειρημένοισι κρίνεται, γνώσῃ τοῖσιν* 24 *αὐτοῖσι σημείοισιν.*

[1] After *πτυέλων* the MSS. have *ὅταν ὁμοῦ πῦόν τε ἀναβήσσῃ καὶ χολήν.* The clause is deleted by Gomperz and Wilamowitz.

[2] After *σημείων* C′M add *καὶ μὴ λανθάνειν.* So apparently Galen.

as I have severally described them in the several kinds of cases. My remarks apply to acute diseases and to all their consequences.

XXV. He who would make accurate forecasts as to those who will recover, and those who will die, and whether the disease will last a greater or less number of days, must understand all the symptoms thoroughly and be able to appreciate them, estimating their powers when they are compared with one another, as I have set forth above, particularly in the case of urine and sputa. It is also necessary promptly to recognize the assaults of the endemic diseases, and not to pass over the constitution of the season. However, one must clearly realize about sure signs and about symptoms generally, that in every year and in every land bad signs indicate something bad, and good signs something favourable, since the symptoms described above prove to have the same significance in Libya, in Delos, and in Scythia. So one must clearly realize that in the same districts it is not strange that one should be right in the vast majority of instances, if one learns them well and knows how to estimate and appreciate them properly. Do not regret the omission from my account of the name of any disease.[1] For it is by the same symptoms in all cases that you will know the diseases that come to a crisis at the times I have stated.

[1] Contrast with this the criticism of the Cnidian physicians in Chapter III of *Regimen in Acute Diseases*, and notice once more the insistence on "general" pathology as contrasted with diagnosis.

[3] χώρῃ C′: ὥρῃ other MSS. and Kühlewein. I adopt this reading (which, as Littré says, is not supported by Galen) because of the ἐπεί-clause which follows.

REGIMEN IN ACUTE DISEASES

INTRODUCTION

The authorship of this work has never been doubted. It is indisputably one of the great Hippocratic group of treatises, being a kind of supplement to *Prognostic*. It has also close affinities with *Ancient Medicine*, the author of which held medicine to be merely a branch of regimen.

In ancient times, besides its usual title, the book was sometimes called *On the Ptisan*, or *Against the Cnidian Sentences*, the former from the chief article of sick food, the latter from the polemic with which the work opens.

The "acute" diseases are those characterized by high fever; they are enumerated in Chapter V.[1] The treatment recommended is supposed in general to apply to any acute disease; the writer is true to the Hippocratic doctrine of "general" pathology. Chest complaints, however, seem to be more in the writer's mind than the other main class of acute diseases.

The Hippocratic treatment is gentle and mild. Little use is made of drugs; those employed are

[1] Pleurisy, pneumonia, phrenitis, causus, and diseases with continuous fever allied to these; *i.e.* chest complaints and remittent malaria. The list is strong proof that the Greeks were ignorant of the zymotic diseases. Unless we bear in mind this peculiarity of Greek endemiology, we can understand neither their medical theory nor their medical practice.

purges and simple herbals. Fomentations and baths are features of Hippocratic regimen, and, did occasion call for them, the enema, suppositories, and venesection were employed. A sparing use was made of water, the drinks recommended being hydromel (honey and water), oxymel (honey and vinegar) and wine. But the great stand-by of the physician in acute diseases was the decoction of barley, "ptisan," which I have translated by "gruel" for the sake of convenience. Great care was bestowed upon its preparation, and the most minute directions were given for its use. Sometimes the pure juice was employed, sometimes more or less of the solid barley was added. Apparently no other nourishment was given, except the things already mentioned, until well after the crisis.[1]

The unpretentious and cautious character of this regimen is in perfect harmony with the modest nature of Greek, particularly of Coan, medicine; no rash promises are made, and no rash experiments attempted.[2]

Galen says that the question of regimen is treated in a confused manner, and his criticism is borne out by a few chapters, which are rather difficult to follow. On the whole, however, the directions for treatment are clearly expressed.

Manuscripts and Editions

The chief manuscripts are A, M and V. The last two generally agree as against A. Of the two classes preference should be given to A, which generally gives the better reading, although its

[1] See Chapter XIII. [2] See p. xxxviii.

excellence is perhaps not so marked as it is in the case of *Ancient Medicine*. R′ and S′ also are occasionally useful. Holkhamensis 282 contains the treatise, but is practically the same as V.

There were many editions during the sixteenth century, the first separate one being apparently that of Haller.[1] In the seventeenth century the chief editions were those of Mercuriali (1602) and Heurnius (1609).

There is a commentary by Galen.

The only English translation, so far as I know, is that of Francis Adams. I have, however, in my possession a MS. English translation, in a late seventeenth-century hand, which is distinctly better than the type of translation fashionable at this period. In a few places it has helped me to make my own translation. The author was a careful scholar, and, to judge from his medical notes, a practitioner. I refer to the translation as "Z."

I have found it hard to translate χυλός. "Barley water" is the natural rendering, but it is not always available. I hope that the word "juice," which I have often employed, will not be thought too strange.

[1] *Liber de Diaeta Acutorum Graece.* Paris, 1530.

ΠΕΡΙ ΔΙΑΙΤΗΣ ΟΞΕΩΝ

I. Οἱ συγγράψαντες τὰς Κνιδίας καλεομένας γνώμας ὁποῖα μὲν πάσχουσιν οἱ κάμνοντες ἐν ἑκάστοισι τῶν νοσημάτων ὀρθῶς ἔγραψαν καὶ ὁποίως ἔνια ἀπέβαινεν· καὶ ἄχρι μὲν τούτων, καὶ ὁ μὴ ἰητρὸς δύναιτ᾽ ἂν ὀρθῶς[1] συγγράψαι, εἰ εὖ παρὰ τῶν καμνόντων ἑκάστου πύθοιτο, ὁποῖα πάσχουσιν· ὁπόσα δὲ προσκαταμαθεῖν δεῖ τὸν ἰητρὸν μὴ λέγοντος τοῦ κάμνοντος, τούτων πολλὰ παρεῖται, ἄλλ᾽ ἐν ἄλλοισιν καὶ ἐπίκαιρα
10 ἔνια ἐόντα ἐς τέκμαρσιν.

II. Ὁπόταν δὲ ἐς τέκμαρσιν λέγηται, ὡς χρὴ ἕκαστα ἰητρεύειν, ἐν τούτοισι πολλὰ ἑτεροίως γινώσκω ἢ ὡς κεῖνοι ἐπεξῄεσαν· καὶ οὐ μοῦνον διὰ τοῦτο οὐκ ἐπαινέω, ἀλλ᾽ ὅτι καὶ ὀλίγοισι τὸν ἀριθμὸν τοῖσιν ἄκεσιν ἐχρέοντο· τὰ γὰρ πλεῖστα αὐτοῖσιν εἴρηται, πλὴν τῶν ὀξειῶν νούσων, φάρμακα ἐλατήρια διδόναι καὶ ὀρὸν καὶ γάλα τὴν
8 ὥρην πιπίσκειν.

III. Εἰ μὲν οὖν ταῦτα ἀγαθὰ ἦν καὶ ἁρμόζοντα τοῖσι νοσήμασιν, ἐφ᾽ οἷσι παρῄνεον διδόναι, πολὺ

[1] A has καὶ ἣν μὴ ἰητρὸς δύναιτ᾽ ἂν ὀρθῶς. The other MSS. omit ἥν. R′ has δύναιτό τις ἂν (with Galen). Kühlewein reads καὶ ἣν μὴ ἰητρός, δύναιτό τις ἂν ὀρθῶς. The reading in the text is that of Wilamowitz.

REGIMEN IN ACUTE DISEASES

I. THE authors of the work entitled *Cnidian Sentences* have correctly described the experiences of patients in individual diseases and the issues of some of them. So much even a layman could correctly describe by carefully inquiring from each patient the nature of his experiences. But much of what the physician should know besides, without the patient's telling him, they have omitted; this knowledge varies in varying circumstances, and in some cases is important for the interpretation of symptoms.

II. And whenever they interpret symptoms with a view to determining the right method of treatment in each case,[1] my judgment in these matters is in many things different from their exposition. And not only on this account do I censure them, but because too the remedies they used were few in number; for most of their prescriptions, except in the case of acute diseases, were to administer purges, and to give to drink, at the proper season, whey and milk.

III. Now were these remedies good, and suited to the diseases for which the Cnidians recommended

[1] I take the ὡς-clause to be epexegetic of τέκμαρσιν.

ἂν ἀξιώτερα ἦν ἐπαίνου, ὅτι ὀλίγα ἐόντα αὐτάρκεά ἐστιν· νῦν δὲ οὐχ οὕτως ἔχει. οἱ μέντοι ὕστερον ἐπιδιασκευάσαντες ἰητρικώτερον δή τι ἐπῆλθον περὶ τῶν προσοιστέων ἑκάστοισιν. ἀτὰρ οὐδὲ περὶ διαίτης οἱ ἀρχαῖοι συνέγραψαν οὐδὲν ἄξιον λόγου· καί τοι μέγα τοῦτο παρῆκαν. τὰς μέντοι πολυτροπίας τὰς ἐν ἑκάστῃ τῶν νούσων καὶ τὴν
10 πολυσχιδίην οὐκ ἠγνόεον ἔνιοι· τοὺς δ' ἀριθμοὺς ἑκάστου τῶν νοσημάτων σάφα ἐθέλοντες φράζειν οὐκ ὀρθῶς ἔγραψαν· μὴ γὰρ οὐκ εὐαρίθμητον ᾖ, εἰ τούτῳ τις σημαίνεται τὴν τῶν καμνόντων νοῦσον, τῷ[1] τὸ ἕτερον τοῦ ἑτέρου διαφέρειν τι, μὴ τωὐτὸ δὲ νόσημα δοκεῖ εἶναι, ἢν μὴ τωὐτὸ ὄνομα
16 ἔχῃ.[2]

(2 L.) IV. Ἐμοὶ δὲ ἀνδάνει μὲν[3] πάσῃ τῇ τέχνῃ προσέχειν τὸν νόον· καὶ γὰρ ὁπόσα ἔργα καλῶς ἔχει ἢ ὀρθῶς, καλῶς ἕκαστα χρὴ ποιεῖν καὶ ὀρθῶς, καὶ ὁπόσα ταχέως, ταχέως, καὶ ὁπόσα καθαρίως, καθαρίως, καὶ ὁπόσα ἀνωδύνως, διαχειρίζεσθαι ὡς ἀνωδυνώτατα καὶ τἄλλα πάντα τοιουτότροπα διαφερόντως τῶν πέλας ἐπὶ τὸ
8 βέλτιον ποιεῖν χρή.

[1] τῷ is not in the MSS., but is added by Gomperz.
[2] Littré reads καὶ ἢν μὴ τωὐτὸ νόσημα δοκῇ εἶναι, μὴ τωὐτὸ ὄνομα ἔχειν.
[3] The MSS. here have ἐν, which is deleted by Gomperz.

[1] The οὐδὲ in this sentence modifies in all probability from περὶ διαίτης to λόγου, and the whole from ἀτὰρ to παρῆκαν is a parenthesis, referring incidentally to the ἀρχαῖοι as similar to the Cnidians in their neglect of regimen. Grammatically it is possible to take οὐδὲ closely with περὶ διαίτης, in which case οἱ ἀρχαῖοι would refer to the earlier Cnidian authors. The translation "Z" identifies the Cnidians and οἱ ἀρχαῖοι.

their use, they would be much more worthy of recommendation, in that though few they were sufficient. But as it is this is not the case. However, the later revisers have showed rather more scientific insight in their discussion of the remedies to be employed in each instance. But in fact regimen received no treatment worth mentioning from the ancient physicians, although this omission is a serious one.[1] Yet the many phases and subdivisions of each disease were not unknown to some; but though they wished clearly to set forth the number of each kind of illness their account was incorrect. For the number will be almost incalculable if a patient's disease be diagnosed as different whenever there is a difference in the symptoms, while a mere variety of name is supposed to constitute a variety of the illness.[2]

IV. The course I recommend is to pay attention to the whole of the medical art. Indeed all acts that are good or correct should be in all cases well or correctly performed; if they ought to be done quickly, they should be done quickly, if neatly, neatly, if painlessly, they should be managed with the minimum of pain; and all such acts ought to be performed excellently, in a manner better than that of one's own fellows.

This view is perhaps unlikely, but, if it be true, *οἱ ἀρχαῖοι* in Chapter V must also refer to the Cnidians, and to them must be attributed the names *πλευρῖτις*, *περιπνευμονία*, *φρενῖτις* and *καῦσος*. We do know that the Cnidians paid special attention to names of diseases.

[2] Littré's emendation would mean that the Cnidians refused to give a disease its usual name whenever a variation occurred in the symptoms. This only repeats the sense of the preceding clause, while H. means that giving a disease another name does not make it another disease.

V. Μάλιστα δ' ἂν ἐπαινέσαιμι ἰητρόν, ὅστις ἐν τοῖσιν ὀξέσι νοσήμασι, ἃ τοὺς πλείστους τῶν ἀνθρώπων κτείνει, ἐν τούτοισι διαφέρων τι τῶν ἄλλων εἴη ἐπὶ τὸ βέλτιον. ἔστιν δὲ ταῦτα ὀξέα, ὁποῖα ὠνόμασαν οἱ ἀρχαῖοι πλευρῖτιν καὶ περιπνευμονίην καὶ φρενῖτιν[1] καὶ καῦσον, καὶ τἄλλα ὅσα τούτων ἐχόμενα, ὧν οἱ πυρετοὶ τὸ ἐπίπαν συνεχέες. ὅταν γὰρ μὴ λοιμώδεος νούσου τρόπος τις κοινὸς ἐπιδημήσῃ, ἀλλὰ σποράδες ἔωσιν αἱ
10 νοῦσοι, καὶ πολλαπλάσιοι[2] ὑπὸ τούτων τῶν νοσημάτων ἀποθνήσκουσι[3] ἢ ὑπὸ τῶν ἄλλων τῶν
12 συμπάντων.

VI. Οἱ μὲν οὖν ἰδιῶται οὐ κάρτα γινώσκουσιν τοὺς ἐς ταῦτα διαφέροντας τῶν πέλας ἑτεροίων τε μᾶλλον ἐπαινέται ἰημάτων καὶ ψέκται εἰσίν· ἐπεί τοι μέγα σημεῖον τόδε, ὅτι οἱ δημόται ἀσυνετώτατοι αὐτοὶ ἑωυτῶν περὶ τούτων τῶν νοσημάτων εἰσίν, ὡς μελετητέα ἐστί· οἱ γὰρ μὴ ἰητροὶ ἰητροὶ δοκέουσιν εἶναι μάλιστα διὰ ταύτας τὰς νούσους· ῥηΐδιον γὰρ τὰ ὀνόματα ἐκμαθεῖν, ὁποῖα νενόμισται προσφέρεσθαι πρὸς τοὺς τὰ τοιάδε κάμνοντας· ἢν
10 γὰρ ὀνομάσῃ τις πτισάνης τε χυλὸν καὶ οἶνον τοῖον ἢ τοῖον καὶ μελίκρητον, πάντα τοῖσι ἰδιώτῃσι δοκέουσιν οἱ ἰητροὶ τὰ αὐτὰ λέγειν, οἵ τε βελτίους καὶ οἱ χείρους. τὰ δὲ οὐχ οὕτως ἔχει, ἀλλ' ἐν τούτοισι καὶ πάνυ διαφέρουσιν ἕτεροι
15 ἑτέρων.

(3 L.) VII. Δοκεῖ δέ μοι ἄξια γραφῆς εἶναι,

[1] After φρενῖτιν M has καὶ λήθαργοι. The case and number seem to indicate a marginal note, and lethargus would certainly be included in ὅσα τούτων ἐχόμενα.

V. I should most commend a physician who in acute diseases, which kill the great majority of patients, shows some superiority. Now the acute diseases are those to which the ancients have given the names of pleurisy, pneumonia, phrenitis, and ardent fever,[1] and such as are akin to these, the fever of which is on the whole continuous. For whenever there is no general type of pestilence prevalent, but diseases are sporadic, acute diseases cause many times more deaths than all others put together.

VI. Now laymen do not accurately distinguish those who are excellent in this respect from their fellows, but rather praise or blame strange remedies. For in very truth there is strong evidence that it is in the proper treatment of these illnesses that ordinary folk show their most stupid side, in the fact that through these diseases chiefly quacks get the reputation of being physicians. For it is an easy matter to learn the names of the remedies usually given to patients in such diseases. If barley-water be mentioned, or such and such a wine, or hydromel,[2] laymen think that physicians, good and bad alike, prescribe all the same things. But it is not so, and there are great differences between physicians in these respects.

VII. And it seems to me worth while to write

[1] For φρενῖτις and καῦσος see General Introduction to Vol. I, pp. lvii, lviii.
[2] A mixture of honey and water.

---

[2] πολλαπλάσιοι Gomperz. V has παραπλήσιοι and M μὴ παραπλήσιοι; A omits (with καὶ).
[3] After ἀποθνῄσκουσι the MSS. have πλείους (AV) or μᾶλλον (M). Deleted by Wilamowitz.

ὁπόσα τε ἀκαταμάθητά ἐστιν τοῖς ἰητροῖς ἐπίκαιρα ἐόντα εἰδέναι[1] καὶ μεγάλας ὠφελείας φέρει ἢ μεγάλας βλάβας. ἀκαταμάθητα οὖν καὶ τάδ' ἐστίν, διὰ τί ἄρα ἐν τῇσιν ὀξείῃσι νούσοισιν οἱ μὲν τῶν ἰητρῶν πάντα τὸν αἰῶνα διατελέουσιν πτισάνας διδόντες ἀδιηθήτους καὶ νομίζουσιν ὀρθῶς ἰητρεύειν, οἱ δέ τινες περὶ παντὸς ποιέονται, ὅπως κριθὴν μηδεμίαν καταπίῃ ὁ κάμνων—μεγάλην 10 γὰρ βλάβην ἡγεῦνται εἶναι—ἀλλὰ δι' ὀθονίου τὸν χυλὸν διηθέοντες διδόασιν· οἱ δ' αὖ τινες αὐτῶν οὔτ' ἂν πτισάνην παχεῖαν δοῖεν οὔτε χυλόν· οἱ μὲν μέχρι ἂν ἑβδομαῖος γένηται, οἱ 14 δὲ καὶ διὰ τέλεος ἄχρι ἂν κριθῇ ἡ νοῦσος.

VIII. Μάλα μὲν οὖν οὐδὲ προβάλλεσθαι τὰ τοιαῦτα ζητήματα εἰθισμένοι εἰσὶν[2] οἱ ἰητροί. ἴσως δὲ οὐδὲ προβαλλόμενα γινώσκεται· καίτοι διαβολήν γε ἔχει ὅλη ἡ τέχνη πρὸς τῶν δημοτέων μεγάλην, ὡς μὴ δοκεῖν ὅλως ἰητρικὴν εἶναι· ὥστ' εἰ ἔν γε τοῖσιν ὀξυτάτοισι τῶν νοσημάτων τοσόνδε διοίσουσιν ἀλλήλων οἱ χειρώνακτες, ὥστε ἃ ὁ ἕτερος προσφέρει ἡγεύμενος ἄριστα εἶναι, ταῦτα νομίζειν τὸν ἕτερον κακὰ εἶναι, σχεδὸν ἂν κατά 10 γε τῶν τοιούτων τὴν τέχνην φαῖεν ὡμοιῶσθαι μαντικῇ, ὅτι καὶ οἱ μάντιες τὸν αὐτὸν ὄρνιθα, εἰ μὲν ἀριστερὸς εἴη, ἀγαθὸν νομίζουσιν εἶναι, εἰ δὲ δεξιός, κακόν—καὶ ἐν ἱεροσκοπίῃ δὲ τοιάδε, ἄλλα ἐπ' ἄλλοις—ἔνιοι δὲ τῶν μαντίων τὰ ἐναντία 15 τούτων.

[1] After εἰδέναι the MSS. have ὁκόσα τε or καὶ ὁπόσα. I have deleted ὁπόσα, on the ground that there are not two classes of points ἄξια γραφῆς, but only one, which contains things that are *both* ἐπίκαιρα εἰδέναι *and* μεγάλας βλάβας φέροντα ἢ μεγάλας ὠφελείας.

on such matters as are not yet ascertained by physicians, though knowledge thereof is important, and on them depend great benefit or great harm. For instance, it has not been ascertained why in acute diseases some physicians think that the correct treatment is to give unstrained barley-gruel throughout the illness; while others consider it to be of first-rate importance for the patient to swallow no particle of barley, holding that to do so is very harmful, but strain the juice through a cloth before they give it. Others again will give neither thick gruel nor yet juice, some not before the seventh day, others at no time until the disease reaches a crisis.

VIII. Now certainly[1] physicians are not at all in the habit of even raising such questions; even when they are raised perhaps nothing is learned. Yet the art as a whole has a very bad name among laymen, so that there is thought to be no art of medicine at all. Accordingly, since among practitioners there will prove to be so much difference of opinion about acute diseases that the remedies which one physician gives in the belief that they are the best are considered by a second to be bad, laymen are likely to object to such that their art resembles divination; for diviners too think that the same bird, which they hold to be a happy omen on the left, is an unlucky one when on the right, while other diviners maintain the opposite. The inspection of entrails shows similar anomalies in its various departments.

[1] *μάλα μὲν οὖν* is a strange phrase with which to begin a sentence. It occurs again at the beginning of Chapter XVIII.

---

[2] *εἰθισμένοι εἰσίν* MV: *εἴθισ(ται) τοῖς ἰητροῖς* A: *εἰθίδαται* Ilberg, followed by Kühlewein.

IX. Φημὶ δὲ πάγκαλον εἶναι τοῦτο τὸ σκέμμα καὶ ἠδελφισμένον τοῖσι πλείστοισι τῶν ἐν τῇ τέχνῃ καὶ ἐπικαιροτάτοισι· καὶ γὰρ τοῖσι νοσέουσι πᾶσιν ἐς ὑγιείην μέγα τι δύναται καὶ τοῖσιν ὑγιαίνουσιν ἐς ἀσφάλειαν καὶ τοῖσιν ἀσκέουσιν
6 ἐς εὐεξίην καὶ ἐς ὅ τι ἕκαστος ἐθέλει.

(4 L.) X. Πτισάνη μὲν οὖν δοκεῖ ὀρθῶς προκεκρίσθαι τῶν σιτηρῶν γευμάτων ἐν τούτοισι τοῖσι νοσήμασιν, καὶ ἐπαινέω τοὺς προκρίναντας. τὸ γὰρ γλίσχρασμα αὐτῆς λεῖον καὶ συνεχὲς καὶ προσηνές ἐστι καὶ ὀλισθηρὸν καὶ πλαδαρὸν μετρίως καὶ ἄδιψον καὶ εὐέκκριτον, εἴ τι καὶ τούτου προσδέοι, καὶ οὔτε στύψιν ἔχον οὔτε ἄραδον κακὸν οὔτε ἀνοιδίσκεται ἐν τῇ κοιλίῃ· ἀνῴδηκε γὰρ ἐν τῇ ἑψήσει, ὅσον πλεῖστον ἐπεφύκει
10 διογκοῦσθαι.

XI. Ὅσοι μὲν πτισάνῃ χρέονται ἐν τούτοισι τοῖσι νοσήμασι, οὐδεμιῇ ἡμέρῃ κενεαγγητέον, ὡς ἔπος εἰρῆσθαι, ἀλλὰ χρηστέον καὶ οὐ διαλειπτέον, ἢν μή τι δέῃ ἢ διὰ φαρμακείην ἢ κλύσιν διαλείπειν. καὶ τοῖσι μέν γε εἰθισμένοισι δὶς σιτεῖσθαι τῆς ἡμέρης δὶς δοτέον· τοῖσι δὲ μονοσιτεῖν εἰθισμένοις ἅπαξ δοτέον τὴν πρώτην· ἐκ προσαγωγῆς[1] δ' ἐνδέχεται καὶ τούτοισιν δὶς διδόναι, ἢν δοκέῃ προσδεῖν. πλῆθος δὲ ἀρκεῖ κατ' ἀρχὰς διδόναι
10 μὴ πολὺ μηδὲ ὑπέρπαχυ, ἀλλ' ὅσον εἵνεκα τοῦ ἔθεος ἐσιέναι τι καὶ κενεαγγίην μὴ γενέσθαι
12 πολλήν.

XII. Περὶ δὲ τῆς ἐπιδόσιος ἐς πλῆθος τοῦ ῥυφήματος, ἢν μὲν ξηρότερον ᾖ τὸ νόσημα ἢ ὡς ἄν τις βούληται, οὐ χρὴ ἐπὶ πλέον διδόναι, ἀλλὰ προπίνειν πρὸ τοῦ ῥυφήματος ἢ μελίκρητον ἢ οἶνον, ὁπότερον

IX. But I am confident that this inquiry is wholly profitable, being bound up with most, and the most important, of the things embraced by the art. In fact, it has great power to bring health in all cases of sickness, preservation of health to those who are well, good condition to athletes in training, and in fact realization of each man's particular desire.

X. Now I think that gruel made from barley has rightly been preferred over other cereal foods in acute diseases, and I commend those who preferred it; for the gluten of it is smooth, consistent, soothing, lubricant, moderately soft, thirst-quenching, easy of evacuation, should this property too be valuable, and it neither has astringency nor causes disturbance in the bowels or swells up in them. During the boiling, in fact, it has expanded to the utmost of its capacity.

XI. Those who use this gruel in acute diseases must not fast, generally speaking, on any day, but they must use it without intermission unless some intermission be called for because of a purge or enema. Those who are wont to eat two meals a day should take gruel twice; those wont to have one meal only should have gruel once on the first day. Gradually, if it be thought that they need it, these also may take a second dose. At first it is sufficient to administer a small quantity, not over-thick, just enough, in fact, to satisfy habit and to prevent severe pangs of hunger.

XII. As to increasing the quantity of the gruel, if the disease be drier than one would wish, you ought not to increase the dose, but to give to drink before the gruel either hydromel or wine, whichever

[1] After προσαγωγῆς the MSS. have ἦν. Deleted by Reinhold.

ἂν ἁρμόζῃ· τὸ δ' ἁρμόζον ἐφ' ἑκάστοισι τῶν τρόπων εἰρήσεται. ἢν δὲ ὑγραίνηται τὸ στόμα καὶ τὰ ἀπὸ τοῦ πνεύμονος ἴῃ ὁποῖα δεῖ, ἐπιδιδόναι χρὴ ἐς πλῆθος τοῦ ῥυφήματος, ὡς ἐν κεφαλαίῳ εἰρῆσθαι· τὰ μὲν γὰρ θᾶσσον καὶ μᾶλλον πλα-
10 δῶντα ταχυτῆτα σημαίνει κρίσιος, τὰ δὲ βραδύτερον πλαδῶντα καὶ ἧσσον βραδυτέρην σημαίνει τὴν κρίσιν. καὶ ταῦτα αὐτὰ μὲν καθ' ἑωυτὰ
13 τοιάδε τὸ ἐπίπαν ἐστί.

XIII. Πολλὰ δὲ καὶ ἄλλα ἐπίκαιρα παρεῖται, οἷσι προσημαίνεσθαι δεῖ, ἃ εἰρήσεται ὕστερον. καὶ ὅσῳ ἂν πλείων ἡ κάθαρσις γίνηται, τοσῷδε χρὴ πλεῖον διδόναι ἄχρι κρίσιος· μάλιστα δὲ κρίσιος ὑπερβολῆς δύο ἡμερέων, οἷσί γε ἢ πεμπταίοισιν ἢ ἑβδομαίοισιν ἢ ἐναταίοισιν δοκεῖ κρίνειν, ὡς καὶ τὸ ἄρτιον καὶ τὸ περισσὸν προμηθήσῃ·[1] μετὰ δὲ τῷ μὲν ῥυφήματι τὸ πρωὶ χρηστέον, ὀψὲ δὲ ἐς σιτία
9 μεταβάλλειν.

XIV. Συμφέρει δὲ τὰ τοιάδε ὡς ἐπὶ τὸ πολὺ τοῖς οὔλῃσι πτισάνῃσιν αὐτίκα χρεωμένοις. αἵ τε γὰρ ὀδύναι ἐν τοῖσι πλευριτικοῖσιν αὐτίκα παύονται αὐτόματοι, ὅταν ἄρξωνται πτύειν τι ἄξιον λόγου καὶ ἐκκαθαίρεσθαι, αἵ τε καθάρσιες πολλὸν τελεώτεραί εἰσι, καὶ ἔμπυοι ἧσσον γίνονται, ἢ εἰ ἀλλοίως τις διαιτῴη, καὶ αἱ κρίσιες ἁπλούστεραι καὶ εὐκριτώτεραι καὶ ἧσσον ὑπο-
9 στροφώδεες.

(5 L.) XV. Τὰς δὲ πτισάνας χρὴ ἐκ κριθέων ὡς βελτίστων εἶναι καὶ κάλλιστα ἑψῆσθαι, καὶ

[1] προμηθήσῃ Littré, the MSS. having προμηθὲς ᾖ. The MS. reading can be kept only if προμηθές be given a passive

is suitable; it will be stated later what is suitable in each form of illness. Should the mouth be moist, and the sputa as they should be, increase as a general rule the quantity of the gruel; for early appearance of abundant moisture indicates an early crisis, while a later appearance of scanty moisture indicates a late crisis. In their essence the facts are on the whole as stated.

XIII. Many other important points have been passed over which must be used in prognosis; these will be discussed later. The more complete the purging of the bowels the more the quantity of gruel administered should be increased until the crisis. In particular, proceed thus for two days after the crisis, in such cases as lead you to suppose that the crisis will be on the fifth, seventh or ninth day, so as to make sure of both the even and the odd day. Afterwards you must administer gruel in the morning, but you may change to solid food in the evening.

XIV. The above rules are on the whole useful to those who administer unstrained gruel from the outset. For in cases of pleurisy the pains at once cease of their own accord, as soon as sputa worth mentioning begin to be brought up and purgings begin to take place; while the purgings are much more complete, and empyema is less likely to occur, than if another regimen were adopted, and the crises are simpler, more decisive, and less liable to relapses.

XV. Gruel should be made from the finest barley, and boiled as well as possible, especially if more

---

meaning ("carefully guarded against"). Not finding a parallel to this I have adopted the reading of Littré.

ἄλλως ἢν μὴ τῷ χυλῷ μούνῳ μέλλῃς χρῆσθαι· μετὰ γὰρ τῆς ἄλλης ἀρετῆς τῆς πτισάνης τὸ ὀλισθηρὸν τὴν κριθὴν καταπινομένην ποιεῖ μὴ βλάπτειν· οὐδαμῇ γὰρ προσίσχει οὐδὲ μένει κατὰ τὴν τοῦ θώρηκος ἴξιν· ὀλισθηροτάτη δὲ καὶ ἀδιψοτάτη καὶ εὐπεπτοτάτη καὶ ἀσθενεστάτη ἐστὶν
9 ἡ κάλλιστα ἑφθή· ὧν πάντων δεῖ.

XVI. Ἢν οὖν μὴ προστιμωρήσῃ τις ὅσων δεῖται αὐτάρκης εἶναι ὁ τρόπος τῆς τοιαύτης πτισανορρυφίης, πολλαχῇ βεβλάψεται. οἷσι γὰρ σῖτος αὐτίκα ἐγκατακέκλεισται, εἰ μή τις ὑποκενώσας δοίη τὸ ῥύφημα, τὴν ὀδύνην ἐνεοῦσαν προσπαροξύνειεν ἂν καὶ μὴ ἐνεοῦσαν ἂν ἐμποιήσειεν, καὶ πνεῦμα πυκνότερον γένοιτ' ἄν· κακὸν δὲ τοῦτο· ξηραντικόν τε γὰρ πνεύμονος καὶ κοπῶδες ὑποχονδρίων καὶ ἤτρου καὶ φρενῶν· τοῦτο
10 δέ, ἢν ἔτι τοῦ πλευροῦ τῆς ὀδύνης συνεχέος ἐούσης καὶ πρὸς τὰ θερμάσματα μὴ χαλώσης καὶ τοῦ πτυάλου μὴ ἀνιόντος, ἀλλὰ καταγλισχραινομένου ἀσαπέως, ἢν μὴ λύσῃ τις τὴν ὀδύνην ἢ κοιλίην μαλθάξας ἢ φλέβα ταμών, ὁπότερον ἂν τούτων σημήνῃ, τὰς δὲ πτισάνας ἢν οὕτως ἔχουσι διδῷ,
16 ταχέες οἱ θάνατοι τῶν τοιούτων γίνονται.

XVII. Διὰ ταύτας οὖν τὰς προφάσιας καὶ ἑτέρας τοιαύτας[1] οἱ οὔλῃσι πτισάνῃσι χρεώμενοι ἑβδομαῖοι καὶ ὀλιγημερώτεροι θνήσκουσιν, οἱ μέν τι καὶ τὴν γνώμην βλαβέντες, οἱ δ' ὑπὸ τῆς ὀρθοπνοίης τε καὶ τοῦ ῥέγχεος ἀποπνιγέντες. μάλα δὲ τοὺς τοιούτους οἱ ἀρχαῖοι βλητοὺς ἐνό

[1] After τοιαύτας A has μᾶλλον and M ἔτι μᾶλλον.

---

[1] αὐτίκα seems to have this sense here.

than the pure juice is going to be used. For one of the virtues of gruel is its lubricant nature, which prevents the barley that is swallowed from doing any harm, since it clings nowhere and does not stick on its way through the chest. In addition to its excellent lubricating qualities the best boiled gruel quenches thirst the most, is the most easily digested, and the least disturbing. All these characteristics are needed.

XVI. The administration of this gruel requires certain aids, if it is to accomplish its purpose; and if they are not given manifold harm will result. When for instance food is at the time[1] confined in the bowels, should the gruel be given without first emptying them, it will increase any pain already existing or cause one if it does not exist already, and the respiration will become more rapid. This is harmful, in that it dries the lungs, besides causing discomfort in the hypochondria, the hypogastrium, and the diaphragm. Moreover, suppose the pain in the side continues and does not yield to the fomentations, while the sputum is not brought up, but becomes viscid without coction; should gruel be administered in these conditions without first relieving the pain, either by loosening the bowels or by venesection, whichever of these courses is indicated, a fatal termination will quickly follow.

XVII. For these reasons, as well as for others like them, those who take unstrained gruel die on the seventh day or earlier, some after being seized with delirium also, others being suffocated by orthopnoea and *râles*. The ancients[2] thought such sufferers "stricken," just because after death the

[2] For these see p. 64.

μιζον εἶναι διὰ τόδε οὐχ ἥκιστα, ὅτι ἀποθανόντων αὐτῶν ἡ πλευρὴ πελιδνὴ εὑρίσκεται, ἴκελόν τι πληγῇ. αἴτιον δὲ τούτου ἐστίν, ὅτι πρὶν λυθῆναι
10 *τὴν ὀδύνην θνῄσκουσιν· ταχέως γὰρ πνευματίαι γίνονται· ὑπὸ δὲ τοῦ πολλοῦ καὶ πυκνοῦ πνεύματος, ὡς ἤδη εἴρηται, καταγλισχραινόμενον τὸ πτύαλον ἀπέπτως κωλύει τὴν ἐπάνοδον γίνεσθαι, ἀλλὰ τὴν ῥέγξιν ποιεῖ ἐνισχόμενον ἐν τοῖσι βρογχίοισι τοῦ πνεύμονος. καὶ ὅταν ἐς τοῦτο ἔλθῃ, θανατῶδες ἤδη ὡς ἐπὶ τὸ πολύ ἐστι· καὶ γὰρ αὐτὸ τὸ πτύαλον ἐνισχόμενον κωλύει μὲν τὸ πνεῦμα ἔσω φέρεσθαι, ἀναγκάζει δὲ ταχέως ἔξω φέρεσθαι· καὶ οὕτως ἐς τὸ κακὸν ἀλλήλοισι συν-*
20 *τιμωρεῖ. τό τε γὰρ πτύαλον ἐνισχόμενον πυκνὸν τὸ πνεῦμα ποιεῖ, τό τε πνεῦμα πυκνὸν ἐὸν ἐπιγλισχραίνει τὸ πτύαλον καὶ κωλύει ἀπολισθάνειν. καταλαμβάνει δὲ ταῦτα οὐ μοῦνον ἢν πτισάνῃ ἀκαίρως χρέωνται, ἀλλὰ πολὺ μᾶλλον, ἤν τι ἄλλο*
25 *φάγωσιν ἢ πίωσι πτισάνης ἀνεπιτηδειότερον.*

(6 L.) XVIII. *Μάλα μὲν οὖν τὰ πλεῖστα παραπλήσιοί εἰσιν αἱ τιμωρίαι τοῖσί τε οὔλῃσι πτισάνῃσι χρεωμένοισι τοῖσί τε χυλῷ αὐτῷ· τοῖσι δὲ μηδετέρῳ τούτων, ἀλλὰ ποτῷ μοῦνον, ἔστιν ὅπῃ καὶ διαφερόντως τιμωρητέον. χρὴ δὲ*
6 *τὸ πάμπαν οὕτω ποιεῖν·*

XIX. *Ἢν μὲν νεοβρῶτι αὐτῷ ἐόντι καὶ κοιλίης μὴ ὑποκεχωρηκυίης ἄρξηται ὁ πυρετός, ἤν τε σὺν ὀδύνῃ ἤν τε ἄνευ ὀδύνης, ἐπισχεῖν τὴν δόσιν τοῦ ῥυφήματος, ἔστ' ἂν οἴηται κεχωρηκέναι ἐς τὸ κάτω μέρος τοῦ ἐντέρου τὸ σιτίον. χρῆσθαι δὲ ποτῷ, ἢν μὲν ἄλγημά τι ἔχῃ, ὀξυμέλιτι, χειμῶνος μὲν θερμῷ, θέρεος δὲ ψυχρῷ· ἢν δὲ πολλὴ δίψα*

side is found to be livid, as if a blow had been received. The reason for this appearance is that death occurs before the pain is relieved. For they quickly suffer from difficulty in breathing. The heavy and rapid respiration, as I have already said, makes the sputum become viscid without coction, and prevents its expulsion, so that it causes the *râles* by being confined in the bronchial passages. At this point death commonly occurs; the mere confinement of the sputum, in fact, while preventing the entrance of breath, forces it out quickly. So one mischief aggravates the other; the confinement of sputum renders respiration rapid, and the rapidity of the respiration makes the sputum viscid, preventing its slipping away. These attacks not only result from unseasonable administration of gruel, but are much more likely to occur if the patient has eaten or drunk something less suitable than gruel.

XVIII. Now the measures necessary to help the administration of the pure juice are practically the same as those required by unstrained gruel; but when neither is given, but only drink, they are in some ways different. In general terms the rules to be observed are the following.

XIX. Should the fever begin when the patient has recently taken food and the bowels have not been emptied, whether pain be present or not, refrain from giving gruel until he thinks that the food has descended to the lower part of the bowel. The drink to be employed, should there be any pain, is oxymel,[1] warm in winter and cold in summer. If there be great thirst, give hydromel

[1] A mixture of vinegar and honey.

ᾖ, καὶ μελικρήτῳ καὶ ὕδατι. ἔπειτα, ἢν μὲν ἄλγημα ἐνῇ ἢ τῶν ἐπικινδύνων τι ἐμφαίνηται,
10 διδόναι τὸ ῥύφημα μήτε πολὺ μήτε παχύ, μετὰ δὲ τὴν ἑβδόμην, ἢν ἰσχύῃ. ἢν δὲ μὴ ὑπεληλύθῃ ὁ παλαιότερος σῖτος νεοβρῶτι ἐόντι, ἢν μὲν ἰσχύῃ τε καὶ ἀκμάζῃ τῇ ἡλικίῃ, κλύσαι, ἢν δὲ ἀσθενέστερος ᾖ, βαλάνῳ προσχρήσασθαι, ἢν μὴ
15 αὐτόματα διεξίῃ καλῶς.

XX. Καιρὸν δὲ τῆς δόσιος τοῦ ῥυφήματος τόνδε μάλιστα φυλάσσεσθαι κατ' ἀρχὰς καὶ διὰ παντὸς τοῦ νοσήματος· ὅταν μὲν οἱ πόδες ψυχροὶ ἔωσιν, ἐπισχεῖν χρὴ τοῦ ῥυφήματος τὴν δόσιν, μάλιστα δὲ καὶ τοῦ ποτοῦ ἀπέχεσθαι· ὅταν δὲ ἡ θέρμη καταβῇ ἐς τοὺς πόδας, τότε διδόναι· καὶ νομίζειν μέγα δύνασθαι τὸν καιρὸν τοῦτον ἐν πάσῃσι τῇσι νούσοισιν, οὐχ ἥκιστα δὲ ἐν τῇσιν ὀξείῃσιν, μάλιστα δ' ἐν τῇσι μᾶλλον πυρε-
10 τώδεσιν καὶ ἐπικινδυνοτάτῃσιν.[1] χρῆσθαι δὲ πρῶτον[2] μὲν χυλῷ, ἔπειτα δὲ πτισάνῃ, κατὰ τὰ
12 τεκμήρια τὰ προγεγραμμένα ἀκριβέως θεωρέων.

(7 L.) XXI. Ὀδύνην δὲ πλευροῦ, ἤν τε κατ' ἀρχὰς γίνηται ἤν θ' ὕστερον, θερμάσμασι μὲν πρῶτον οὐκ ἀπὸ τρόπου χρησάμενον πειρηθῆναι διαλῦσαι. θερμασμάτων δὲ κράτιστον μὲν ὕδωρ θερμὸν ἐν ἀσκῷ ἢ ἐν κύστει ἢ ἐν χαλκῷ ἀγγείῳ ἢ ἐν ὀστρακίνῳ. προϋποτιθέναι δὲ χρὴ μαλθακόν τι πρὸς τὴν πλευρὴν προσηνείης εἵνεκεν. ἀγαθὸν δὲ καὶ σπόγγος μαλθακὸς μέγας ἐξ ὕδατος θερμοῦ ἐκπεπιεσμένος προστίθεσθαι· περιστέγειν δὲ ἄνω

[1] MV have ἐπικινδυνοτάτῃσι. A omits καὶ ἐπικιν. altogether. Possibly the words are a gloss.

and water. Later, should there be any pain or should any dangerous symptom appear, let the gruel given be neither much nor thick, and give it only after the seventh day, and if the strength be maintained. If the previous food which the patient has recently eaten should not have gone down, give an enema if the patient be strong and in the prime of life, but if he be too weak use a suppository, should the bowels be not well moved of their own accord.

XX. This is the time for administering gruel that must be most carefully observed both at the beginning of the illness and throughout its course. When the feet are cold you must refrain from giving gruel, and especially from giving drinks; give the gruel when the heat descends to the feet. Consider this time of great importance in all diseases, particularly in acute diseases, and most of all in those where the fever is high and the danger very great. Use first the pure juice, then the gruel, keeping a sharp eye for the signs already described.

XXI. When there is pain in the side, whether at the beginning or later, it is not amiss to try to dissipate it first by hot fomentations. The best fomentation is hot water in a skin, or bladder, or bronze or earthen vessel. Apply something soft to the side first to prevent discomfort. A good thing also to apply is a big, soft sponge dipped in hot water and squeezed out. You must, however, cover up the heat on the upper part,[1] for doing so will

[1] *I. e.* on the part of the sponge not next to the skin.

---

[2] πρῶτον is my reading. MV have πρῶτον μάλιστα μὲν and A has μάλιστα μὲν only. μάλιστα is omitted by the Paris MS. 2276 (S′).

10 τὴν θάλψιν χρή. πλείω τε γὰρ χρόνον ἀρκέσει καὶ παραμενεῖ, καὶ ἅμα ὡς μὴ ἡ ἀτμὶς πρὸς τὸ πνεῦμα τοῦ κάμνοντος φέρηται, ἢν ἄρα μὴ δοκῇ καὶ τοῦτο χρήσιμον πρός τι εἶναι· ἔστι γὰρ ὅτε δεῖ πρός τι. ἔτι δὲ καὶ κριθαὶ ἢ ὄροβοι· ἐν ὄξει κεκρημένῳ σμικρῷ ὀξυτέρῳ ἢ ὡς ἂν πίοι τις διέντα καὶ ἀναζέσαντα ἐς μαρσίππια καταρράψαντα προστιθέναι. καὶ πίτυρα τὸν αὐτὸν τρόπον. ξηραὶ δὲ πυρίαι, ἅλες, κέγχροι πεφρυγμένοι ἐν εἰρινέοισι μαρσιππίοισιν ἐπιτηδειότατοι·
20 καὶ γὰρ κοῦφον καὶ προσηνὲς ὁ κέγχρος.

XXII. Λύει δὲ μάλθαξις ἡ τοιήδε καὶ τὰς πρὸς κληῖδα περαινούσας ἀλγηδόνας· τομὴ μέντοι οὐχ ὁμοίως λύει ὀδύνην, ἢν μὴ πρὸς τὴν κληῖδα περαίνῃ ἡ ὀδύνη· ἢν δὲ μὴ λύηται πρὸς τὰ θερμάσματα ὁ πόνος, οὐ χρὴ πολὺν χρόνον θερμαίνειν· καὶ γὰρ ξηραντικὸν τοῦ πνεύμονος τοῦτο καὶ ἐμπυητικόν· ἀλλ' ἢν μὲν σημαίνῃ ἡ ὀδύνη ἐς κληῖδα ἢ ἐς βραχίονα βάρος ἢ περὶ μαζὸν ἢ ὑπὲρ τῶν φρενῶν, τάμνειν χρὴ τὴν ἐν τῷ ἀγκῶνι φλέβα τὴν ἔσω
10 καὶ μὴ ὀκνεῖν συχνὸν ἀφαιρεῖν, ἔστ' ἂν ἐρυθρότερον πολλῷ ῥυῇ ἢ ἀντὶ καθαροῦ τε καὶ ἐρυθροῦ
12 πελιδνόν· ἀμφότερα γὰρ γίνεται.

XXIII. Ἢν δ' ὑπὸ φρένας ᾖ τὸ ἄλγημα, ἐς δὲ τὴν κληῖδα μὴ σημαίνῃ, μαλθάσσειν χρὴ τὴν κοιλίην ἢ μέλανι ἐλλεβόρῳ ἢ πεπλίῳ, μέλανι μὲν δαῦκος ἢ σέσελι ἢ κύμινον ἢ ἄνησον ἢ ἄλλο τι τῶν εὐωδέων μίσγοντα, πεπλίῳ δὲ ὀπὸν σιλφίου. ἀτὰρ καὶ μισγόμενα ἀλλήλοισιν ὁμοιότροπα ταῦτ'

[1] *Helleborus niger.*
[2] *Euphorbia peplus.*
[3] *Athamanta cretensis.*
[4] *Laserpitium latifolium.*

make it hold out and last for a longer time; besides, it will prevent the steam being carried towards the breath of the patient—unless indeed the patient's breathing it be considered an advantage, as in fact it occasionally is. Barley too or vetches: soak in vinegar that is slightly stronger than could be drunk, boil, sew up in bags and then apply. Bran may be used in like manner. For dry fomentations, salt or toasted millet in woollen bags is most suitable; millet is also light and soothing.

XXII. A soft fomentation like this relieves the pains too that extend to the collar-bone. Venesection, however, does not relieve the pain so well unless it extends to the collar-bone. If the pain does not give way before the hot applications, do not continue them for long; continued heat dries the lungs and is apt to cause empyema. Should, however, the pain show signs of extending to the collar-bone, or should there be a weight in the fore-arm, or in the region of the breast, or above the diaphragm, you must open the inner vein at the elbow, and not hesitate to take away much blood until it flows much redder, or until it becomes livid instead of clear and red. Either of these changes may occur.

XXIII. If the pain be under the diaphragm, and does not declare itself towards the collar-bone, soften the bowels with black hellebore[1] or peplium,[2] mixing with the black hellebore daucus,[3] seseli,[4] cumin, anise or some other fragrant herb, and with the peplium juice of silphium.[5] In fact the blending

[5] A sort of assafoetida.

ἐστιν. ἄγει δὲ μέλας μὲν καλλίω καὶ κρισιμώτερα πεπλίου, πέπλιον δὲ μέλανος φυσέων καταρρηκτικώτερόν ἐστιν. ἄμφω δὲ ταῦτα ὀδύνην παύει·
10 παύει δὲ καὶ ἄλλα συχνὰ τῶν ὑπηλάτων· κράτιστα δὲ ταῦτα ὧν ἐγὼ οἶδα ἐστίν· ἐπεὶ καὶ τὰ ἐν τοῖσι ῥυφήμασι διδόμενα ὑπήλατα ἀρήγει, ὅσα μὴ ἄγαν ἐστὶν ἀηδέα ἢ διὰ πικρότητα ἢ δι' ἄλλην τινὰ ἀηδίην, ἢ διὰ πλῆθος ἢ διὰ χροιὴν ἢ
15 ὑποψίην τινά.

XXIV. Τῆς μέντοι πτισάνης, ὅταν πίῃ τὸ φάρμακον, ἐπιρρυφεῖν αὐτίκα χρὴ διδόναι μηδὲν ἔλασσον ἀξίως λόγου ἢ ὅσον εἴθιστο· ἐπεὶ καὶ κατὰ λόγον ἐστὶ μεσηγὺ τῆς καθάρσιος μὴ διδόναι ῥυφεῖν· ὅταν δὲ λήξῃ ἡ κάθαρσις, τότε ἔλασσον ῥυφείτω ἢ ὅσον εἴθιστο. μετὰ δὲ ταῦτα ἀναγέτω ἐπὶ τὸ πλεῖον, ἢν ἥ τε ὀδύνη πεπαυμένη ᾖ καὶ
8 μηδὲν ἄλλο ἐναντιῶται.

XXV. Ωὑτὸς δέ μοι λόγος ἐστίν, κἢν χυλῷ δέῃ πτισάνης χρῆσθαι. φημὶ γὰρ ἄμεινον εἶναι αὐτίκα ἄρξασθαι ῥυφεῖν τὸ ἐπίπαν μᾶλλον ἢ προκενεαγγήσαντα ἄρξασθαι τοῦ ῥυφήματος τριταῖον ἢ τεταρταῖον ἢ πεμπταῖον ἢ ἑκταῖον ἢ ἑβδομαῖον, ἤν γε μὴ προκριθῇ ἡ νοῦσος ἐν τούτῳ τῷ χρόνῳ. αἱ δὲ προπαρασκευαὶ καὶ τούτοισι
8 παραπλήσιοι ποιητέαι, ὁποῖαι εἴρηνται.

(8 L.) XXVI. Περὶ μὲν οὖν ῥυφήματος προσάρσιος οὕτω γινώσκω. ἀτὰρ καὶ περὶ ποτοῦ, ὁποῖον ἄν τις μέλλῃ πίνειν, τῶν προσγραφησομένων ωὑτὸς λόγος τὸ ἐπίπαν ἐστίν. οἶδα δὲ τοὺς ἰητροὺς τὰ ἐναντιώτατα ἢ ὡς δεῖ ποιέοντας· βούλονται γὰρ πάντες ὑπὸ τὰς ἀρχὰς τῶν νούσων προταριχεύσαντες τοὺς ἀνθρώπους ἢ δύο ἢ τρεῖς

of these constituents gives a harmonious compound. Black hellebore causes evacuations that are better and more favourable to the crisis than does peplium; but peplium breaks flatulence better than black hellebore. Both, however, stop pain, as do also many other evacuants; but these are the best I know of, though evacuants given in the gruel help, if they are not too unpleasant owing to bitterness or other unpleasant taste, or owing to quantity, colour, or some quality that arouses the patient's suspicion.

XXIV. Immediately after he has taken the purge, give the patient a quantity of gruel not appreciably less than usual, though it is reasonable to suspend giving it while the purge is acting. When the purging has ceased, give less gruel than usual, afterwards increasing it gradually, if the pain have ceased and nothing else indicate the contrary.

XXV. I recommend the same rule if it be necessary to use the pure juice of barley. For I hold it to be better on the whole to begin giving it at once rather than to starve the patient and then to begin giving the gruel on the third, fourth, fifth, sixth or seventh day, should the disease not reach a crisis in the interval. In this case too the preparations to be made are similar to those I have described.

XXVI. Such are my recommendations for the administration of gruel; and as to drink, whatever be the nature of that to be given, the directions that I shall set forth are in general the same. I am convinced that the practice of physicians is the exact opposite of what it should be; for they all wish at the beginning of a disease to reduce the patient by

ἢ καὶ πλείους ἡμέρας οὕτω προσφέρειν τὰ ῥυφήματα καὶ τὰ πόματα· καὶ ἴσως τι καὶ εἰκὸς δοκεῖ 10 *αὐτοῖσι εἶναι μεγάλης μεταβολῆς γινομένης*[1] *τῷ* 11 *σώματι μέγα τι κάρτα καὶ ἀντιμεταβάλλειν.*

XXVII. Τὸ δὲ *μεταβάλλειν μὲν εὖ ἔχει μὴ ὀλίγον· ὀρθῶς μέντοι ποιητέη καὶ βεβαίως ἡ μεταβολὴ καὶ ἔκ γε τῆς μεταβολῆς ἡ πρόσαρσις τῶν γευμάτων ἔτι μᾶλλον. μάλιστα μὲν οὖν ἂν βλάπτοιντο, εἰ μὴ ὀρθῶς μεταβάλλοιεν, οἱ οὔλῃσι τῇσι πτισάνῃσι χρεώμενοι· βλάπτοιντο δ' ἂν καὶ οἱ μούνῳ τῷ ποτῷ χρεώμενοι, βλάπτοιντο δ' ἂν καὶ οἱ μούνῳ τῷ χυλῷ χρεώμενοι,*[2] *ἥκιστα δ' ἂν* 9 *οὗτοι.*

(9 L.) XXVIII. Χρὴ δὲ *καὶ τὰ μαθήματα ποιεῖσθαι ἐν τῇ διαίτῃ τῶν ἀνθρώπων ἔτι ὑγιαινόντων, οἷα συμφέροι.*[3] *εἰ γὰρ δὴ τοῖσί γε ὑγιαίνουσι φαίνεται διαφέροντα μεγάλα τὰ τοῖα ἢ τοῖα διαιτήματα καὶ ἐν ἄλλῳ τινὶ καὶ ἐν τῇσι μεταβολῇσι, πῶς οὐχὶ καὶ ἐν τῇσι νούσοισι διαφέρει μέγα καὶ τούτων ἐν τῇσιν ὀξυτάτῃσι μέγιστα; ἀλλὰ μὴν εὐκαταμάθητόν γέ ἐστιν, ὅτι φαύλη δίαιτα βρώσιος καὶ πόσιος αὐτὴ ἑωυτῇ* 10 *ἐμφερὴς αἰεὶ ἀσφαλεστέρη ἐστὶν τὸ ἐπίπαν ἐς ὑγιείην, ἢ εἴ τις ἐξαπίνης μέγα μεταβάλλοι ἐς ἄλλα. ἐπεὶ καὶ τοῖσι δὶς σιτεομένοισι τῆς ἡμέρης καὶ τοῖσι μονοσιτέουσιν αἱ ἐξαπιναῖοι μεταβολαὶ βλάβας καὶ ἀρρωστίην παρέχουσιν. καὶ τοὺς*

[1] MV read *γενομένης*.

[2] A omits the second clause (*βλάπτοιντο . . . χρεώμενοι*).

[3] *συμφέροι* A: *συμφέρει* other MSS. I have kept the reading of A, as the "vague" optative without *ἂν* is common in the *Corpus*. See Vol. I., p. 59 (footnote).

starvation for two, three, or even more days before administering gruel and drink. Perhaps they consider it natural, when a violent change is taking place[1] in the body, to counteract it by another violent change.

XXVII. Now to bring about a change is no small gain, but the change must be carried out correctly and surely, a remark which applies even more to the administration of food after the change. Now those will be most harmed, should the change not be correct, who take unstrained gruel. Those too will be harmed who take drink only, as well as those who take the juice of barley only, but the last least of all.

XXVIII. A physician's studies should include a consideration of what is beneficial in a patient's regimen while he is yet in health. For surely, if men in health find that one regimen produces very different results from another, especially when the regimen is changed, in disease too there will be great differences, and the greatest in acute diseases. But it is easily discovered that a simple[2] diet of food and drink, if it be persevered in without a break, is on the whole safer for health than a sudden, violent change. For example, sudden changes cause harm and weakness, both to those who take one, and to those who take two full meals a day. Those too who are not in the habit of lunching, if they have taken lunch, immediately become feeble, heavy in all

[1] Or, reading γενομένης, "has taken place."

[2] So apparently is the meaning of φαῦλος here; Galen comments on its meaning. See *e. g.* xv. 341. But it may be "bad," "poor."

μέν γε μὴ μεμαθηκότας ἀριστᾶν, ἢν ἀριστήσωσιν, εὐθέως ἀρρώστους ποιεῖ καὶ βαρέας ὅλον τὸ σῶμα καὶ ἀσθενέας καὶ ὀκνηρούς· ἢν δὲ καὶ ἐπιδειπνήσωσιν, ὀξυρεγμιώδεας. ἐνίοισι δ' ἂν καὶ σπατίλη γένοιτο, ὅτι παρὰ τὸ ἔθος ἠχθοφόρηκεν ἡ κοιλίη 20 εἰθισμένη ἐπιξηραίνεσθαι καὶ μὴ δὶς διογκοῦσθαι 21 μηδὲ δὶς ἕψειν τὰ σιτία.

XXIX. Ἀρήγει οὖν τούτοισιν ἀνασηκῶσαι τὴν μεταβολήν· ἐγκοιμηθῆναι γὰρ χρή, ὥσπερ νύκτα ἄγοντα μετὰ τὸ δεῖπνον, τοῦ μὲν χειμῶνος ἀρριγέως, τοῦ δὲ θέρεος ἀθαλπέως· ἢν δὲ καθεύδειν μὴ δύνηται, βραδεῖαν, συχνὴν ὁδὸν περιπλανηθέντα, μὴ στασίμως, δειπνῆσαι μηδὲν ἢ ὀλίγα μηδὲ βλαβερά· ἔτι δὲ ἔλασσον πιεῖν καὶ μὴ ὑδαρές. ἔτι δὲ μᾶλλον ἂν πονήσειεν ὁ τοιοῦτος, εἰ τρὶς φάγοι τῆς ἡμέρης ἐς κόρον· ἔτι δὲ μᾶλλον, 10 εἰ πλεονάκις· καίτοι γε πολλοί εἰσιν οἳ εὐφόρως φέρουσι τρὶς σιτεόμενοι τῆς ἡμέρης ἐς πλῆθος, 12 οἳ ἂν οὕτως ἐθισθῶσιν.

XXX. Ἀλλὰ μὴν καὶ οἱ μεμαθηκότες δὶς σιτεῖσθαι τῆς ἡμέρης, ἢν μὴ ἀριστήσωσιν, ἀσθενέες καὶ ἄρρωστοί εἰσιν καὶ δειλοὶ ἐς πᾶν ἔργον καὶ καρδιαλγέες· κρέμασθαι γὰρ αὐτοῖσι δοκεῖ τὰ σπλάγχνα, καὶ οὐρέουσι θερμὸν καὶ χλωρόν, καὶ ἡ ἄφοδος συγκαίεται. ἔστι δ' οἷσι καὶ τὸ στόμα πικραίνεται καὶ οἱ ὀφθαλμοὶ κοιλαίνονται καὶ οἱ κρόταφοι πάλλονται καὶ τὰ ἄκρα διαψύχεται,[1] καὶ οἱ μὲν πλεῖστοι ἀνηριστηκότες[2] οὐ

[1] διαψύχονται MSS. : διαψύχεται Galen.

[2] A has πλεῖστοι οἳ ἂν ἠριστηκότες, M πλεῖστοι τῶν ἀνηριστηκότων οὐ δύνανται τὸ δεῖπνον, V πλεῖστοι τῶν ἀνηριστηκότων οὐ δύνανται κατεσθίειν τὸ δεῖπνον. I read (with Kühlewein)

the body, weak and sluggish. Should they also dine, they suffer from acid eructations. Diarrhœa too may occur in some cases, because the digestive organs have been loaded, contrary to habit, when they are accustomed to a period of dryness, and not to be twice distended with food and to digest food twice.

XXIX. It is beneficial, then, in these cases to counterbalance the change. Thus one should sleep off the meal,[1] as one passes the night after dinner,[2] avoiding cold in winter and heat in summer. If sleep be impossible, a slow, long walk should be taken, without stopping; then no dinner should be eaten, or at least only a little light food; still less should be drunk, and that not diluted. Such a man will suffer yet more if he eat three times a day to surfeit, and still more if he eat more often. Yet there are many who, if accustomed to it, can easily bear three full meals a day.

XXX. But, indeed, those too who have the habit of taking two meals a day, should they omit lunch, find themselves weak, feeble, averse to all exertion, and the victims of heart-burn. Their bowels seem to hang, the urine is hot and yellow, and the stools are parched. In some cases the mouth is bitter, the eyes are hollow, the temples throb, and the extremities are chilled; most men who have missed

[1] Such I take to be the force of the preposition in *ἐγκοιμηθῆναι*.

[2] Galen says that we must either change *τὸ δεῖπνον* to *τὸν ἄριστον*, or understand *μετὰ τὸν ἄριστον* after *χρή*. The latter suggestion is the simpler. The text of Galen appears to be corrupt, but the drift of the passage is clear.

ἀνηριστηκότες from A, and omit the *οἱ* of A as a repetition of the preceding syllable. Kühlewein puts *ἀνηριστηκότες* after *κατεσθίειν*.

10 δύνανται κατεσθίειν τὸ δεῖπνον, δειπνήσαντες δὲ βαρύνουσι τὴν κοιλίην καὶ δυσκοιτέουσι πολὺ
12 μᾶλλον ἢ εἰ προηριστήκεσαν.

XXXI. Ὁπότε οὖν ταῦτα τοιαῦτα γίνεται τοῖσιν ὑγιαίνουσιν εἵνεκεν ἡμίσεος ἡμέρης διαίτης μεταβολῆς, παρὰ τὸ ἔθος οὔτε προσθεῖναι λυσι-
4 τελεῖν φαίνεται οὔτε ἀφελεῖν.

XXXII. Εἰ τοίνυν οὗτος ὁ παρὰ τὸ ἔθος μονοσιτήσας ὅλην τὴν ἡμέρην κενεαγγήσας δειπνήσειεν ὁπόσον εἴθιστο, εἰκὸς αὐτόν, εἰ τότε ἀνάριστος ἐὼν ἐπόνει καὶ ἠρρώστει, δειπνήσας δὲ τότε βαρὺς ἦν, πολὺ μᾶλλον βαρύνεσθαι· εἰ δέ γε ἔτι πλείω χρόνον κενεαγγήσας ἐξαπίνης
7 μεταδειπνήσειεν, ἔτι μᾶλλον βαρυνεῖται.[1]

XXXIII. Τὸν οὖν παρὰ τὸ ἔθος κενεαγγήσαντα συμφέρει ταύτην τὴν ἡμέρην ἀντισηκῶσαι ὧδε· ἀρριγέως καὶ ἀθαλπέως καὶ ἀταλαιπώρως—ταῦτα γὰρ πάντα βαρέως ἂν ἐνέγκαι—τὸ δὲ δεῖπνον

[1] In this chapter there are two noticeable variants. MV omit τὴν after ὅλην, and A for εἰ τότε ἀνάριστος reads εἰ ὅτε ἀνάριστος. Littré, however, building on Galen's comment (ὃ λέγει τοιοῦτον ἐστίν· εἰ ὁ παρὰ τὸ ἔθος ἀναρίστητος μείνας, εἶτα δειπνήσας τῶν εἰθισμένων ἐλάττω, τῆς νυκτὸς ἐβαρύνθη, πολὺ μειζόνως βαρυνθήσεται ὁ πλείω ἢ ὅσα εἴθιστο δειπνήσας) reads as follows: εἰ τοίνυν οὗτος . . . ὅλην ἡμέρην κενεαγγήσας, δειπνήσειεν ὁκόσον εἴθιστο, δειπνήσας δέ, τότε βαρὺς ἦν, εἰκὸς αὐτόν, εἰ, ὅτι ἀνάριστος ἐὼν ἐπόνεε καὶ ἠῤῥώστει, δειπνήσειε πλείω ἢ ὁκόσον εἴθιστο, πουλὺ μᾶλλον βαρύνεσθαι.

---

[1] There is a remarkable likeness between Chapters XXVIII–XXX and *Ancient Medicine*, Chapters X–XII. The similarity is verbal, and can hardly be due to chance. Littré thinks the likeness proves that the author of *Ancient Medicine* was Hippocrates. I confess that I feel the force of his argument more now than I did when I was translating

lunch cannot eat their dinner; and if they do dine their bowels are heavy, and they sleep much worse than if they had previously taken lunch.[1]

XXXI. Since then men in health suffer in this way through a change in regimen for half a day, it is plainly beneficial neither to increase nor yet to decrease what is customary.

XXXII. If then this man, who contrary to custom took only one meal, should fast strictly the whole day and then eat his usual quantity of dinner, it is likely that—since on the other occasion he suffered from pain and weakness after taking no lunch, and was heavy after dinner—he will feel much heavier. And if he keep a strict fast for a still longer period, and then suddenly eat a dinner, he will feel heavier still.[2]

XXXIII. He therefore who has fasted strictly contrary to his custom is benefited if he compensate for the day of starvation in the following manner. He should avoid cold, heat and fatigue—all of which will distress him—and his dinner should be consider-

Ancient Medicine, but one treatise may contain a passage appearing in another without the author of the two being the same. One may be copying the other, or both may be copying a third. The truth probably is that the writer of *Regimen in Acute Diseases* imitated *Ancient Medicine.*

[2] In this chapter I follow Kühlewein, but with no confidence. Our MS. tradition seems to make the severity of the change depend upon the length of the fast (*ἡμίσεος ἡμέρης, ὅλην τὴν ἡμέρην, ἔτι πλείω χρόνον*). Furthermore, the grammatical confusion of Ch. XXXII, with its strange *τότε* before *βαρύς*, suggests corruption. Galen's comment points to a text now lost, although *ὁπόσον εἴθιστο* is a part of it, in which the severity of the change was made to depend upon the quantity of food taken. It is easy to suggest possible restorations, but none are likely.

συχνῷ ἔλασσον ποιήσασθαι ἢ ὅσον εἴθιστο καὶ μὴ ξηρόν, ἀλλὰ τοῦ πλαδαρωτέρου τρόπου· καὶ πιεῖν μὴ ὑδαρὲς μηδὲ ἔλασσον ἢ κατὰ λόγον τοῦ βρώματος· καὶ τῇ ὑστεραίῃ ὀλίγα ἀριστῆσαι, ὡς
9 *ἐκ προσαγωγῆς ἀφίκηται ἐς τὸ ἔθος.*

XXXIV. *Αὐτοὶ μέντοι σφέων αὐτῶν δυσφορώτερον δὴ τὰ τοιαῦτα φέρουσιν οἱ πικρόχολοι τὰ ἄνω· τὴν δέ γε ἀσιτίην τὴν παρὰ τὸ ἔθος οἱ φλεγματίαι τὰ ἄνω εὐφορώτερον τὸ ἐπίπαν, ὥστε καὶ τὴν μονοσιτίην τὴν παρὰ τὸ ἔθος εὐφορώτερον*
6 *ἂν οὗτοι ἐνέγκαιεν.*

XXXV. *Ἱκανὸν μὲν οὖν καὶ τοῦτο σημεῖον, ὅτι αἱ μέγισται μεταβολαὶ τῶν περὶ τὰς φύσιας ἡμέων καὶ τὰς ἕξιας συμβαινόντων μάλιστα νοσοποιέουσιν. οὐ δὴ οἷόν τε παρὰ καιρὸν οὔτε σφοδρὰς τὰς κενεαγγίας ποιεῖν οὔτε ἀκμαζόντων τῶν νοσημάτων καὶ ἐν φλεγμασίῃ ἐόντων προσφέρειν οὔτε ἐξαπίνης οἷόν τε ὅλῳ τῷ πρήγματι*
8 *μεταβάλλειν οὔτε ἐπὶ τὰ οὔτε ἐπὶ τά.*

(10 L.) XXXVI. Π*ολλὰ δ' ἄν τις ἠδελφισμένα τούτοισι τῶν ἐς κοιλίην καὶ ἄλλα εἴποι, ὡς εὐφόρως μὲν φέρουσι τὰ βρώματα, ἃ εἰθίδαται, ἢν καὶ μὴ ἀγαθὰ ᾖ φύσει· ὡσαύτως δὲ καὶ τὰ ποτά· δυσφόρως δὲ φέρουσι τὰ βρώματα, ἃ μὴ εἰθίδαται, κἢν μὴ κακὰ ᾖ· ὡσαύτως δὲ καὶ τὰ*
7 *ποτά.*

XXXVII. Κ*αὶ ὅσα μὲν κρεηφαγίη πολλὴ παρὰ τὸ ἔθος βρωθεῖσα ποιεῖ ἢ σκόροδα ἢ σίλφιον ἢ ὀπὸς ἢ καυλὸς ἢ ἄλλα ὅσα τοιουτότροπα μεγάλας δυνάμιας ἰδίας ἔχοντα, ἧσσον ἄν τις θαυμάσειεν, εἰ τὰ τοιαῦτα πόνους ἐμποιεῖ ἐν τῇσι κοιλίῃσι μᾶλλον ἄλλων· ἀλλὰ εἰ δὴ καταμάθοι, ὅσον μᾶζα*

ably less than usual, not dry but of rather a liquid character. His drink must not be watery nor out of proportion to the quantity of the food. On the next day he should take a light lunch, and so by degrees return to his usual practice.

XXXIV. The people who bear these changes with more than usual distress are those who are bilious in the upper digestive tract. Those who bear unaccustomed fasting better are generally the phlegmatic in the upper tract, so that these will also bear better the unaccustomed taking of one meal only.

XXXV. Now this too is adequate proof that the chief causes of diseases are the most violent changes in what concerns our constitutions and habits. Therefore it is not possible unseasonably to produce utter starvation, nor to give food while a disease is at its height and an inflammation remains, nor is it possible suddenly to make a complete change either in this direction or in that.

XXXVI. There are many other things akin to these that one might say about the digestive organs, to show that people readily bear the food to which they are accustomed, even though it be not naturally good. It is the same also with drinks. Men with difficulty bear the food to which they are unaccustomed, even though it be not bad. It is the same also with drinks.

XXXVII. If it were a question of eating much meat contrary to custom, or garlic, or silphium, juice or stalk, or anything else of the same kind possessing powerful qualities of its own, one would be less surprised at its producing more pains in the bowels than do other things. But it is surprising to learn

ὄχλον καὶ ὄγκον καὶ φῦσαν καὶ στρόφον κοιλίῃ παρέχει παρὰ τὸ ἔθος βρωθεῖσα τῷ ἀρτοφαγεῖν εἰθισμένῳ ἢ οἷον ἄρτος βάρος καὶ στάσιν κοιλίης
10 τῷ μαζοφαγεῖν εἰθισμένῳ ἢ αὐτός γε ὁ ἄρτος θερμὸς βρωθεὶς οἵην δίψαν παρέχει καὶ ἐξαπιναίην πληθώρην διὰ τὸ ξηραντικόν τε καὶ βραδύπορον, καὶ οἱ ἄγαν καθαροί τε καὶ συγκομιστοὶ παρὰ τὸ ἔθος βρωθέντες οἷα διαφέροντα ἀλλήλων ποιεῦσι καὶ μᾶζά γε ξηρὴ παρὰ τὸ ἔθος ἢ ὑγρὴ ἢ γλίσχρη, καὶ τὰ ἄλφιτα οἷόν τι ποιεῖ τὰ ποταίνια τοῖσι μὴ εἰωθόσι καὶ τὰ ἑτεροῖα τοῖσι τὰ ποταίνια εἰωθόσι· καὶ οἰνοποσίη καὶ ὑδροποσίη παρὰ τὸ ἔθος ἐς θάτερα μεταβληθέντα ἐξαπίνης καὶ ὑδαρής
20 τε οἶνος καὶ ἄκρητος παρὰ τὸ ἔθος ἐξαπίνης ποθείς—ὁ μὲν γὰρ πλάδον τε ἐν τῇ ἄνω κοιλίῃ ἐμποιήσει καὶ φῦσαν ἐν τῇ κάτω, ὁ δὲ παλμόν τε φλεβῶν καὶ καρηβαρίην καὶ δίψαν—καὶ λευκός τε καὶ μέλας παρὰ τὸ ἔθος μεταβάλλουσιν,[1] εἰ καὶ ἄμφω οἰνώδεες εἶεν, ὅμως πολλὰ ἂν ἑτεροιώσειαν κατὰ τὸ σῶμα· ὡς δὴ γλυκύν τε καὶ οἰνώδεα ἧσσον ἄν τις φαίη θαυμαστὸν εἶναι μὴ
28 τωὐτὸ δύνασθαι ἐξαπίνης μεταβληθέντα.

(11 L.) Τιμωρητέον μὲν δὴ τοιόνδε τι μέρος τῷ ἐναντίῳ λόγῳ· ὅτι ἡ μεταβολὴ τῆς διαίτης τούτοισιν ἐγένετο οὐ μεταβάλλοντος τοῦ σώματος οὔτ᾽ ἐπὶ τὴν ῥώμην, ὥστε προσθέσθαι δεῖν σιτία,
5 οὔτ᾽ ἐπὶ τὴν ἀρρωστίην, ὥστ᾽ ἀφαιρεθῆναι.

XXXVIII. Προστεκμαρτέα δὴ καὶ ἡ ἰσχὺς καὶ ὁ τρόπος τοῦ νοσήματος ἑκάστου καὶ τῆς φύσιος

[1] μεταβάλλουσι A (A² adding -ν): μεταβάλλοντι MV: μεταβληθείς Kühlewein. I retain the reading of A, taking it to be a dative of disadvantage.

the trouble, distension, flatulence and tormina produced in the digestive organs by barley-cake eaten by one used to eating bread, or the heaviness and stagnation in digestive organs caused by bread eaten by one accustomed to eat barley-cake, or the thirst and sudden fulness produced by bread itself, when eaten hot, because of its drying and indigestible qualities; and the different effects caused by over-fine and over-coarse bread when partaken of contrary to custom; and by barley-cake unusually dry, or moist, or viscid; the effect of new barley-bread on those not used to it, and of old on those accustomed to new. Again, the drinking of wine or the drinking of water, when one habit is suddenly changed to the other, diluted wine or neat wine drunk with a sudden break of habit; the former produces water-brash in the upper bowels and flatulence in the lower, while the second causes throbbing of the veins, heaviness of the head, and thirst. Again, an exchange of white and dark wine, although both are vinous, if contrary to habit will cause many alterations in the body. So that one should express less surprise that the sudden exchange of a sweet wine for a vinous, and of a vinous for a sweet, should have the same effect.

Let me now say what may be said in favour of the opposite reasoning; in these cases the change of regimen took place without any change in the body, either towards strength, so as to render necessary an increase of food, or towards weakness, so as to require a diminution of it.

XXXVIII. Account too must certainly be taken of the strength and character of each illness, of the

τοῦ[1] ἀνθρώπου καὶ τοῦ ἔθεος τῆς διαίτης τοῦ κάμνοντος, οὐ μοῦνον σιτίων, ἀλλὰ καὶ ποτῶν. πολλῷ δ' ἧσσον ἐπὶ τὴν πρόσθεσιν ἰτέον· ἐπεί γε τὴν ἀφαίρεσιν ὅλως ἀφελεῖν πολλαχοῦ λυσιτελεῖ, ὅπου διαρκεῖν μέλλει ὁ κάμνων, μέχρι ἂν τῆς νούσου ἡ ἀκμὴ πεπανθῇ. ἐν ὁποίοισι δὲ τὸ
9 τοιόνδε ποιητέον, γεγράψεται.

XXXIX. Πολλὰ δ' ἄν τις καὶ ἄλλα ἠδελφισμένα τοῖς εἰρημένοισι γράφοι· τόδε γε μὴν κρέσσον μαρτύριον· οὐ γὰρ ἠδελφισμένον μοῦνόν ἐστι τῷ πρήγματι, περὶ οὗ μοι ὁ πλεῖστος λόγος εἴρηται, ἀλλ' αὐτὸ τὸ πρῆγμα ἐπικαιρότατόν ἐστιν διδακτήριον· οἱ γὰρ ἀρχόμενοι τῶν ὀξέων νοσημάτων ἔστιν ὅτε οἱ μὲν σιτία ἔφαγον αὐθημερὸν ἠργμένοι ἤδη, οἱ δὲ καὶ τῇ ὑστεραίῃ, οἱ δὲ καὶ ἐρρύφεον τὸ προστυχόν, οἱ δὲ καὶ κυκεῶνα
10 ἔπιον. ἅπαντα δὲ ταῦτα κακίω μέν ἐστιν, ἢ εἰ ἑτεροίως διαιτηθείη· πολλῷ μέντοι ἐλάσσω βλάβην φέρει ἐν τούτῳ τῷ χρόνῳ ἁμαρτηθέντα, ἢ εἰ τὰς μὲν πρώτας δύο ἡμέρας ἢ τρεῖς κενεαγγήσειε τελέως, τεταρταῖος δὲ ἐὼν τοιάδε διαιτηθείη ἢ πεμπταῖος· ἔτι μέντοι κάκιον, εἰ ταύτας πάσας τὰς ἡμέρας προκενεαγγήσας ἐν τῇσιν ὕστερον ἡμέρῃσιν οὕτω διαιτηθείη, πρὶν ἢ πέπειρον τὴν νοῦσον γενέσθαι· οὕτω μὲν γὰρ θάνατον φέρει φανερῶς τοῖς πλείστοις, εἰ μὴ πάμπαν ἡ νοῦσος
20 εὐήθης εἴη. αἱ δὲ κατ' ἀρχὰς ἁμαρτάδες οὐχ ὁμοίως ταύτῃσιν ἀνήκεστοι, ἀλλὰ πολλῷ εὐακε-

[1] After τοῦ the MSS. have τε. It is omitted by Littré after Galen.

constitution of the individual, and of the habitual regimen of the patient, of his drink as well as of his food. Much the greater caution should be shown in increasing the quantities, since it is often beneficial to enforce total abstinence until the disease reaches its height and coction has taken place, should the patient be likely to hold out. The circumstances in which such a course ought to be adopted I shall state later.

XXXIX. There are many other remarks, closely related to what has been already said, that might be made; the following, however, is a stronger piece of evidence, for it is not merely closely related to the matter which I have mostly been discussing, but it is the matter itself, and so its teaching is of the first importance. Cases have occurred where patients at the beginning of acute diseases have eaten solid food on the very first day when the onset has already taken place, others on the next day; others again have taken the first gruel that came to hand, while some have even drunk cyceon.[1] Another regimen, no doubt, would have been an improvement on any of these courses; yet mistakes at this time cause much less harm than if the patient had completely starved for the first two or three days, and then adopted this regimen on the fourth or fifth. It would be still worse, however, if he were first to starve for all these days and then to adopt such a regimen in the following days, before the disease became concocted. The consequence is plainly death in most cases, unless the disease be very mild indeed. But mistakes at the beginning are not so irremediable,

[1] A mixed food, usually containing cheese, honey and wine.

στότεραι. τοῦτο οὖν ἡγεῦμαι μέγιστον διδακτήριον, ὅτι οὐ στερητέαι αἱ πρῶται ἡμέραι τοῦ ῥυφήματος ἢ τοίου ἢ τοίου τοῖσι μέλλουσιν ὀλίγον
25 *ὕστερον ῥυφήμασιν ἢ σιτίοισι χρῆσθαι.*

XL. *Πυθμενόθεν μὲν οὖν οὐκ ἴσασιν οὔθ' οἱ τῇσι κριθώδεσι πτισάνῃσι χρεώμενοι, ὅτι αὐτῇσι κακοῦνται, ὅταν ῥυφεῖν ἄρξωνται, ἢν προκενεαγγήσωσιν δύο ἢ τρεῖς ἡμέρας ἢ πλείους, οὔτ' αὖ οἱ τῷ χυλῷ χρεώμενοι γινώσκουσιν ὅτι τοιούτοισι βλάπτονται ῥυφέοντες, ὅταν μὴ ὀρθῶς ἄρξωνται τοῦ ῥυφήματος. τόδε γε μὴν καὶ φυλάσσουσι καὶ γινώσκουσιν, ὅτι μεγάλην τὴν βλάβην φέρει, ἤν, πρὶν πέπειρον τὴν νοῦσον γενέσθαι, κριθώδεα*
10 *πτισάνην ῥυφῇ ὁ κάμνων, εἰθισμένος χυλῷ*
11 *χρῆσθαι.*

XLI. *Πάντα οὖν ταῦτα μεγάλα μαρτύρια, ὅτι οὐκ ὀρθῶς ἄγουσιν ἐς τὰ διαιτήματα οἱ ἰητροὶ τοὺς κάμνοντας· ἀλλ' ἐν ᾗσί τε νούσοισιν οὐ χρὴ κενεαγγεῖν τοὺς μέλλοντας ῥυφήμασι διαιτᾶσθαι, κενεαγγέουσιν, ἐν ᾗσί τε οὐ χρὴ μεταβάλλειν ἐκ κενεαγγίης ἐς ῥυφήματα, ἐν ταύτῃσι μεταβάλλουσι. καὶ ὡς ἐπὶ τὸ πολὺ ἀπαρτὶ*[1] *ἐν τούτοισι τοῖσι καιροῖσι μεταβάλλουσιν ἐς τὰ ῥυφήματα ἐκ τῆς κενεαγγίης, ἐν οἷσι πολλάκις ἀρήγει ἐκ*
10 *τῶν ῥυφημάτων πλησιάζειν τῇ κενεαγγίῃ, ἢν οὕτω*
11 *τύχῃ παροξυνομένη ἡ νοῦσος.*

[1] ἀπαρτὶ is the reading of Littré, found in Galen and also in R′. A has ἁμαρτάνει, M and V have ἁμαρτάνουσιν, followed by ὁτὲ δ' in A and by ἐνίοτε in M and V. A[2] changed ἁμαρτάνει to ἁμαρταίνει, and Littré thinks that ἀπαρτὶ ἐν became ἁμαρτηνη and ἁμαρταινει, which was corrected to ἁμαρτάνει and ἁμαρτάνουσι.

but are much more easy to counteract. This fact, then, I consider to be very strong testimony that during the first days there should not be abstinence from gruel of one kind or another, if the patient is going to be given gruel or solid food a little later on.

XL. So there is radical ignorance among both those who use unstrained gruel and those who use only the juice; the former do not know that injury is done if a fast of two, three, or more days precede the commencement of taking gruel, the latter do not know that harm comes from taking their gruel when the commencement is not correctly made.[1] They do know, however, and regulate the treatment accordingly, that great injury is done if a patient, used to taking barley-water, take unstrained gruel before the disease is concocted.

XLI. All these things are strong testimony that physicians do not correctly guide their patients in the matter of regimen. They make them fast when the disease is one where fasting before taking gruel is wrong, and they change from fasting to gruel when the disease is one where such a change is wrong. And generally they make the change from fasting to gruel exactly at those times at which often it is profitable to exchange gruel for what is virtually fasting, should for instance an exacerbation of the disease occur during a gruel diet.

[1] There is some confusion in this sentence owing to the grammatical subject being uncertain. What is the subject of ἴσασιν, the physicians or the patients? The sense requires the former, but χρεώμενοι, βλάπτονται and other words point to the latter. Perhaps the explanation is that the true subject is an indefinite "they," a blank cheque to be filled up by "physicians" in some cases and by "patients" in others.

XLII. Ἐνίοτε δὲ καὶ ὠμὰ ἐπισπῶνται ἀπὸ τῆς κεφαλῆς καὶ τοῦ περὶ θώρηκα τόπου χολώδεα· ἀγρυπνίαι τε συνεμπίπτουσιν αὐτοῖσι, δι' ἃς οὐ πέσσεται ἡ νοῦσος, περίλυποι δὲ καὶ πικροὶ γίνονται καὶ παραφρονέουσι, καὶ μαρμαρυγώδεά σφεων τὰ ὄμματα καὶ αἱ ἀκοαὶ ἤχου μεσταὶ καὶ τὰ ἀκρωτήρια κατεψυγμένα καὶ οὖρα ἄπεπτα καὶ πτύσματα λεπτὰ καὶ ἁλυκὰ καὶ κεχρωσμένα ἀκρήτῳ χρώματι σμικρὰ καὶ ἱδρῶτες περὶ τράχη-
10 λον καὶ διαπορήματα καὶ πνεῦμα προσπταῖον ἐν τῇ ἄνω φορῇ πυκνὸν ἢ μέγα λίην, ὀφρύες δεινώσιος μετέχουσαι, λειποψυχώδεα πονηρὰ καὶ τῶν ἱματίων ἀπορρίψιες ἀπὸ τοῦ στήθεος καὶ χεῖρες τρομώδεες, ἐνίοτε δὲ καὶ χεῖλος τὸ κάτω σείεται. ταῦτα δ' ἐν ἀρχῇσι παραφαινόμενα παραφροσύνης δηλωτικά ἐστι σφοδρῆς, καὶ ὡς ἐπὶ τὸ πολὺ θνήσκουσιν· οἱ δὲ διαφεύγοντες ἢ μετὰ ἀποστήματος ἢ αἵματος ῥύσιος ἐκ τῆς ῥινὸς ἢ πῦον παχὺ
19 πτύσαντες διαφεύγουσιν, ἄλλως δὲ οὔ.

XLIII. Οὐδὲ γὰρ τῶν τοιούτων ὁρῶ ἐμπείρους τοὺς ἰητρούς, ὡς χρὴ διαγινώσκειν τὰς ἀσθενείας ἐν τῇσι νούσοισιν, αἵ τε διὰ κενεαγγίην ἀσθενεῦνται, αἵ τε δι' ἄλλον τινὰ ἐρεθισμόν, αἵ τε διὰ πόνον καὶ ὑπὸ ὀξύτητος τῆς νούσου, ὅσα τε ἡμέων ἡ φύσις καὶ ἡ ἕξις ἑκάστοισιν ἐκτεκνοῖ πάθεα καὶ εἴδεα παντοῖα· καίτοι σωτηρίην ἢ θάνατον φέρει
8 γινωσκόμενα καὶ ἀγνοεύμενα τὰ τοιάδε.

XLIV. Μέζον μὲν γὰρ κακόν ἐστιν, ἢν διὰ τὸν πόνον καὶ τὴν ὀξύτητα τῆς νούσου ἀσθενέοντι

---

[1] "Unrelieved," "pure."

XLII. Sometimes such treatment draws crude matters from the head and bilious matters from the region of the chest. The patient is afflicted with sleeplessness, in consequence of which the disease is not concocted, and he becomes depressed, peevish and delirious; flashes of light come to the eyes; the ears are full of noise; the extremities are chilled; urine is unconcocted; sputa thin, salt, slightly tinged with an unmixed[1] colour; sweats about the neck; disquietude[2]; respiration, interrupted in the ascent of the breath, rapid or very deep; eye-brows dreadful[3]; distressing faints; casting away of the clothes from the chest; trembling of the hands; in some cases there is also shaking of the lower lip. These symptoms, when manifesting themselves at the beginning, are indications of violent delirium, and usually the patient dies. Those who recover do so with an abscession, or a flow of blood from the nose, or by expectoration of thick pus; otherwise they do not recover at all.

XLIII. Nor indeed do I see that physicians are experienced in the proper way to distinguish the kinds of weakness that occur in diseases, whether it be caused by starving, or by some other irritation, or by pain, or by the acuteness of the disease; the affections again, with their manifold forms, that our individual constitution and habit engender—and that though a knowledge of such things brings safety and ignorance brings death.

XLIV. For example, it is one of the more serious blunders, when the patient is weak through the pain

[2] Restlessness; the patient "does not know what to do with himself."

[3] Probably "frowning."

προσαίρῃ τις ποτὸν ἢ ῥύφημα πλέον ἢ σιτίον, οἰόμενος διὰ κενεαγγίην ἀσθενεῖν. ἀεικὲς δὲ καὶ διὰ κενεαγγίην ἀσθενέοντα μὴ γνῶναι καὶ πιέζειν τῇ διαίτῃ· φέρει μὲν γάρ τινα κίνδυνον καὶ αὕτη ἡ ἁμαρτάς, πολλῷ δὲ ἥσσονα τῆς ἑτέρης· καταγελαστοτέρη δὲ πολλῷ αὕτη μᾶλλον ἡ ἁμαρτὰς τῆς ἑτέρης· εἰ γὰρ ἄλλος ἰητρὸς ἢ καὶ δημότης 10 ἐσελθὼν καὶ γνοὺς τὰ συμβεβηκότα δοίη καὶ φαγεῖν καὶ πιεῖν, ἃ ὁ ἕτερος ἐκώλυεν, ἐπιδήλως ἂν δοκέοι ὠφεληκέναι. τὰ δὲ τοιάδε μάλιστα καθυβρίζεται τῶν χειρωνακτέων ὑπὸ τῶν ἀνθρώπων· δοκεῖ γὰρ αὐτοῖσιν ὁ ἐπεσελθὼν ἰητρὸς ἢ ἰδιώτης ὡσπερεὶ τεθνεῶτα ἀναστῆσαι. γεγράψεται οὖν καὶ περὶ τούτων σημεῖα, οἷσι χρὴ ἕκαστα 17 τούτων διαγινώσκειν.

(12 L.) XLV. Παραπλήσια μέντοι τοῖσι κατὰ κοιλίην ἐστὶ καὶ ταῦτα· καὶ γὰρ ἢν ὅλον τὸ σῶμα ἀναπαύσηται πολὺ παρὰ τὸ ἔθος, οὐκ αὐτίκα ἔρρωται μᾶλλον· ἢν δὲ δὴ καὶ πλείω χρόνον διελινύσαν ἐξαπίνης ἐς τοὺς πόνους ἔλθῃ, φαῦλόν τι πρήξει ἐπιδήλως. οὕτω δὲ καὶ ἓν ἕκαστον τοῦ σώματος· καὶ γὰρ οἱ πόδες τοιόνδε τι πρήξειαν[1] καὶ τἄλλα ἄρθρα, μὴ εἰθισμένα πονεῖν, ἢν διὰ χρόνου ἐξαπίνης ἐς τὸ πονεῖν ἔλθῃ· ταὐτὰ δ' ἂν 10 καὶ οἱ ὀδόντες καὶ οἱ ὀφθαλμοὶ πάθοιεν, καὶ οὐδὲν ὅ τι οὔ·[2] ἐπεὶ καὶ κοίτη παρὰ τὸ ἔθος μαλθακὴ

[1] Gomperz here adds ἄν, which might easily fall out after πρήξειαν. The scribe of A has πρήξει. Gomperz is probably right, but the optative without ἂν is often found in the Hippocratic writings where we should expect the ἂν to be added.

[2] οὐδὲν ὅ τι οὔ Cobet: οὐδὲν ὅτιαν A[1] (A[2] changes -αν to -ουν): οὐθὲν ὁτιοῦν MV. Littré reads πᾶν ὁτιοῦν (R′ has πάν).

or the acuteness of the disease, to administer drink, or more gruel, or food, under the impression that the weakness is due to want of nourishment. It is a shame too not to recognise weakness that is due to such want, and to aggravate it by the regimen; for this mistake too carries with it some danger, though far less than the other mistake. It is, however, much more likely to make the physician a laughing-stock; for if another physician or a layman were to come in, and, recognising what had taken place, were to give to eat and drink things contrary to the doctor's orders, he would show himself a manifest helper of the patient. It is especially such mistakes of practitioners that are regarded with contempt by the public[1]; for they think that the physician or layman who came in later raised up the patient as it were from the dead. So the symptoms in these cases also shall be described, whereby each kind can be discriminated.

XLV. I will now give some facts that are analogous to those already given about the bowels. If the whole body have a long and unusual rest, it does not gain strength all at once; and should it have a yet longer period of idleness, and then suddenly undergo fatigue, it will manifestly fare somewhat badly. Similarly too with the several parts of the body; the feet, and the other limbs, will suffer in a like manner, if, when not accustomed to fatigue for a long time, they suddenly undergo it. The teeth too, the eyes, and everything else would fare in the same way. For even a bed that is soft, or

[1] Possibly; "by their patients."

πόνον ἐμποιεῖ καὶ σκληρὴ παρὰ τὸ ἔθος, καὶ ὑπαίθριος κοίτη[1] *παρὰ τὸ ἔθος σκληρύνει τὸ*
14 *σῶμα.*

XLVI. *Ἀτὰρ τῶν τοιῶνδε πάντων ἀρκέσει παράδειγμά τι γράψαι· εἰ γάρ τις ἕλκος λαβὼν ἐν κνήμῃ μήτε λίην ἐπίκαιρον μήτε λίην εὔηθες, μήτε ἄγαν εὐελκὴς ἐὼν μήτε ἄγαν δυσελκής, αὐτίκα ἀρξάμενος ἐκ πρώτης κατακείμενος ἰητρεύοιτο καὶ μηδαμῇ μετεωρίζοι τὸ σκέλος, ἀφλέγμαντος μὲν ἂν εἴη οὗτος μᾶλλον καὶ ὑγιὴς πολλῷ θᾶσσον ἂν γένοιτο, ἢ εἰ περιπλανώμενος ἰητρεύοιτο· εἰ μέντοι πεμπταῖος ἢ ἑκταῖος ἐών, ἢ καὶ*
10 *ἔτι ἀνωτέρω, ἀναστὰς ἐθέλοι προβαίνειν, μᾶλλον ἂν πονέοι τότε, ἢ εἰ αὐτίκα ἐξ ἀρχῆς πλανώμενος ἰητρεύοιτο· εἰ δὲ καὶ πολλὰ ταλαιπωρήσειεν ἐξαπίνης, πολλῷ ἂν μᾶλλον πονήσειεν, ἢ εἰ κείνως ἰητρευόμενος τὰ αὐτὰ ταῦτα ταλαιπωρήσειεν ἐν ταύτῃσιν τῇσιν ἡμέρῃσιν. διὰ τέλεος οὖν μαρτυρεῖ ταῦτα πάντα ἀλλήλοισιν, ὅτι πάντα ἐξαπίνης μέζον πολλῷ τοῦ μετρίου μεταβαλλόμενα καὶ*
18 *ἐπὶ τὰ καὶ ἐπὶ τὰ βλάπτει.*

XLVII. *Πολλαπλασίη μὲν οὖν κατὰ κοιλίην ἡ βλάβη ἐστίν, ἢν ἐκ πολλῆς κενεαγγίης ἐξαπίνης πλέον τοῦ μετρίου προσαίρηται—καὶ κατὰ τὸ ἄλλο σῶμα, ἢν ἐκ πολλῆς ἡσυχίης ἐξαίφνης ἐς πλείω πόνον ἔλθῃ, πολλῷ πλείω βλάψει—ἢ εἰ ἐκ πολλῆς ἐδωδῆς ἐς κενεαγγίην μεταβάλλοι· δεῖ μέντοι καὶ τὸ σῶμα τούτοισιν ἐλινύειν· κἢν ἐκ πολλῆς ταλαιπωρίης ἐξαπίνης ἐς σχολήν τε καὶ ῥᾳθυμίην ἐμπέσῃ, δεῖ δὲ καὶ τούτοισι τὴν κοιλίην*

[1] *κοίτη*: Gomperz would delete.

hard, contrary to what a man is used to, produces fatigue, and sleeping contrary to habit in the open air stiffens the body.

XLVI. A single example of all these things will suffice. Take the case of a man on whose leg appears a sore that is neither very serious nor very slight, and suppose he is neither a very good nor a very bad subject. If from the very first day he undergo treatment while lying on his back and never raise his leg at all, he will suffer less from inflammation, and will recover much more quickly than if he walk about while being treated. If, however, on the fifth or sixth day, or later still, he were to get up and move about, he would then suffer more pain[1] than if he were to walk about under treatment from the very first. And if he should suddenly undertake many exertions, he would suffer much more pain[1] than if with the other treatment he undertook the same exertions on these days. So in all cases all the evidence concurs in proving that all sudden changes, that depart widely from the mean in either direction, are injurious.

XLVII. So the harm to the bowels, if the patient after long fasting suddenly take more than a moderate quantity—the body too in general, if after long rest it suddenly undergo an extra amount of fatigue, will receive far greater harm therefrom—is many times greater than that which results from a change from full diet to strict fasting. However, the body also must rest in this case; and if after great exertion the body suddenly indulge in idleness and ease, the bowels in this case too must

[1] *πόνος* is "pain" here, but "fatigue," "tired aches," in the preceding chapter.

10 ἐλινύειν ἐκ πλήθεος βρώμης· εἰ δὲ μή, πόνον ἐν τῷ σώματι ἐμποιήσει καὶ βάρος ὅλου τοῦ
12 σώματος.

(13 L.) XLVIII. Ὁ οὖν πλεῖστός μοι λόγος γέγονεν περὶ τῆς μεταβολῆς τῆς ἐπὶ τὰ καὶ ἐπὶ τά. ἐς πάντα μὲν οὖν εὔχρηστον ταῦτ' εἰδέναι· ἀτὰρ καί, περὶ οὗ ὁ λόγος ἦν, ὅτι ἐν τῇσιν ὀξείῃσι νούσοισιν ἐς τὰ ῥυφήματα μεταβάλλουσιν ἐκ τῆς κενεαγγίης· μεταβλητέον γὰρ ὡς ἐγὼ κελεύω· ἔπειτα οὐ χρηστέον ῥυφήμασιν, πρὶν ἡ νοῦσος πεπανθῇ ἢ ἄλλο τι σημεῖον φανῇ ἢ κατὰ ἔντερον, κενεαγγικὸν ἢ ἐρεθιστικόν, ἢ κατὰ τὰ ὑποχόνδρια,
10 οἷα γεγράψεται.

XLIX. Ἀγρυπνίη ἰσχυρὴ πόμα καὶ σιτίον ἀπεπτότερα ποιεῖ, καὶ ἡ ἐπὶ θάτερα αὖ μεταβολὴ λύει τὸ σῶμα καὶ ἑφθότητα καὶ καρηβαρίην
4 ἐμποιεῖ.

(14 L.) L. Γλυκὺν δὲ οἶνον καὶ οἰνώδεα, καὶ λευκὸν καὶ μέλανα, καὶ μελίκρητον καὶ ὕδωρ καὶ ὀξύμελι τοισίδε σημαινόμενον χρὴ διορίζειν ἐν τῇσιν ὀξείῃσι νούσοισι· ὁ μὲν γλυκὺς ἧσσόν ἐστιν καρηβαρικὸς τοῦ οἰνώδεος καὶ ἧσσον φρενῶν ἁπτόμενος καὶ διαχωρητικώτερος δή τι τοῦ ἑτέρου κατὰ ἔντερον, μεγαλόσπλαγχνος δὲ σπληνὸς καὶ ἥπατος· οὐκ ἐπιτήδειος δὲ οὐδὲ τοῖσι πικροχόλοισι· καὶ γὰρ οὖν καὶ διψώδης τοῖσί γε τοιού-
10 τοις· ἀτὰρ καὶ φυσώδης τοῦ ἐντέρου τοῦ ἄνω, οὐ μὴν πολέμιός γε τῷ ἐντέρῳ τῷ κάτω κατὰ λόγον

---

[1] According to Galen, ἑφθότης means here a heated state connected with the humours, a sort of flabbiness akin to the condition produced by boiling.

rest from abundance of food, otherwise pain will occur in the body and heaviness in every part of it.

XLVIII. So most of my account has dealt with change in one direction or another. Now while this knowledge is useful for all purposes, it is especially important because in acute diseases there is a change, the subject of our discussion, from strict fasting to gruels. This change should be made in accordance with my instructions; and then gruels must not be employed before the disease is concocted, or some other symptom, either of inanition or of irritation, appear in the intestine, or in the hypochondria, according to the description I shall give later.

XLIX. Obstinate sleeplessness makes food and drink less digestible, while a change to the opposite extreme relaxes the body, and causes flabbiness [1] and heaviness of the head.

L. The following criteria enable us to decide when in acute diseases we should administer sweet wine, vinous wine, white wine and dark wine, hydromel, water and oxymel.[2] Sweet wine causes less heaviness in the head than the vinous, goes to the brain less,[3] evacuates the bowels more than the other, but causes swelling of the spleen and liver. It is not suited either to the bilious [4]; in fact it also makes them thirsty. Moreover it causes flatulence in the upper intestine, without, however, disagreeing with the lower intestine proportionately to the

[2] Hydromel (honey and water) and oxymel (honey and vinegar) were, with wine, the chief drinks given in serious diseases.

[3] Is less apt to cause delirium, or (perhaps) semi-intoxication.

[4] See Vol. I, p. 255, note 2.

τῆς φύσης· καίτοι οὐ πάνυ πορίμη ἐστὶν ἡ ἀπὸ τοῦ γλυκέος οἴνου φῦσα, ἀλλ' ἐγχρονίζει περὶ ὑποχόνδριον. καὶ γὰρ οὖν οὗτος ἧσσον διουρητικός ἐστιν τὸ ἐπίπαν τοῦ οἰνώδεος λευκοῦ· πτυάλου δὲ μᾶλλον ἀναγωγὸς τοῦ ἑτέρου ὁ γλυκύς. καὶ οἷσι μὲν διψώδης ἐστὶν πινόμενος, ἧσσον ἂν τούτοις ἀνάγοι ἢ ὁ ἕτερος οἶνος, οἷσι δὲ μὴ διψώδης, μᾶλλον ἀνάγοι ἂν τοῦ
20 *ἑτέρου.*

LI. *Ὁ δὲ λευκὸς οἰνώδης οἶνος ἐπῄνηται μὲν καὶ ἔψεκται τὰ πλεῖστα καὶ τὰ μέγιστα ἐν τῇ τοῦ γλυκέος οἴνου διηγήσει· ἐς δὲ κύστιν μᾶλλον πόριμος ἐὼν τοῦ ἑτέρου καὶ διουρητικὸς καὶ καταρρηκτικὸς ἐὼν αἰεὶ πολλὰ προσωφελεῖ ἐν ταύτῃσι τῇσι νούσοισι· καὶ γὰρ εἰ πρὸς ἄλλα ἀνεπιτηδειότερος τοῦ ἑτέρου πέφυκεν, ἀλλ' ὅμως κατὰ κύστιν ἡ κάθαρσις ὑπ' αὐτοῦ γινομένη ῥύεται, ἣν προτρέπηται ὁποῖα δεῖ. καλὰ δὲ ταῦτα*
10 *τεκμήρια*[1] *περὶ τοῦ οἴνου καὶ ὠφελείης καὶ βλάβης· ἅσσα ἀκαταμάθητα ἦν τοῖσιν ἐμεῦ*
12 *γεραιτέροισιν.*

LII. Κιρρῷ *δ' αὖ*[2] *οἴνῳ καὶ μέλανι αὐστηρῷ ἐν ταύτῃσι τῇσι νούσοισιν ἐς τάδε ἂν χρήσαιο· εἰ καρηβαρίη μὲν μὴ ἐνείη μηδὲ φρενῶν ἅψις μηδὲ τὸ πτύαλον κωλύοιτο τῆς ἀνόδου μηδὲ τὸ οὖρον ἴσχοιτο, διαχωρήματα δὲ πλαδαρώτερα καὶ ξυσματωδέστερα εἴη, ἐν δὴ τοῖσι τοιούτοισι πρέποι ἂν μάλιστα μεταβάλλειν ἐκ τοῦ λευκοῦ καὶ ὅσα*

[1] V has here *τῆς*, the other MSS. *τά*. Omitted by Kühlewein.

[2] *αὖ* Reinhold and Kühlewein: *ἂν* A. Omitted by MV.

flatulence produced. And yet flatulence from sweet wine is not at all transient,[1] but stays in the region of the hypochondrium. In fact it is on the whole less diuretic than vinous white wine; but sweet wine is more expectorant than the other. In persons who are made thirsty by drinking it, it proves less expectorant than the other; but when it does not produce thirst it is the more expectorant.

LI. As to white vinous wine, most and the most important of its virtues and bad effects have already been given in my account of sweet wine. Passing more readily than the other into the bladder, being diuretic and laxative, it always is in many ways beneficial in acute diseases. For although in some respects its nature is less suitable than the other, nevertheless the purging through the bladder that it causes is helpful, if it be administered[2] as it should be. These are good testimonies to the advantages and disadvantages of the wine, and they were left undetermined by my predecessors.

LII. A pale wine, again, and an astringent, dark wine, may be used in acute diseases for the following purposes. If there be no heaviness of the head, if the brain be not affected,[3] nor the sputum checked, nor the urine stopped, and if the stools be rather loose and like shavings, in these and in similar circumstances it will be very suitable to change

[1] πορίμη is a most difficult word to translate. "Transient" is the translation of Adams, and is only partially satisfactory. The word means "easily moving itself," "apt to shift."

[2] προτρέπηται is a difficult word. It suggests that the λευκὸς οἰνώδης οἶνος must be "encouraged" by careful precautions in administering it, if the effects are to be the best.

[3] See note on p. 105.

τούτοισιν ἐμφερέα. προσσυνιέναι δὲ χρὴ ὅτι τὰ μὲν ἄνω πάντα καὶ τὰ κατὰ κύστιν ἧσσον βλάψει,
10 ἢν ὑδαρέστερος ᾖ, τὰ δὲ κατ' ἔντερον μᾶλλον
11 ὀνήσει, ἢν ἀκρητέστερος ᾖ.

(15 L.) LIII. Μελίκρητον δὲ πινόμενον διὰ πάσης τῆς νούσου ἐν τῇσιν ὀξείῃσι νούσοισιν τὸ ἐπίπαν μὲν τοῖσι πικροχόλοισι καὶ μεγαλοσπλάγχνοις ἧσσον ἐπιτήδειον ἢ τοῖσι μὴ τοιούτοισι· διψῶδές γε μὴν ἧσσον τοῦ γλυκέος οἴνου· πνεύμονός τε γὰρ μαλθακτικόν ἐστιν καὶ πτυάλου ἀναγωγὸν μετρίως καὶ βηχὸς παρηγορικόν· ἔχει γὰρ σμηγματῶδές τι, ὃ οὐ μᾶλλον τοῦ καιροῦ[1] καταγλισχραίνει τὸ πτύαλον. ἔστι δὲ καὶ
10 διουρητικὸν μελίκρητον ἱκανῶς, ἢν μή τι τῶν ἀπὸ σπλάγχνων κωλύῃ· καὶ διαχωρητικὸν δὲ κάτω χολωδέων, ἔστι μὲν ὅτε καλῶν, ἔστι δ' ὅτε κατακορεστέρων μᾶλλον τοῦ καιροῦ καὶ ἀφρωδεστέρων. μᾶλλον δὲ τὸ τοιοῦτο τοῖσι χολώδεσί
15 τε καὶ μεγαλοσπλάγχνοισι γίνεται.

LIV. Πτυάλου μὲν οὖν ἀναγωγὴν καὶ πνεύμονος μάλθαξιν τὸ ὑδαρέστερον μελίκρητον ποιεῖ μᾶλλον· τὰ μέντοι ἀφρώδεα διαχωρήματα καὶ μᾶλλον τοῦ καιροῦ κατακορέως χολώδεα καὶ μᾶλλον θερμὰ τὸ ἄκρητον μᾶλλον τοῦ ὑδαρέος ἄγει· τὸ δὲ τοιόνδε διαχώρημα ἔχει μὲν καὶ ἄλλα σίνεα μεγάλα· οὔτε γὰρ ἐξ ὑποχονδρίου καῦμα σβεννύει, ἀλλ' ὁρμᾷ, δυσφορίην τε καὶ ῥιπτασμὸν

[1] Coray was the first to give a simple explanation of this difficult passage by adding *οὐ* before *μᾶλλον*. See the note of Littré for the views of earlier commentators.

---

[1] The phrase *μᾶλλον τοῦ καιροῦ* occurs several times in this

from white wine. It must further be understood that the wine under consideration will do less harm to all the upper parts and to the bladder, if it be more diluted, but will benefit the bowels the more if it be less so.

LIII. Hydromel, drunk throughout the course of an acute disease, is less suited on the whole to the bilious, and to those with enlarged bellies, than to those who are not such. It causes less thirst than does sweet wine, for it softens the lungs, is mildly expectorant, and relieves a cough. It has, in fact, a detergent quality, which makes the sputum viscid, but not more so than is seasonable.[1] Hydromel is also considerably diuretic, unless some condition of the bowels prove a hindrance. It also promotes the evacuation downwards of bilious matters, that are sometimes favourable, sometimes more intense and frothy than is seasonable. This effect, however, happens rather to those who are bilious and have enlarged bellies.

LIV. Now the bringing up of sputum, and the softening of the lungs, are effected rather by hydromel which has been considerably diluted with water. Frothy stools, however, that are more intensely bilious, and hotter, than is seasonable,[2] are more provoked by neat hydromel than by that which is diluted. Such stools cause besides serious mischiefs; they intensify, rather than extinguish, the heat in the hypochondrium, cause distress and

part of the book—a good instance of the psychological truth that a phrase once used is apt to suggest itself subconsciously. It means "abnormal," "more than is usual in the circumstances."

[2] See previous note.

τῶν μελέων ἐμποιεῖ ἑλκῶδές τ' ἐστὶ καὶ ἐντέρου
10 καὶ ἕδρης· ἀλεξητήρια δὲ τούτων γεγράψεται.

LV. Ἄνευ μὲν ῥυφημάτων μελικρήτῳ χρεώμενος ἀντ' ἄλλου ποτοῦ ἐν ταύτῃσι τῇσι νούσοισι πολλὰ ἂν εὐτυχοίης καὶ οὐκ ἂν πολλὰ ἀτυχοίης· οἷσι δὲ δοτέον καὶ οἷσιν οὐ δοτέον, τὰ μέγιστα
5 εἴρηται, καὶ δι' ὃ οὐ δοτέον.

LVI. Κατέγνωσται δὲ μελίκρητον ὑπὸ τῶν ἀνθρώπων, ὡς καταγυιοῖ τοὺς πίνοντας, καὶ διὰ τοῦτο ταχυθάνατον εἶναι νενόμισται. ἐκλήθη δὲ τοῦτο διὰ τοὺς ἀποκαρτερέοντας· ἔνιοι γὰρ μελικρήτῳ ποτῷ χρέονται ὡς τοιούτῳ δῆθεν ἐόντι. τὸ δὲ οὐ παντάπασιν ὧδε ἔχει, ἀλλὰ ὕδατος μὲν πολλῷ ἰσχυρότερόν ἐστιν πινόμενον μοῦνον, εἰ μὴ ταράσσοι τὴν κοιλίην· ἀτὰρ καὶ οἴνου λεπτοῦ καὶ ὀλιγοφόρου καὶ ἀνόδμου ᾗ μὲν ἰσχυρότερον,
10 ᾗ[1] δὲ ἀσθενέστερον. μέγα μὴν διαφέρει καὶ οἴνου καὶ μέλιτος ἀκρητότης ἐς ἰσχύν· ἀμφοτέρων δ' ὅμως τούτων, εἰ καὶ διπλάσιον μέτρον οἴνου ἀκρήτου πίνοι τις, ἢ ὅσον μέλι ἐκλείχοι, πολλὸν ἂν δήπου ἰσχυρότερος εἴη ὑπὸ τοῦ μέλιτος, εἰ μοῦνον μὴ ταράσσοιτο τὴν κοιλίην· πολλαπλάσιον γὰρ ἂν καὶ τὸ κόπριον διεξίοι ἂν αὐτῷ. εἰ μέντοι ῥυφήματι χρέοιτο πτισάνῃ, ἐπιπίνοι δὲ μελίκρητον, ἄγαν πλησμονῶδες ἂν εἴη καὶ φυσῶδες καὶ τοῖσι κατὰ ὑποχόνδριον σπλάγχνοις
20 ἀσύμφορον· προπινόμενον μέντοι πρὸ ῥυφημάτων

[1] ᾗ A: ἐνείῃ MV (V has also ἐνείῃ for the former ᾗ). Galen recognises two readings, ᾗ and ἐνίῃ.

---

[1] I cannot make sense out of this passage if διαφέρει means "is different," as Littré and Adams take it. The word ὅμως

agitation of the limbs, and ulcerate the intestines and the seat. I shall, however, write afterwards remedies for these troubles.

LV. The use of hydromel, without gruel, instead of other drink in acute diseases will cause many successes and few failures. I have already given the most important directions as to whom it should, and to whom it should not, be administered, as well as the reason why it should not be administered.

LVI. Hydromel has been condemned by the public on the ground that it weakens those who drink it, and for this reason it has the reputation of hastening death. This reputation it has won through those who starve themselves to death, some of whom use hydromel as a drink, under the impression that it will hasten their end. But it by no means has this character, being much more nutritive, when drunk alone, than water is, unless it deranges the digestive organs. Moreover, it is in some respects more, and in some respects less nourishing than wine that is thin, weak and odourless. Both neat wine and neat honey are indeed strong[1] in nutritive power, but if a man were to take both, even though he took twice as much neat wine as he swallowed honey, he would, I think, get from the honey much more strength, if only his digestive organs were not disordered, as the quantity of the stools also would be multiplied. If, however, he use barley gruel, and then drink hydromel, it will cause fullness, flatulence, and trouble in the bowels about the hypochondrium. Drunk before the gruel,

in the next sentence suggests that though both honey and wine are nutritive, yet honey is much more so. Hence I take διαφέρει to mean "is pre-eminent."

μελίκρητον οὐ βλάπτει ὡς μεταπινόμενον, ἀλλά
22 *τι καὶ ὠφελεῖ.*

LVII. *Ἑφθὸν δὲ μελίκρητον ἐσιδεῖν μὲν πολλῷ κάλλιον τοῦ ὠμοῦ· λαμπρόν τε γὰρ καὶ λεπτὸν καὶ λευκὸν καὶ διαφανὲς γίνεται. ἀρετὴν δὲ ἥντινα αὐτῷ προσθέω διαφέρουσάν τι τοῦ ὠμοῦ οὐκ ἔχω· οὐδὲ γὰρ ἥδιόν ἐστιν τοῦ ὠμοῦ, ἢν τυγχάνῃ γε καλὸν τὸ μέλι ἐόν· ἀσθενέστερον μέντοι τοῦ ὠμοῦ καὶ ἀκοπρωδέστερόν ἐστιν· ὧν οὐδετέρης τιμωρίης προσδεῖται μελίκρητον. ἄγχιστα δὲ χρηστέον αὐτῷ τοιῷδε ἐόντι, εἰ τὸ μέλι*
10 *τυγχάνοι πονηρὸν ἐὸν καὶ ἀκάθαρτον καὶ μέλαν καὶ μὴ εὐῶδες· ἀφέλοιτο γὰρ ἂν ἡ ἕψησις τῶν*
12 *κακοτήτων αὐτοῦ τὸ πλεῖον τοῦ αἴσχεος.*

(16 L.) LVIII. *Τὸ δὲ ὀξύμελι καλεύμενον ποτὸν πολλαχοῦ εὔχρηστον ἐν ταύτῃσι τῇσι νούσοισιν εὑρήσεις ἐόν· πτυάλου γὰρ ἀναγωγόν ἐστιν καὶ εὔπνοον. καιροὺς μέντοι τοιούσδε ἔχει· τὸ μὲν κάρτα ὀξὺ οὐδὲν ἂν μέσον*[1] *ποιήσειεν πρὸς τὰ πτύαλα τὰ μὴ ῥηϊδίως ἀνιόντα· εἰ γὰρ ἀναγάγοι μὲν τὰ ἐγκέρχνοντα καὶ ὄλισθον ἐμποιήσειεν καὶ ὥσπερ διαπτερώσειε τὸν βρόγχον. παρηγορήσειεν ἄν τι τὸν πνεύμονα· μαλθακτικὸν*
10 *γάρ. καὶ εἰ μὲν ταῦτα συγκυρήσειε, μεγάλην ὠφελείην ἐμποιήσει. ἔστι δ' ὅτε τὸ κάρτα ὀξὺ οὐκ ἐκράτησε τῆς ἀναγωγῆς τοῦ πτυάλου, ἀλλὰ προσεγλίσχρηνε καὶ ἔβλαψε· μάλιστα δὲ τοῦτο πάσχουσιν οἵπερ καὶ ἄλλως ὀλέθριοί εἰσι καὶ ἀδύνατοι βήσσειν τε καὶ ἀποχρέμπτεσθαι τὰ ἐνεχόμενα. ἐς μὲν οὖν τόδε προστεκμαίρεσθαι χρὴ τὴν ῥώμην τοῦ ἀνθρώπου καί, ἢν ἐλπίδα ἔχῃ, διδόναι· διδόναι δέ, ἢν διδῷς, ἀκροχλίαρον καὶ*
19 *κατ' ὀλίγον τὸ τοιόνδε καὶ μὴ λάβρως.*

however, it does not harm as it does if drunk after—nay, it is even somewhat beneficial.

LVII. Boiled hydromel is much more beautiful in appearance than is unboiled, being bright, thin, white and transparent, but I know of no virtue to attribute to it which the unboiled does not possess equally. It is not more pleasant either, provided that the honey be good. It is, however, less nutritious than the unboiled, and causes less bulky stools, neither of which properties are of any use to hydromel. Boil it by all means before use if the honey should be bad, impure, black and not fragrant, as the boiling will take away most of the unpleasantness of these defects.

LVIII. You will find the drink called oxymel often useful in acute diseases, as it brings up sputum and eases respiration. The occasions, however, for it are the following. When very acid it has no slight effect on sputum that will not easily come up; for if it will bring up the sputa that cause hawking, promote lubrication, and so to speak sweep out the windpipe, it will cause some relief to the lungs by softening them. If it succeed in effecting these things it will prove very beneficial. But occasionally the very acid does not succeed in bringing up the sputum, but merely makes it viscid, so causing harm. It is most likely to produce this result in those who are mortally stricken, and have not the strength to cough and bring up the sputa that block the passages. So with an eye to this take into consideration the patient's strength, and give acid oxymel only if there be hope. If you do give it, give it tepid and in small doses, never much at one time.

---

[1] μέσον A and some other MSS.: μέζον M. Galen refers to both readings.

LIX. Τὸ μέντοι ὀλίγον ὕποξυ ὑγραίνει μὲν στόμα καὶ φάρυγγα ἀναγωγόν τε πτυάλου ἐστὶ καὶ ἄδιψον· ὑποχονδρίῳ δὲ καὶ σπλάγχνοισιν τοῖσι ταύτῃ εὐμενές· καὶ τὰς ἀπὸ μέλιτος βλάβας κωλύει· τὸ γὰρ ἐν μέλιτι χολῶδες κολάζεται. ἔστι δὲ καὶ φυσέων καταρρηκτικὸν καὶ ἐς οὔρησιν προτρεπτικόν· ἐντέρου μέντοι τῷ κάτω μέρει πλαδαρώτερον καὶ ξύσματα ἐμποιεῖ· ἔστι δ' ὅτε καὶ φλαῦρον τοῦτο ἐν τῇσιν ὀξείῃσιν
10 τῶν νούσων γίνεται, μάλιστα μὲν ὅτι φύσας κωλύει περαιοῦσθαι, ἀλλὰ παλινδρομεῖν ποιεῖ. ἔτι δὲ καὶ ἄλλως γυιοῖ καὶ ἀκρωτήρια ψύχει· ταύτην καὶ οἶδα μούνην τὴν βλάβην δι' ὀξυ-
14 μέλιτος γινομένην, ἥτις ἀξίη γραφῆς.

LX. Ὀλίγον δὲ τὸ τοιόνδε ποτὸν νυκτὸς μὲν καὶ νήστει πρὸ ῥυφήματος ἐπιτήδειον προπίνεσθαι· ἀτὰρ καὶ ὅταν πολὺ μετὰ ῥύφημα ᾖ, οὐδὲν κωλύει πίνειν. τοῖσι δὲ ποτῷ μοῦνον διαιτωμένοισιν ἄνευ ῥυφημάτων διὰ τόδε οὐκ ἐπιτήδειόν ἐστιν αἰεὶ διὰ παντὸς[1] χρῆσθαι τούτῳ. μάλιστα μὲν διὰ ξύσιν καὶ τρηχυσμὸν τοῦ ἐντέρου· ἀκόπρῳ γὰρ ἐόντι μᾶλλον ἐμποιοίη ἂν ταῦτα κενεαγγίης παρεούσης· ἔπειτα δὲ καὶ τὸ
10 μελίκρητον τῆς ἰσχύος ἀφαιρέοιτ' ἄν. ἢν μέντοι ἀρήγειν φαίνηται πρὸς τὴν σύμπασαν νοῦσον πολλῷ ποτῷ τούτῳ χρῆσθαι, ὀλίγον χρὴ τὸ ὄξος παραχεῖν, ὅσον μοῦνον γινώσκεσθαι· οὕτω γὰρ καὶ ἃ φιλεῖ βλάπτειν, ἥκιστα ἂν βλάπτοι, καὶ ἃ
15 δεῖται ὠφελείης, προσωφελοίη ἄν.

[1] διὰ παντὸς MV : μοῦνον A.

[1] Oxymel in general, not the particular kind discussed in the previous chapter.

LIX. But slightly acid oxymel moistens the mouth and throat, brings up sputum and quenches thirst. It is soothing to the hypochondrium and to the bowels in that region. It counteracts the ill effects of honey, by checking its bilious character. It also breaks flatulence and encourages the passing of urine. In the lower part of the intestines, however, it tends to produce moisture in excess and discharges like shavings. Occasionally in acute diseases this character does mischief, especially because it prevents flatulence from passing along, forcing it to go back. It has other weakening effects as well, and chills the extremities. This is the only ill effect worth writing about that I know can be produced by this oxymel.

LX. It is beneficial to give a little drink of this kind[1] at night and when the patient is fasting before taking gruel. Moreover, there is nothing to prevent its being drunk a long time after the gruel. But those who are restricted to drink alone without gruels are harmed by a constant use of it throughout the illness for the following reasons. The chief is that it scrapes and roughens[2] the intestine, which effects are intensified by the absence of excreta due to the fasting. Then it will also take away from the hydromel its nutritive power. If, however, it appear helpful to the disease as a whole to use this drink in large quantity, reduce the amount of the vinegar so that it can just be tasted. In this way the usual bad effects of oxymel will be reduced to a minimum, and the help required will also be rendered.

[2] Or, as we should say, "irritates."

LXI. Ἐν κεφαλαίῳ δὲ εἰρῆσθαι, αἱ ἀπὸ ὄξεος ὀξύτητες πικροχόλοισι μᾶλλον ἢ μελαγχολικοῖσι συμφέρουσι· τὰ μὲν γὰρ πικρὰ διαλύεται καὶ ἐκφλεγματοῦται ὑπ' αὐτοῦ οὐ μετεωριζόμενα·[1] τὰ δὲ μέλανα ζυμοῦται καὶ μετεωρίζεται[2] καὶ πολλαπλασιοῦται· ἀναγωγὸν γὰρ μελάνων ὄξος. γυναιξὶ δὲ τὸ ἐπίπαν πολεμιώτερον ἢ ἀνδράσιν 8 ὄξος· ὑστεραλγὲς γάρ ἐστιν.

(17 L.) LXII. Ὕδατι δὲ ποτῷ ἐν τῇσιν ὀξείῃσι νούσοισιν ἄλλο μὲν οὐδὲν ἔχω ἔργον ὅ τι προσθέω· οὔτε γὰρ βηχὸς παρηγορικόν ἐστιν ἐν τοῖσι περιπνευμονικοῖσιν οὔτε πτυάλου ἀναγωγόν, ἀλλ' ἧσσον τῶν ἄλλων, εἴ τις διὰ παντὸς ποτῷ ὕδατι χρέοιτο· μεσηγὺ μέντοι ὀξυμέλιτος καὶ μελικρήτου ὕδωρ ἐπιρρυφεόμενον ὀλίγον πτυάλου ἀναγωγόν ἐστι διὰ τὴν μεταβολὴν τῆς ποιότητος τῶν ποτῶν· πλημμυρίδα γάρ τινα ἐμποιεῖ. ἄλλως 10 δὲ οὐδὲ δίψαν παύει, ἀλλ' ἐπιπικραίνει· χολῶδες

[1] ὑπ' αὐτοῦ οὐ μετεωριζόμενα is my conjecture: μετεωριζόμενα ὑπ' αὐτοῦ all MSS.
[2] μετεωρίζεται MV: μερίζεται A.

---

[1] This sentence is a puzzle, owing to the difficulty of getting the required contrast between μετεωριζόμενα and μετεωρίζεται if the MS. reading be retained. Littré translates the former "met en mouvement," the latter "soulève." Adams has "suspended" and "swells up." The translations are plainly impossible; surely μετεωρίζομαι must mean the same thing in both clauses. The verb μετεωρίζω ("I raise," "lift up") is mostly used of fermenting food inflating the bowels. It is therefore just possible that μετεωριζόμενα should be transposed, and placed after πικρά. "Bitter humours, when inflated, are dissolved by it into phlegm; black humours are fermented, inflated and multiplied." The chief objection to this version is that ἀναγωγὸν γὰρ μελάνων ὄξος is pointless,

LXI. To put it briefly, acidities from vinegar benefit those who suffer from bitter bile more than those who suffer from black. For the bitter humours are dissolved and turned into phlegm by it, not being brought up;[1] but the black are fermented, brought up and multiplied, vinegar being apt to raise black humours. Women on the whole are more liable to be hurt by vinegar than are men, as it causes pain in the womb.

LXII. Water as a drink in acute diseases has no particular quality I can attribute to it, as it neither sooths a cough in pneumonia nor brings up sputum, having in these respects less effect than other things, if it be used throughout as a drink. If however it be swallowed between the giving of oxymel and that of hydromel it slightly[2] favours the bringing up of sputum, owing to the change in the quality of the drinks, as it causes a kind of flood. Apart from this it is of no use, not even quenching thirst, but adding a bitterness to it; for it increases the

for ἀναγωγὸν must mean "bring up into the mouth," as this is the sense of ἀναγωγὸς throughout this treatise. The same objection applies to the otherwise attractive reading of A, μερίζεται for μετεωρίζεται.

I once thought that μετεωρίζεται had displaced some verb of the opposite meaning to μετεωριζόμενα, but once more ἀναγωγὸν γὰρ μελάνων ὄξος is against this. I therefore suggest the reading in the text, though with no great confidence. It allows ἀναγωγὸν γὰρ κ.τ.λ. to have its full and proper meaning, but it gives a rare meaning to μετεωρίζω as used in the medical writers. Still in *Regimen in Health* 5 (Littré VI. 78), τὰ μετεωριζόμενα κάτω ὑπάγειν, it almost certainly has the sense of movement towards the mouth from the stomach.

[2] ὀλίγον is perhaps an adjective agreeing with ὕδωρ, "A little water favours, etc."

γὰρ φύσει χολώδει καὶ ὑποχονδρίῳ κακόν. κάκιστον δ' ἑωυτοῦ καὶ χολωδέστατον καὶ φιλαδυναμώτατον, ὅταν ἐς κενεότητα ἐσέλθῃ. καὶ σπληνὸς δὲ αὐξητικὸν καὶ ἥπατός ἐστιν, ὁπόταν πεπυρωμένον ᾖ, καὶ ἐγκλυδαστικόν τε καὶ ἐπιπολαστικόν· βραδύπορον γὰρ διὰ τὸ ὑπόψυχρον εἶναι καὶ ἄπεπτον, καὶ οὔτε διαχωρητικὸν οὔτε διουρητικόν. προσβλάπτει δέ τι[1] καὶ διὰ τόδε, ὅτι ἄκοπρόν ἐστι φύσει. ἢν δὲ δὴ καὶ
20 ποδῶν ποτε ψυχρῶν ἐόντων ποθῇ, πάντα ταῦτα πολλαπλασίως βλάπτει, ἐς ὅ τι ἂν αὐτῶν
22 ὁρμήσῃ.

LXIII. Ὑποπτεύσαντι μέντοι ἐν ταύτῃσι τῇσι νούσοισι καρηβαρίην ἰσχυρὴν ἢ φρενῶν ἅψιν παντάπασιν οἴνου ἀποσχετέον. ὕδατι δ' ἐν τῷ τοιῷδε χρηστέον ἢ ὑδαρέα καὶ κιρρὸν οἶνον παντελῶς δοτέον καὶ ἄνοδμον παντάπασι, καὶ μετὰ τὴν πόσιν αὐτοῦ ὕδωρ μεταποτέον ὀλίγον· ἧσσον γὰρ ἂν οὕτω τὸ ἀπὸ τοῦ οἴνου μένος ἅπτοιτο κεφαλῆς καὶ γνώμης. ἐν οἷσι δὲ μάλιστα αὐτῷ ὕδατι ποτῷ χρηστέον καὶ ὁπότε πολλῷ κάρτα
10 καὶ ὅπου μετρίῳ, καὶ ὅπου θερμῷ καὶ ὅπου ψυχρῷ, τὰ μέν που πρόσθεν εἴρηται, τὰ δ' ἐν
12 αὐτοῖσι τοῖσι καιροῖσι ῥηθήσεται.

LXIV. Κατὰ ταῦτα δὲ καὶ περὶ τῶν ἄλλων ποτῶν, οἷον κρίθινον καὶ τὰ ἀπὸ χλοιῆς ποιεύμενα καὶ τὰ ἀπὸ σταφίδος καὶ στεμφύλων καὶ πυρῶν καὶ κνήκου καὶ μύρτων καὶ ῥοιῆς καὶ τῶν ἄλλων, ὅταν τινὸς αὐτῶν καιρὸς ᾖ χρῆσθαι, γεγράψεται

[1] δέ τι MSS.: δ' ἔτι Coray and Reinhold.

bile of the naturally bilious and is injurious to the hypochondrium. Its bad qualities are at their worst, it is most bilious, and most weakening, when it is drunk during a fast. It enlarges the spleen, and the liver, when inflamed; it causes a gurgling inside without penetrating downwards.[1] For it travels slowly owing to its being cool and difficult of digestion, while it is neither laxative nor diuretic. It also causes some harm because by nature it does nothing to increase faeces. If furthermore it be drunk while the feet are cold, all its harmful effects are multiplied, no matter which of them it happens to aggravate.

LXIII. Should you suspect, however, in these diseases an overpowering heaviness of the head, or that the brain is affected, there must be a total abstinence from wine. In such cases use water, or at most give a pale-yellow wine, diluted and entirely without odour. After each draft of it give a little water to drink, for so the strength of the wine will affect less the head and the reason. As to the principal cases in which water alone must be employed as a drink, when it should be used in abundance and when in moderation, when it should be warm and when cold, I have in part discussed these things already, and shall do so further when the occasions arise.

LXIV. Similarly with the other kinds of drink, barley-water for instance, herbal drinks, those made from raisins, grape-skins, wheat, bastard saffron, myrtle, pomegranates and so forth, along with the proper times for their use, a discussion will be

[1] ἐπιπολαστικὸν means literally "remaining on the surface"; hence "not going downwards."

παρ' αὐτῷ τῷ νοσήματι ὅπωσπερ καὶ τἄλλα τῶν
7 συνθέτων φαρμάκων.

(18 L.) LXV. Λουτρὸν δὲ συχνοῖσι τῶν νοσημάτων[1] ἀρήγοι ἂν χρεωμένοισιν ἐς τὰ μὲν συνεχέως, ἐς τὰ δ' οὔ. ἔστι δ' ὅτε ἧσσον χρηστέον διὰ τὴν ἀπαρασκευασίην τῶν ἀνθρώπων· ἐν ὀλίγῃσι γὰρ οἰκίῃσι παρεσκεύασται τὰ ἄρμενα καὶ οἱ θεραπεύσοντες[2] ὡς δεῖ. εἰ δὲ μὴ παγκάλως λούοιτο, βλάπτοιτο ἂν οὐ σμικρά· καὶ γὰρ σκέπης ἀκάπνου δεῖ καὶ ὕδατος δαψιλέος καὶ τοῦ λουτροῦ συχνοῦ καὶ μὴ λίην λάβρου, ἢν μὴ οὕτω
10 δέῃ. καὶ μᾶλλον μὲν μὴ σμήχεσθαι· ἢν δὲ σμήχηται, θερμῷ χρῆσθαι αὐτῷ καὶ πολλαπλασίῳ ἢ ὡς νομίζεται σμήγματι, καὶ προσκαταχεῖσθαι μὴ ὀλίγῳ, καὶ ταχέως μετακαταχεῖσθαι. δεῖ δὲ καὶ τῆς ὁδοῦ βραχείης ἐς τὴν πύαλον, καὶ ἐς εὐέμβατόν τε καὶ εὐέκβατον. εἶναι δὲ καὶ τὸν λουόμενον κόσμιον καὶ σιγηλὸν καὶ μηδὲν αὐτὸν προσεργάζεσθαι, ἀλλ' ἄλλους καὶ καταχεῖν καὶ σμήχειν· καὶ μετακέρασμα πολλὸν ἡτοιμάσθαι καὶ τὰς ἐπαντλήσιας ταχείας ποιεῖσθαι· καὶ
20 σπόγγοισι χρῆσθαι ἀντὶ στεγγίδος, καὶ μὴ ἄγαν ξηρὸν χρίεσθαι τὸ σῶμα. κεφαλὴν μέντοι ἀνεξηράνθαι χρὴ ὡς οἷόν τε μάλιστα ὑπὸ σπόγγου

[1] νοσημάτων MSS.: νοσεύντων Kühlewein.

[2] θεραπεύσοντες my suggestion: θεραπεύσαντες A: θεραπέοντες V.

---

[1] It should be noticed that these promises are not fulfilled. Perhaps the author wrote, or intended to write, a book on particular diseases to supplement his "general" pathology.

given together with the particular disease in question;[1] similarly too with the rest of the compound medicines.

LXV. The bath will be beneficial to many patients, sometimes when used continuously, sometimes at intervals. Occasionally its use must be restricted, because the patients have not the necessary accommodation, for few houses have suitable apparatus and attendants to manage the bath properly. Now if the bath be not carried out thoroughly well, no little harm will be done. The necessary things include a covered place free from smoke, and an abundant supply of water, permitting bathings that are frequent but not violent, unless violence is necessary.

If rubbing with soap be avoided, so much the better; but if the patient be rubbed, let it be with soap[2] that is warm, and many times greater in amount than is usual, while an abundant affusion should be used both at the time and immediately afterwards. A further necessity is that the passage to the basin should be short, and that the basin should be easy to enter and to leave. The bather must be quiet and silent; he should do nothing himself, but leave the pouring of water and the rubbing to others. Prepare a copious supply of tepid[3] water, and let the affusions be rapidly made. Use sponges instead of a scraper, and anoint the body before it is quite dry. The head, however, should be rubbed with a sponge until it is as dry

[2] *σμῆγμα*, the Greek equivalent for soap, usually consisted of olive oil and an alkali mixed into a paste.

[3] *μετακέρασμα*, a mixture of hot and cold water, to enable the bather to "cool down" by degrees.

ἐκμασσομένην. καὶ μὴ διαψύχεσθαι τὰ ἄκρα μηδὲ τὴν κεφαλὴν μηδὲ τὸ ἄλλο σῶμα· καὶ μήτε νεορρύφητον μήτε νεόποτον λούεσθαι μηδὲ ῥυφεῖν
26 μηδὲ πίνειν ταχὺ μετὰ τὸ λουτρόν.

LXVI. Μέγα μὲν δὴ μέρος χρὴ νέμειν τῷ κάμνοντι, ἢν ὑγιαίνων ᾖ φιλόλουτρος ἄγαν καὶ εἰθισμένος λούεσθαι· καὶ γὰρ ποθέουσι μᾶλλον οἱ τοιοίδε καὶ ὠφελέονται λουσάμενοι καὶ βλάπτονται μὴ λουσάμενοι. ἁρμόζει δ' ἐν περιπνευμονίῃσι μᾶλλον ἢ ἐν καύσοισι τὸ ἐπίπαν· καὶ γὰρ ὀδύνης τῆς κατὰ πλευρῶν καὶ στήθεος καὶ μεταφρένου παρηγορικόν ἐστιν λουτρὸν καὶ πτυάλου πεπαντικὸν καὶ ἀναγωγὸν καὶ εὔπνοον καὶ ἄκο-
10 πον· μαλθακτικὸν γὰρ καὶ ἄρθρων καὶ τοῦ ἐπιπολαίου δέρματος· καὶ οὐρητικὸν δὲ καὶ
12 κάρηβαρίην λύει καὶ ῥῖνας ὑγραίνει.

LXVII. Ἀγαθὰ μὲν οὖν λουτρῷ τοσαῦτα πάρεστιν, ὧν πάντων δεῖ. εἰ μέντοι τῆς παρασκευῆς ἔνδειά τις ἔσται ἑνὸς ἢ πλειόνων, κίνδυνος μὴ λυσιτελεῖν τὸ λουτρόν, ἀλλὰ μᾶλλον βλάπτειν· ἓν γὰρ ἕκαστον αὐτῶν μεγάλην φέρει βλάβην μὴ προπαρασκευασθὲν ὑπὸ τῶν ὑπουργῶν ὡς δεῖ. ἥκιστα δὲ λούειν καιρὸς τούτους, οἷσιν ἡ κοιλίη ὑγροτέρη τοῦ καιροῦ ἐν τῇσι νούσοισιν· ἀτὰρ οὐδ' οἷσιν ἕστηκε μᾶλλον τοῦ καιροῦ καὶ μὴ προδιελή-
10 λυθεν. οὐδὲ δὴ τοὺς γεγυιωμένους χρὴ λούειν οὐδὲ τοὺς ἀσώδεας ἢ ἐμετικοὺς οὐδὲ τοὺς ἐπανερευγομένους χολῶδες οὐδὲ τοὺς αἱμορραγέοντας ἐκ ῥινῶν, εἰ μὴ ἔλασσον τοῦ καιροῦ ῥέοι· τοὺς δὲ καιροὺς οἶδας. εἰ δὲ ἔλασσον τοῦ καιροῦ ῥέοι, λούειν, ἤν τε ὅλον τὸ σῶμα πρὸς τὰ ἄλλα ἀρήγῃ,
16 ἤν τε τὴν κεφαλὴν μοῦνον.

as possible. Keep chill from the extremities and the head, as well as from the body generally. The bath must not be given soon after gruel or drink has been taken, nor must these be taken soon after a bath.

LXVI. Let the habits of the patient carry great weight—whether he is very fond of his bath when in health, or is in the habit of bathing. Such people feel the need of a bath more, are more benefited by its use and more harmed by its omission. On the whole, bathing suits pneumonia rather than ardent fevers, for it soothes pain in the sides, chest and back; besides, it concocts and brings up sputum, eases respiration, and removes fatigue, as it softens the joints and the surface of the skin. It is diuretic, relieves heaviness of the head, and moistens the nostrils.

LXVII. Such are the benefits from bathing, and they are all needed. If, however, one or more requisites be wanting, there is a danger that the bath will do no good, but rather harm. For each neglect of the attendants to make proper preparations brings great harm. It is a very bad time to bathe when the bowels are looser than they ought to be[1] in acute diseases, likewise too when they are more costive than they ought to be, and have not previously been moved. Do not bathe the debilitated, those affected by nausea or vomiting, those who belch up bile, nor yet those who bleed from the nose, unless the hemorrhage be less than normal, and you know what the normal is. If the hemorrhage be less than normal, bathe either the whole body, if that be desirable for other considerations, or else the head only.

[1] Or, "normal"; see note on p. 108.

LXVIII. Ἢν οὖν αἵ τε παρασκευαὶ ἔωσιν ἐπιτήδειοι καὶ ὁ κάμνων μέλλῃ εὖ δέξασθαι τὸ λουτρόν, λούειν χρὴ ἑκάστης ἡμέρης· τοὺς δὲ φιλολουτρέοντας, οὐδ' εἰ δὶς τῆς ἡμέρης λούοις, οὐδὲν ἂν βλάπτοις. χρῆσθαι δὲ λουτροῖσι τοῖσι οὔλῃσι πτισάνῃσι χρεωμένοισι παρὰ πολὺ μᾶλλον ἐνδέχεται, ἢ τοῖσι χυλῷ μοῦνον χρεωμένοισιν· ἐνδέχεται δὲ καὶ τούτοισιν ἐνίοτε· ἥκιστα δὲ[1] τοῖσι ποτῷ μοῦνον χρεωμένοις· ἔστι δ' οἷσι καὶ
10 τούτων ἐνδέχεται. τεκμαίρεσθαι δὲ χρὴ τοῖσι προγεγραμμένοισιν, οὕς τε μέλλει λουτρὸν ὠφελεῖν ἐν ἑκάστοισι τῶν τρόπων τῆς διαίτης οὕς τε μή· οἷσι μὲν γὰρ προσδεῖ[2] τινος κάρτα τούτων, ὅσα λουτρὸν ἀγαθὰ ποιεῖ, λούειν καθ'[3] ὅσα ἂν λουτρῷ ὠφελῆται· οἷσι δὲ τούτων μηδενὸς προσδεῖ καὶ πρόσεστιν αὐτοῖσί τι τῶν σημείων,
17 ἐφ' οἷς λούεσθαι οὐ συμφέρει, οὐ δεῖ λούειν.

[1] After δὲ the MSS have καὶ which Ermerins deletes.

[2] προσδεῖ Kühlewein for προσδεῖταί (A) or προσδέεταί (MV) of the MSS.

LXVIII. If the preparations be adequate, and the patient likely to benefit by the bath, bathe every day. Those who are fond of bathing will not be harmed even by two baths a day. Patients taking unstrained gruel are much more capable of using the bath than those taking juice only, though these too can use it sometimes. Those taking nothing but drink are the least capable, though some even of these can bathe. Judge by means of the principles given above who are likely and who are unlikely to profit by the bath in each kind of regimen. Those who really need one of the benefits given by the bath you should bathe as far as they are profited by the bath. Those should not be bathed who have no need of these benefits, and who furthermore show one of the symptoms that bathing is not suitable.

---

[3] καθ' Kühlewein: καὶ MSS.

THE SACRED DISEASE

INTRODUCTION

This book was apparently known to Bacchius,[1] and is referred to by Galen [2] without his mentioning the author's name. It is in Erotian's list of the genuine works of Hippocrates.

Modern critics are by no means agreed about either its authorship or its merits. Littré [3] has very little to say about it. Ermerins regards it as the patchwork composition of a second-rate sophist much later than Hippocrates. Gomperz [4] speaks of the "wonderfully suggestive formula" invented by its author, and calls him pugnacious and energetic. Wilamowitz [5] rates it very highly indeed, and considers that it was written by the author of *Airs Waters Places.* Wellmann [6] believes it was written in opposition to the Sicilian school, including Diocles, who believed in incantations. An English writer [7] speaks of it as "a masterpiece of scientific sanity; broad in outlook, keen and ironical in argument and humane in spirit."

One point at least is certain—*The Sacred Disease* cannot be independent of *Airs Waters Places.* It will be convenient to quote the parallel passages side by side.

[1] See Littré, I. 137.

[2] XVII. pt. 2., 341 and XVIII. pt. 2., 18.

[3] VI. 350 foll.

[4] *Greek Thinkers*, I. 311–313.

[5] *Griechisches Lesebuch*, 269, 270.

[6] *Fragmentensammlung*, I. 30, 31.

[7] John Naylor in *Hibbert Journal* (Oct., 1909), *Luke the Physician and Ancient Medicine.*

Airs Waters Places	*The Sacred Disease*
τούς τε ἀνθρώπους τὰς κεφαλὰς ὑγρὰς ἔχειν καὶ φλεγματώδεας, τάς τε κοιλίας αὐτῶν πυκνὰ ἐκταράσσεσθαι ἀπὸ τῆς κεφαλῆς τοῦ φλέγματος ἐπικαταρρέοντος. III. κατάρροοι ἐπιγενόμενοι ἐκ τοῦ ἐγκεφάλου παραπληκτικοὺς ποιέουσι τοὺς ἀνθρώπους, ὁκόταν ἐξαίφνης ἡλιωθέωσι τὴν κεφαλὴν ἢ ῥιγώσωσι. III.	See Chapters VIII.–XII.
ἐξ ἁπάντων ἐν ὁκόσοισι ὑγρόν τι ἔνεστιν. ἔνεστι δὲ ἐν παντὶ χρήματι. VIII.	ὅσα φύεται καὶ ἐν οις τι ὑγρόν ἐστιν· ἔστι δὲ ἐν παντί. XVI.
φλέγματος ἐπικαταρρυέντος ἀπὸ τοῦ ἐγκεφάλου. X.	ὁ ἐγκέφαλος . . . ὥστε οὐκ ἐπικαταρρεῖ. XIII.
ὁ γὰρ γόνος πανταχόθεν ἔρχεται τοῦ σώματος, ἀπό τε τῶν ὑγιηρῶν ὑγιηρὸς ἀπό τε τῶν νοσερῶν νοσερός. εἰ οὖν γίνονται ἔκ τε φαλακρῶν φαλακροὶ καὶ ἐκ γλαυκῶν γλαυκοὶ καὶ ἐκ διεστραμμένων στρεβλοὶ ὡς ἐπὶ τὸ πλῆθος, καὶ περὶ τῆς ἄλλης μορφῆς ὁ αὐτὸς λόγος, τί κωλύει καὶ ἐκ μακροκεφάλου μακροκέφαλον γίνεσθαι; XIV.	ὁ γόνος ἔρχεται πάντοθεν τοῦ σώματος, ἀπό τε τῶν ὑγιηρῶν ὑγιηρός, καὶ ἀπὸ τῶν νοσερῶν νοσερός. V. εἰ γὰρ ἐκ φλεγματώδεος φλεγματώδης, καὶ ἐκ χολώδεος χολώδης γίνεται, καὶ ἐκ φθινώδεος φθινώδης, καὶ ἐκ σπληνώδεος σπληνώδης, τί κωλύει κ.τ.λ. V.

ἐμοὶ δὲ καὶ αὐτῷ δοκεῖ ταῦτα τὰ πάθεα θεῖα εἶναι καὶ τἄλλα πάντα καὶ οὐδὲν ἕτερον ἑτέρου θειότερον οὐδὲ ἀνθρωπινώτερον, ἀλλὰ πάντα ὁμοῖα καὶ πάντα θεῖα. ἕκαστον δὲ αὐτῶν ἔχει φύσιν τὴν ἑωυτοῦ καὶ οὐδὲν ἄνευ φύσιος γίνεται. XXII.	οὐδέν τί μοι δοκεῖ τῶν ἄλλων θειοτέρη εἶναι νούσων οὐδὲ ἱερωτέρη, ἀλλὰ φύσιν μὲν ἔχει καὶ πρόφασιν. I. ἀλλὰ πάντα θεῖα καὶ πάντα ἀνθρώπινα· φύσιν δὲ ἕκαστον ἔχει καὶ δύναμιν ἐφ' ἑωυτοῦ. XXI.
καὶ ἐχρῆν, ἐπεὶ θειότερον τοῦτο τὸ νόσευμα τῶν λοιπῶν ἐστιν, οὐ . . . προσπίπτειν μούνοις, ἀλλὰ τοῖς ἅπασιν ὁμοίως. XXII.	καίτοι εἰ θειότερόν ἐστι τῶν ἄλλων, τοῖσιν ἅπασιν ὁμοίως ἔδει γίνεσθαι τὴν νοῦσον ταύτην. V.
ἀλλὰ γάρ, ὥσπερ καὶ πρότερον ἔλεξα, θεῖα μὲν καὶ ταῦτά ἐστιν ὁμοίως τοῖς ἄλλοις· γίνεται δὲ κατὰ φύσιν ἕκαστα. XXII.	τὸ δὲ νόσημα τοῦτο οὐδέν τί μοι δοκεῖ θειότερον εἶναι τῶν λοιπῶν, ἀλλὰ φύσιν ἔχει ἣν καὶ τὰ ἄλλα νοσήματα, καὶ πρόφασιν. V.

Besides these special passages, both treatises lay stress upon moistening of the brain as a cause of disease, and upon the purging and drying of that organ by "catarrhs"; both insist upon supposed functions of veins, upon the importance of winds and the change of the seasons; both too have much the same "pet" words, ἐκκρίνειν, ἀποκρίνειν, κοιλίαι and so on. In one occurs the phrase ὧδε ἢ ὅτι τούτων ἐγγύτατα, in the other οὕτω ἢ ὅτι τούτων ἐγγυτάτω.

So much for the similarities. There are also dissimilarities. *Airs Waters Places* is free from sophistic rhetoric, but the author of *The Sacred Disease* is not above such artifices as this: κατὰ μὲν τὴν ἀπορίην αὐτοῖσι τοῦ μὴ γινώσκειν τὸ θεῖον διασῴζεται, κατὰ δὲ *τὴν εὐπορίην τοῦ τρόπου τῆς ἰήσιος ᾧ ἰῶνται, ἀπόλλυται.* *A. W. P.* seems to be dominated by no postulates of philosophy; *S. D.* is eclectic, laying stress now upon air, as the element which makes the brain intelligent,[1] now upon the four traditional "opposites," the wet, the dry, the hot and the cold.[2] Above all, *A. W. P.* is more dignified, more reserved, and more compact in style.

Wilamowitz may possibly be right in his contention that both works are by the same writer. If this be so, the writer was almost certainly not the author of *Epidemics*. The latter would never have said that cures can be effected by creating at the proper seasons the dry or the moist, the hot or the cold.

A confident verdict would be rash, but I am inclined to believe that the writer of *S. D.* was a pupil of the writer of *A. W. P.* Perhaps the master set his pupil a thesis on a subject which was a favourite of his—"Superstition and Medicine." It would be natural in the circumstances for the student to borrow without acknowledgment from his master not only arguments but also verbal peculiarities, but he would not hide his own characteristics either of thought or of style.

Although the work is generally supposed to refer to epilepsy,[3] other seizures, including certain forms of insanity, must not be excluded. Epilepsy generally conforms to a regular type, and scarcely corresponds to the elaborate classification in Chapter IV.

[1] *τὴν δὲ φρόνησιν ὁ ἀὴρ παρέχεται.* XIX.

[2] See Chapter XXI.

[3] It should be noticed that the usual term employed is "this disease." The word ἐπίληψις occurs once only (Chapter XIII.), where it means "seizure."

INTRODUCTION

In opposition to popular opinion, the writer maintains that these seizures are not due to "possession" by a god but to a natural cause. He insists upon the uniformity of Nature, and protests against the unscientific dualism which characterizes some phenomena as natural and others as divine. All phenomena, he says, are both natural and divine. He holds that epilepsy is curable by natural means, intending, apparently, to imply that it can be cured if the right remedies are discovered, and not that cures actually did occur.

The "cause" of epilepsy is said to be the stoppage of life-giving air in the veins[1] by a flow of phlegm from the head into them. The crude and mistaken physiology of this part of the work need not detain us,[2] but the function assigned to air is important, and shows the influence of Diogenes of Apollonia.

Far more interesting is the function attributed to the brain, which, in opposition to the popular view, is regarded as the seat of consciousness, and not the heart or the midriff. The view was not novel, and can be traced back to Alcmaeon.[3] It was accepted by Plato and rejected by Aristotle.[4]

[1] I have translated φλέβες by "veins" and φλέβια by "minor veins," though I do not think that the writer always maintained a distinction between the two words. Of course φλέβες includes what are now called "arteries," but as the difference between veins and arteries was not known in the author's time "veins" must be the normal translation.

[2] The confident assurance with which the writer enunciates his views on phlegm and air is in sharp contrast with the extreme caution of the writer of *Epidemics I.* and *III.*

[3] See Beare *Greek Theories of Elementary Cognition*, 93 and 160.

[4] See Beare *op. cit.* index *s.v.* "brain."

INTRODUCTION

The date of the work can be fixed with tolerable certainty. Nobody would put it before *Airs Waters Places*, unless indeed with Wilamowitz one holds that the two were written by the same author, in which case *The Sacred Disease* might be a youthful composition. But even on this supposition the difference between the dates of the two would not be great. On the other hand the work was known to Bacchius, early in the third century, and apparently regarded as genuine. There are in the vulgate two places where μὴ has displaced οὐ (a sure sign of late date) but an examination of the best manuscript shows that in both οὐ is the true reading. Here and there occur touches of sophistic rhetoric which make a fourth-century date unlikely, and an impartial reader feels that the writer, whoever he was, was a contemporary, probably a younger contemporary, of Socrates.[1] There is no internal sign of the part of Greece in which the author lived, except that the list of gods given in Chapter IV. seems to be Ionian.[2]

The more often *The Sacred Disease* is read, the more it attracts the reader, particularly if it be realized that the sequence of thought is sometimes impaired by glosses, which must be removed if a fair judgment on the writer is to be given. At first one notices the crudities, the slight logical lapses, the unwarranted assumptions, all of which are natural enough if the writer was a pupil writing a set thesis for his teacher.

[1] The writer is even more vigorously opposed to superstition than the great Socrates himself, with his δαιμόνιον and faith in oracles.

[2] See the writer in *Pauly-Wissowa*, "Hippokrates," p. 1827.

Then little by little the grandeur of the main theme, the uniformity of Nature, every aspect of which is equally divine, grips the attention. We realise that we are in contact with a great mind, whether the words in front of us are the direct expression of that mind, or only the indirect expression through the medium of a pupil's essay.

Manuscripts and Editions

The chief MSS. are θ and M, supplemented by (*a*) some Paris MSS. of an inferior class and by (*b*) those MSS. which Littré called ι, κ, λ and μ.[1]

Of these the best is θ, a tenth-century MS. at Vienna, for which see Ilberg in the *Prolegomena* to the Teubner edition of Hippocrates. If θ be closely followed it produces on the text of *The Sacred Disease* much the same effect as following A produces on *Ancient Medicine*; there is greater simplicity, while the dialect is much improved. By its help the editor is often able to remove the faults which so disfigure the text of Littré and even that of Reinhold.

The Sacred Disease is included in Reinhold's edition, while a great part appears in the *Lesebuch* of Wilamowitz-Moellendorff.[2] It is translated into English in the second volume of Adams.

I have myself collated both θ and M for the present edition. The collation of θ used by Littré was very accurate, but he appears to have known but little about M. Many of the Paris manuscripts,

[1] See Littré, VI., 351.

[3] See also *Die hippokratische Schrift* περὶ ἱρῆς νούσου in *Sitzungsberichte der Berliner Akademie*, 1901. In 1827 there was published in Leipzig an edition by Fr. Dietz.

however, are so similar to M that they supplied him with nearly all, if not quite all, of its readings.

The printed text follows θ closely, but on several occasions I have preferred M. I believe that I have given in the footnotes the reading of θ on at any rate the most important of these occasions. So the reader will find the critical notes to this treatise more elaborate than usual. As no full edition exists, perhaps this novelty will not be unwelcome.

The scribe of M appears to have been a fairly good Greek scholar, and his text is on the whole smoother and more regular than that of θ. He prefers the pronominal forms in ὁκ- to those in ὁπ-, and he uses the long forms ποιέειν, etc. Punctuation and accents are fairly correct. His marginal notes sometimes run into verse. Thus on fol. 85r (bottom) we have:—

Ἱππόκρατες, τὸ θεῖον ἵλεων ἔχοις,

a pious wish that the author may not be punished for "denying divinely the divine." On 91r he has this note on the last sentence of the treatise:—

ἰητρὲ πρόσσχες, γνῶθι τῶν καιρῶν ὅρους.

On the whole, the readings of M in *Sacred Disease* are rather better than they are in the treatises already translated.

The manuscript called θ is written in a very clear and beautiful script. The scribe, however, seems to have been a poor Greek scholar. The punctuation is hopeless, and the accentuation far from good. He writes ἄνωι, κάτωι, διεφθάρηι, πλείωι, ἐπάγηι and κεφαλῆι (nominative). On the other hand we have τῶ χρόνω. He occasionally slips into Attic forms, *e.g.* θαλάττης, and μεταβολαῖς with μεταβολῇσι immediately following. Vagaries such as these, combined with the fact that he cannot make up his mind whether to write ἱρὸς or ἱερός, show how little we can hope to regain exactly the spelling of the Hippocratic writers. We must be content with very approximate knowledge.

The most interesting point brought out by a comparison between M and θ is the great number of trivial differences, chiefly in the order of the words. There are also many little words and phrases in M which are not found in θ. In many cases it almost seems that a rough text has been purposely made smoother. For instance, M has μὲν γὰρ on at least two occasions when θ has μὲν only. But there are many differences which are in no way corrections or improvements, and it is therefore difficult, if not impossible, to say always which manuscript is to be preferred. Fortunately these differences do not affect the general sense; they do, however, tend to show that at some period (or periods) in the history of the text the Hippocratic writings were copied with much more attention to the meaning than to verbal faithfulness.

ΠΕΡΙ ΙΕΡΗΣ ΝΟΥΣΟΥ

I. Περὶ τῆς ἱερῆς νούσου καλεομένης ὧδ' ἔχει. οὐδέν τί μοι δοκεῖ τῶν ἄλλων θειοτέρη εἶναι νούσων οὐδὲ ἱερωτέρη, ἀλλὰ φύσιν μὲν ἔχει καὶ πρόφασιν, οἱ δ' ἄνθρωποι[1] ἐνόμισαν θεῖόν τι πρῆγμα[2] εἶναι ὑπὸ ἀπειρίης καὶ θαυμασιότητος, ὅτι οὐδὲν ἔοικεν ἑτέροισι· καὶ κατὰ μὲν τὴν ἀπορίην αὐτοῖσι τοῦ μὴ γινώσκειν τὸ θεῖον διασῴζεται, κατὰ δὲ τὴν εὐπορίην τοῦ τρόπου τῆς ἰήσιος ᾧ ἰῶνται,[3] ἀπόλλυται, ὅτι 10 καθαρμοῖσί τε ἰῶνται καὶ ἐπαοιδῇσιν. εἰ δὲ διὰ τὸ θαυμάσιον θεῖον νομιεῖται, πολλὰ τὰ ἱερὰ νοσήματα ἔσται καὶ οὐχὶ ἕν,[4] ὡς ἐγὼ ἀποδείξω ἕτερα οὐδὲν ἧσσον ἐόντα θαυμάσια οὐδὲ τερα-

[1] φύσιν μὲν ἔχει καὶ πρόφασιν, οἱ δ' ἄνθρωποι κ.τ.λ. my emendation: φύσιν μὲν ἔχει καὶ τὰ λοιπὰ νοσήματα ὅθεν γίνεται φύσιν τε αὐτὴ (αὐτὴν ι) καὶ πρόφασιν οἱ δ' ἄνθρωποι θι: M has δὲ for τε and omits δ'. The punctuation of θ is very erratic here. φύσιν μὲν ἔχει ἣν καὶ τὰ λοιπὰ νοσήματα, ὅθεν γίνεται. φύσιν δὲ αὐτῇ καὶ πρόφασιν οἱ ἄνθρωποι κ.τ.λ. Littré: φύσιν μὲν ἔχειν, ἣν καὶ τὰ λοιπὰ νοσήματα, ὅθεν γίγνεται. φύσιν δὲ αὐτῇ καὶ πρόφασιν οἱ ἄνθρωποι κ.τ.λ. Ermerins: φύσιν μὲν ἔχει καὶ τἄλλα νοσήματα καὶ πρόφασιν ἕκαστα ὅθεν γίγνεται, φύσιν δὲ καὶ τοῦτο καὶ πρόφασιν· οἱ δ' ἄνθρωποι κ.τ.λ. Reinhold: φύσιν μὲν ἔχει καὶ τἆλλα νοσήματα, ὅθεν γίνεται, φύσιν δὲ καὶ αὕτη καὶ πρόφασιν· οἱ δ' Wilamowitz.

[2] M omits τι πρῆγμα.

[3] ὠπῶνται θ: M has ἰήσιος ἰῶνται· ἀπολύονται γὰρ ἢ καθαρμοῖσιν ἢ ἐπαοιδῇσιν (the final ν is very faint).

THE SACRED DISEASE

I. I AM about to discuss the disease called "sacred." It is not, in my opinion, any more divine or more sacred than other diseases, but has a natural cause, and its supposed divine origin is due to men's inexperience, and to their wonder at its peculiar character.[1] Now while men continue to believe in its divine origin because they are at a loss to understand it, they really disprove its divinity by the facile method of healing which they adopt, consisting as it does of purifications and incantations. But if it is to be considered divine just because it is wonderful, there will be not one sacred disease but many, for I will show that other diseases are no less

[1] I am by no means satisfied that the text I have given is correct, but I am sure that the received text is wrong. However, as our best manuscript has δ' before ἄνθρωποι, probably φύσιν μὲν ἔχει is answered by οἱ δ' ἄνθρωποι ἐνόμισαν, in which case the intervening words are a gloss, or parts of a gloss. The fact is that φύσιν μὲν ἔχει, even without πρόφασιν, is enough to make clear the writer's meaning, as we can see from the passage in *Airs Waters Places*, XXII. (Vol. I. p. 126), which was certainly in his mind: ἕκαστον δὲ αὐτῶν ἔχει φύσιν τὴν ἑωυτοῦ καὶ οὐδὲν ἄνευ φύσιος γίνεται. But a scholiast would be very tempted to round off the sentence, and in particular to explain πρόφασιν. Hence arose, I think, καὶ τὰ λοιπὰ νοσήματα and ὅθεν γίνεται. Whatever the correct reading may be, and this is uncertain, the sense of the passage is perfectly clear.

[4] So M: τούτου εἵνεκεν θ.

τώδεα,[1] ἃ οὐδεὶς νομίζει ἱερὰ εἶναι. τοῦτο μὲν οἱ πυρετοὶ οἱ ἀμφημερινοὶ καὶ οἱ τριταῖοι καὶ οἱ τεταρταῖοι οὐδὲν ἧσσόν μοι δοκέουσιν ἱεροὶ εἶναι καὶ ὑπὸ θεοῦ γίνεσθαι ταύτης τῆς νούσου, ὧν οὐ θαυμασίως ἔχουσιν· τοῦτο δὲ ὁρῶ μαινομένους ἀνθρώπους καὶ παραφρονέοντας ἀπὸ οὐδεμιῆς
20 προφάσιος ἐμφανέος, καὶ πολλά τε καὶ ἄκαιρα ποιέοντας, ἔν τε τῷ ὕπνῳ οἶδα πολλοὺς οἰμώζοντας καὶ βοῶντας, τοὺς δὲ πνιγομένους, τοὺς δὲ καὶ ἀναΐσσοντάς τε καὶ φεύγοντας ἔξω καὶ παραφρονέοντας μέχρι ἐπέγρωνται,[2] ἔπειτα δὲ ὑγιέας ἐόντας καὶ φρονέοντας ὥσπερ καὶ πρότερον, ἐόντας τ' αὐτοὺς ὠχρούς τε καὶ ἀσθενέας, καὶ ταῦτα οὐχ ἅπαξ, ἀλλὰ πολλάκις. ἄλλα τε πολλά ἐστι καὶ παντοδαπὰ ὧν περὶ ἑκάστου λέγειν
29 πολὺς ἂν εἴη λόγος.

II. Ἐμοὶ δὲ δοκέουσιν οἱ πρῶτοι τοῦτο τὸ νόσημα ἱερώσαντες τοιοῦτοι εἶναι ἄνθρωποι οἷοι καὶ νῦν εἰσι μάγοι τε καὶ καθάρται καὶ ἀγύρται καὶ ἀλαζόνες, οὗτοι δὲ καὶ[3] προσποιέονται σφόδρα θεοσεβέες εἶναι καὶ πλέον τι εἰδέναι. οὗτοι τοίνυν παραμπεχόμενοι καὶ προβαλλόμενοι τὸ θεῖον τῆς ἀμηχανίης τοῦ μὴ ἔχειν ὅ τι προσενέγκαντες ὠφελήσουσι, καὶ ὡς μὴ κατάδηλοι ἔωσιν οὐδὲν ἐπιστάμενοι, ἱερὸν ἐνόμισαν τοῦτο
10 τὸ πάθος εἶναι· καὶ λόγους ἐπιλέξαντες ἐπιτηδείους τὴν ἴησιν κατεστήσαντο ἐς τὸ ἀσφαλὲς σφίσιν αὐτοῖσι, καθαρμοὺς προσφέροντες καὶ ἐπαοιδάς, λουτρῶν τε ἀπέχεσθαι κελεύοντες[4] καὶ ἐδεσμάτων πολλῶν καὶ ἀνεπιτηδείων ἀνθρώποισι

[1] θ omits οὐδὲ τερατώδεα.

wonderful and portentous, and yet nobody considers them sacred. For instance, quotidian fevers, tertians and quartans seem to me to be no less sacred and god-sent than this disease,[1] but nobody wonders at them. Then again one can see men who are mad and delirious from no obvious cause, and committing many strange acts; while in their sleep, to my knowledge, many groan and shriek, others choke, others dart up and rush out of doors, being delirious until they wake, when they become as healthy and rational as they were before, though pale and weak; and this happens not once but many times. Many other instances, of various kinds, could be given, but time does not permit us to speak of each separately.

II. My own view is that those who first attributed a sacred character to this malady were like the magicians, purifiers, charlatans and quacks of our own day, men who claim great piety and superior knowledge. Being at a loss, and having no treatment which would help, they concealed and sheltered themselves behind superstition, and called this illness sacred, in order that their utter ignorance might not be manifest. They added a plausible story, and established a method of treatment that secured their own position. They used purifications and incantations; they forbade the use of baths, and of many foods that are unsuitable for sick folk—of sea

[1] Because of the regularity of the attacks of fever, which occur every day (quotidians), every other day (tertians), or with intermissions of two whole days (quartans).

---

[2] So θ: *μέχρις ἐξεγρέωνται* M.

[3] M has *ἀφιερώσαντες αὐτοὶ τοιοῦτοι*, and *ὁκόσοι* for *οὗτοι δὲ καί*.

[4] *ἀπέχεσθαι κελεύοντες* M: *ἀπέχοντες* θ.

νοσέουσιν ἐσθίειν· θαλασσίων μὲν τρίγλης, μελανούρου, κεστρέος, ἐγχέλυος (οὗτοι γὰρ ἐπικηρότατοί εἰσιν),[1] κρεῶν δὲ αἰγείων[2] καὶ ἐλάφων καὶ χοιρίων καὶ κυνὸς (ταῦτα γὰρ κρεῶν ταρακτικώτατά ἐστι τῆς κοιλίης), ὀρνίθων δὲ ἀλεκτρυόνος[3]
20 καὶ τρυγόνος καὶ ὠτίδος, ἔτι δὲ ὅσα[4] νομίζεται ἰσχυρότατα εἶναι, λαχάνων δὲ μίνθης, σκορόδου καὶ κρομμύων (δριμὺ γὰρ ἀσθενέοντι οὐδὲν συμφέρει), ἱμάτιον δὲ μέλαν μὴ ἔχειν (θανατῶδες γὰρ τὸ μέλαν), μηδὲ ἐν αἰγείῳ κατακεῖσθαι δέρματι μηδὲ φορεῖν, μηδὲ[5] πόδα ἐπὶ ποδὶ ἔχειν, μηδὲ χεῖρα ἐπὶ χειρὶ (πάντα γὰρ ταῦτα κωλύματα εἶναι). ταῦτα δὲ τοῦ θείου εἵνεκα προστιθέασιν, ὡς πλέον τι εἰδότες, καὶ ἄλλας προφάσιας λέγοντες, ὅπως, εἰ μὲν ὑγιὴς γένοιτο,[6] αὐτῶν ἡ δόξα
30 εἴη καὶ ἡ δεξιότης, εἰ δὲ ἀποθάνοι, ἐν ἀσφαλεῖ καθισταῖντο αὐτῶν αἱ ἀπολογίαι καὶ ἔχοιεν πρόφασιν ὡς οὐδὲν αἴτιοί εἰσιν, ἀλλ' οἱ θεοί· οὔτε γὰρ φαγεῖν οὔτε πιεῖν ἔδοσαν φάρμακον οὐδέν, οὔτε λουτροῖσι καθήψησαν, ὥστε δοκεῖν αἴτιοι εἶναι. ἐγὼ δὲ δοκέω Λιβύων ἂν τῶν τὴν μεσόγειον οἰκεόντων οὐδέν' ἂν[7] ὑγιαίνειν, ὅτι ἐπ' αἰγείοισι δέρμασι κατακέονται καὶ κρέασιν αἰγείοισι χρέονται,[8] ἐπεὶ οὐκ ἔχουσιν οὔτε στρῶμα οὔτε ἱμάτιον οὔτε ὑπόδημα ὅ τι μὴ αἴγειόν ἐστιν· οὐ
40 γάρ ἐστιν αὐτοῖς ἄλλο προβάτιον οὐδὲν ἢ αἶγες

[1] ἐπικηρότατοι θι: ἐπικαιρότατοι M, Littré, Ermerins, Reinhold. Some MSS. have οἱ ἰχθύες after γάρ.

[2] After αἰγείων θ adds καὶ τύρου αἰγείου. The MSS. vary at this point between adjectives and nouns, but the sense is quite plain.

[3] ἀλεκτρυόνος M: ἀλεκτόριδος θ.

[4] ἔτι δὲ ὅσα M: ἃ θ.

fishes: red mullet, black-tail, hammer and the eel (these are the most harmful sorts); the flesh of goats, deer, pigs and dogs (meats that disturb most the digestive organs); the cock, pigeon and bustard, with all birds that are considered substantial foods; mint, leek and onion among the vegetables, as their pungent character is not at all suited to sick folk; the wearing of black (black is the sign of death); not to lie on or wear goat-skin, not to put foot on foot or hand on hand (all which conduct is inhibitive).[1] These observances they impose because of the divine origin of the disease, claiming superior knowledge and alleging other causes, so that, should the patient recover, the reputation for cleverness may be theirs; but should he die, they may have a sure fund of excuses, with the defence that they are not at all to blame, but the gods. Having given nothing to eat or drink, and not having steeped their patients in baths, no blame can be laid, they say, upon them. So I suppose that no Libyan dwelling in the interior can enjoy good health, since they lie on goat-skins and eat goats' flesh, possessing neither coverlet nor cloak nor footgear that is not from the goat; in fact they possess no cattle save

[1] Here is probably a reference to "binding" by sorcery. So Wilamowitz. But may not κωλύματα mean that if the patient follows the advice of the quacks an attack (so it is said) will be "prevented"?

---

[5] θ omits μηδέ.

[6] θ has the plural throughout this sentence.

[7] M has ἂν after Λιβύων but not after οὐδενα. θι have οὐδὲν ἄν. It is therefore probable that it should be in both places.

[8] The MSS. are here unintelligible. The text is Littré's.

εἰ δὲ ταῦτα ἐσθιόμενα καὶ προσφερομενα τὴν νοῦσον τίκτει τε καὶ αὔξει καὶ μὴ ἐσθιόμενα ἴηται, οὐκέτι ὁ θεὸς αἴτιος ἐστίν, οὐδὲ οἱ καθαρμοὶ ὠφελέουσιν, ἀλλὰ τὰ ἐδέσματα τὰ ἰώμενά ἐστι καὶ τὰ βλάπτοντα, τοῦ δὲ θεοῦ ἀφανίζεται ἡ
46 δύναμις.

III. Οὕτως οὖν ἔμοιγε δοκέουσιν οἵτινες τῷ τρόπῳ τούτῳ ἐγχειρέουσιν ἰῆσθαι ταῦτα τὰ νοσήματα οὔτε ἱερὰ νομίζειν εἶναι οὔτε θεῖα· ὅπου γὰρ ὑπὸ καθαρμῶν τοιούτων μετάστατα γίνεται καὶ ὑπὸ θεραπείης τοιῆσδε, τί κωλύει καὶ ὑφ' ἑτέρων τεχνημάτων ὁμοίων τούτοισιν ἐπιγίνεσθαί τε τοῖσιν ἀνθρώποισι καὶ προσπίπτειν; ὥστε τὸ θεῖον μηκέτι αἴτιον εἶναι, ἀλλά τι ἀνθρώπινον. ὅστις γὰρ οἷός τε περικαθαίρων ἐστὶ καὶ μαγεύων
10 ἀπάγειν τοιοῦτον πάθος, οὗτος κἂν ἐπάγοι ἕτερα τεχνησάμενος, καὶ ἐν τούτῳ τῷ λόγῳ τὸ θεῖον ἀπόλλυται.[1] τοιαῦτα λέγοντες καὶ μηχανώμενοι προσποιέονται πλέον τι εἰδέναι, καὶ ἀνθρώπους ἐξαπατῶσι προστιθέμενοι αὐτοῖς ἁγνείας τε καὶ καθάρσιας, ὅ τε πολὺς αὐτοῖς τοῦ λόγου ἐς τὸ θεῖον ἀφήκει καὶ τὸ δαιμόνιον. καίτοι ἔμοιγε οὐ περὶ εὐσεβείης τοὺς λόγους δοκέουσι ποιεῖσθαι, ὡς οἴονται, ἀλλὰ περὶ ἀσεβείης μᾶλλον, καὶ ὡς οἱ θεοὶ οὐκ εἰσί, τὸ δὲ εὐσεβὲς αὐτῶν καὶ τὸ θεῖον
20 ἀσεβές ἐστι καὶ ἀνόσιον, ὡς ἐγὼ διδάξω.

IV. Εἰ γὰρ σελήνην καθαιρεῖν[2] καὶ ἥλιον ἀφανίζειν καὶ χειμῶνά τε καὶ εὐδίην ποιεῖν καὶ ὄμβρους καὶ αὐχμοὺς καὶ θάλασσαν ἄπορον καὶ γῆν ἄφορον[3] καὶ τἄλλα τὰ τοιουτότροπα πάντα

[1] Both θ and M have ἀπολύεται.

[2] κατάγειν θι and Wilamowitz (perhaps rightly).

goats. But if to eat or apply these things engenders and increases the disease, while to refrain works a cure, then neither is godhead[1] to blame nor are the purifications beneficial; it is the foods that cure or hurt, and the power of godhead disappears.

III. Accordingly I hold that those who attempt in this manner to cure these diseases cannot consider them either sacred or divine; for when they are removed by such purifications and by such treatment as this, there is nothing to prevent the production of attacks in men by devices that are similar. If so, something human is to blame, and not godhead. He who by purifications and magic can take away such an affection can also by similar means bring it on, so that by this argument the action of godhead is disproved. By these sayings and devices they claim superior knowledge, and deceive men by prescribing for them purifications and cleansings, most of their talk turning on the intervention of gods and spirits. Yet in my opinion their discussions show, not piety, as they think, but impiety rather, implying that the gods do not exist, and what they call piety and the divine is, as I shall prove, impious and unholy.

IV. For if they profess to know how to bring down the moon, to eclipse the sun, to make storm and sunshine, rain and drought, the sea impassable and the earth barren, and all such wonders, whether

[1] ὁ θεός does not imply any sort of monotheism. The article is generic, and the phrase therefore means "*a* god" rather than "*the* god." See my article on the vague use of ὁ θεὸς in *Classical Review*, Dec. 1913.

---

[3] θάλασσαν ἄπορον καὶ γῆν ἄφορον Lobeck (*Aglaophamus* I. 634), Ermerins: θάλασσαν εὔφορον καὶ γῆν ἄφορον Reinhold: θάλασσαν ἄφορον καὶ γῆν MSS.

ὑποδέχονται ἐπίστασθαι, εἴτε καὶ ἐκ τελετέων εἴτε καὶ ἐξ ἄλλης τινὸς γνώμης καὶ μελέτης φασὶ ταῦτα οἷόν τ' εἶναι γενέσθαι οἱ ταῦτ' ἐπιτηδεύοντες, δυσσεβεῖν ἔμοιγε δοκέουσι καὶ θεοὺς οὔτε εἶναι νομίζειν οὔτε ἰσχύειν οὐδὲν οὔτε εἴργεσθαι
10 ἂν οὐδενὸς τῶν ἐσχάτων. ἃ[1] ποιέοντες πῶς οὐ δεινοὶ αὐτοῖς εἰσίν; εἰ γὰρ ἄνθρωπος μαγεύων καὶ θύων σελήνην καθαιρήσει καὶ ἥλιον ἀφανιεῖ καὶ χειμῶνα καὶ εὐδίην ποιήσει, οὐκ ἂν ἔγωγέ τι θεῖον νομίσαιμι τούτων εἶναι οὐδέν, ἀλλ' ἀνθρώπινον, εἰ δὴ τοῦ θείου ἡ δύναμις ὑπὸ ἀνθρώπου γνώμης κρατεῖται καὶ δεδούλωται. ἴσως δὲ οὐχ οὕτως ἔχει ταῦτα, ἀλλ' ἄνθρωποι βίου δεόμενοι πολλὰ καὶ παντοῖα τεχνῶνται καὶ ποικίλλουσιν ἔς τε τἄλλα πάντα καὶ ἐς τὴν νοῦσον ταύτην,
20 ἑκάστῳ εἴδει τοῦ πάθεος θεῷ τὴν αἰτίην προστιθέντες.[2] καὶ ἢν μὲν γὰρ αἶγα μιμῶνται, καὶ ἢν[3] βρύχωνται, ἢ τὰ δεξιὰ σπῶνται, μητέρα θεῶν φασὶν αἰτίην εἶναι. ἢν δὲ ὀξύτερον καὶ εὐτονώτερον φθέγγηται, ἵππῳ εἰκάζουσι,[4] καὶ φασὶ Ποσειδῶνα αἴτιον εἶναι. ἢν δὲ καὶ τῆς κόπρου τι παρῇ, ὅσα πολλάκις γίνεται ὑπὸ τῆς νούσου βιαζομένοισιν, Ἐνοδίῃ πρόσκειται ἡ ἐπωνυμίη· ἢν δὲ πυκνότερον καὶ λεπτότερον, οἷον ὄρνιθες, Ἀπόλλων νόμιος. ἢν δὲ ἀφρὸν ἐκ τοῦ στόματος ἀφίῃ καὶ τοῖσι ποσὶ λα-

[1] ἃ my emendation (anticipated by Ermerins): ποιέοντες ἕνεκά γε· πῶς M: ποιέοντες ὡς θ: τῶν ἐσχάτων ποιέοντες, ἕνεκά γε τῶν θεῶν. δεινοὶ ἄρ' αὐτοῖσίν εἰσιν Reinhold: τῶν ἐσχάτων ποιέοντες ἕνεκά γε θεῶν. εἰ γὰρ κ.τ.λ. Wilam. See *Postscript*.

[2] After προστιθέντες the MSS., with many variations, have a sentence which in Littré appears as οὐ γὰρ καθάπαξ ἀλλὰ πλεονάκις ταῦτα μέμνηνται. θι omit καθάπαξ and add γε μὴν after πλεονάκις. M has ἓν for καθάπαξ, and so have two other

it be by rites or by some cunning or practice that they can, according to the adepts, be effected, in any case I am sure that they are impious, and cannot believe that the gods exist or have any strength, and that they would not refrain from the most extreme actions. Wherein surely they are terrible in the eyes of the gods. For if a man by magic and sacrifice will bring the moon down, eclipse the sun, and cause storm and sunshine, I shall not believe that any of these things is divine, but human, seeing that the power of godhead is overcome and enslaved by the cunning of man. But perhaps what they profess is not true, the fact being that men, in need of a livelihood, contrive and devise many fictions of all sorts, about this disease among other things, putting the blame, for each form of the affection, upon a particular god.[1] If the patient imitate a goat, if he roar, or suffer convulsions in the right side, they say that the Mother of the Gods is to blame. If he utter a piercing and loud cry, they liken him to a horse and blame Poseidon. Should he pass some excrement, as often happens under the stress of the disease, the surname Enodia is applied. If it be more frequent and thinner, like that of birds, it is Apollo Nomius. If he foam at the mouth and kick, Ares has the

[1] If the sentence be retained which I have deleted as a gloss the general meaning will be: "Again and again do they bethink themselves of this trick."

MSS. M and θ have *μεμίμηνται*. Ermerins reads *οὐ γὰρ ἓν ἀλλὰ πολλὰ ταῦτα μέμνηνται*: Reinhold *οὐ γὰρ καθάπαξ ἐνί, ἀλλὰ πλεόνεσι ταῦτα νενέμηται*. The last reading is the most intelligible, but I reject the whole sentence as a gloss. So apparently Wilamowitz.

[3] θ omits *γὰρ* to *καὶ ἥν*.

[4] *ἰκάζουσι* (or *ἱκάζουσι*) θ.

30 κτίζῃ, Ἄρης τὴν αἰτίην ἔχει. οἷσι δὲ νυκτὸς δείματα παρίσταται καὶ φόβοι καὶ παράνοιαι καὶ ἀναπηδήσιες ἐκ τῆς κλίνης[1] καὶ φεύξιες ἔξω, Ἑκάτης φασὶν εἶναι ἐπιβολὰς καὶ ἡρώων ἐφόδους. καθαρμοῖσί τε χρέονται καὶ ἐπαοιδῇσι, καὶ ἀνοσιώτατόν τε καὶ ἀθεώτατον πρῆγμα ποιέουσιν, ὥς ἔμοιγε δοκεῖ· καθαίρουσι γὰρ τοὺς ἐχομένους τῇ νούσῳ αἵματί τε καὶ ἄλλοισι τοιούτοις ὥσπερ μίασμά τι ἔχοντας, ἢ ἀλάστορας, ἢ πεφαρμακευμένους[2] ὑπὸ ἀνθρώπων, ἤ τι ἔργον 40 ἀνόσιον εἰργασμένους, οὓς ἐχρῆν τἀναντία τούτων ποιεῖν, θύειν[3] τε καὶ εὔχεσθαι καὶ ἐς τὰ ἱερὰ φέροντας ἱκετεύειν τοὺς θεούς· νῦν δὲ τούτων μὲν ποιέουσιν οὐδέν, καθαίρουσι δέ. καὶ τὰ μὲν τῶν καθαρμῶν[4] γῇ κρύπτουσι, τὰ δὲ ἐς θάλασσαν ἐμβάλλουσι, τὰ δὲ ἐς τὰ ὄρεα ἀποφέρουσιν,[5] ὅπῃ μηδεὶς ἅψεται μηδὲ ἐμβήσεται· τὰ δ' ἐχρῆν ἐς τὰ ἱερὰ φέροντας τῷ θεῷ ἀποδοῦναι, εἰ δὴ ὁ θεός ἐστιν αἴτιος· οὐ μέντοι ἔγωγε ἀξιῶ ὑπὸ θεοῦ ἀνθρώπου σῶμα μιαίνεσθαι, τὸ ἐπικηρότατον ὑπὸ 50 τοῦ ἁγνοτάτου· ἀλλὰ καὶ ἢν τυγχάνῃ ὑπὸ ἑτέρου μεμιασμένον ἤ τι πεπονθός, ὑπὸ τοῦ θεοῦ καθαίρεσθαι ἂν αὐτὸ καὶ ἁγνίζεσθαι μᾶλλον ἢ μιαίνεσθαι. τὰ γοῦν μέγιστα τῶν ἁμαρτημάτων καὶ ἀνοσιώτατα τὸ θεῖόν ἐστι τὸ καθαῖρον καὶ ἁγνίζον καὶ ῥύμμα[6] γινόμενον ἡμῖν, αὐτοί τε ὅρους τοῖσι

[1] After κλίνης some MSS. have καὶ φόβητρα, which the editors retain. θι omit.

[2] πεφαρμακευμένους θι: πεφαρμαγμένους most MSS. and editors.

[3] θύειν omitted by θ.

[4] καθαρμῶν. Should not this be καθαρμάτων?

[5] φέρουσιν θ.

blame. When at night occur fears and terrors, delirium, jumpings from the bed and rushings out of doors, they say that Hecate is attacking or that heroes are assaulting.[1] In making use, too, of purifications and incantations they do what I think is a very unholy and irreligious thing. For the sufferers from the disease they purify with blood and such like, as though they were polluted, blood-guilty, bewitched by men, or had committed some unholy act. All such they ought to have treated in the opposite way; they should have brought them to the sanctuaries, with sacrifices and prayers, in supplication to the gods. As it is, however, they do nothing of the kind, but merely purify them. Of the purifying objects[2] some they hide in the earth, others they throw into the sea, others they carry away to the mountains, where nobody can touch them or tread on them. Yet, if a god is indeed the cause, they ought to have taken them to the sanctuaries and offered them to him. However, I hold that a man's body is not defiled by a god, the one being utterly corrupt the other perfectly holy. Nay, even should it have been defiled or in any way injured through some different agency, a god is more likely to purify and sanctify it than he is to cause defilement. At least it is godhead that purifies, sanctifies and cleanses us from the greatest and most impious of our sins; and we ourselves fix

[1] The person is "possessed," as we say.

[2] If *καθαρμάτων* be right, the translation will be "refuse," "off-scourings." I am not sure that my emendation is right, because what are *καθαρμοὶ* *before* the process of purification become *καθάρματα* afterwards.

---

[6] *ῥύμμα θι*: *ἔρυμα* M: *ῥῦμα* Reinhold.

θεοῖσι τῶν ἱερῶν καὶ τῶν τεμενέων ἀποδείκνυμεν,[1] ὡς ἂν μηδεὶς ὑπερβαίνῃ ἢν μὴ ἁγνεύῃ, ἐσιόντες τε ἡμεῖς περιρραινόμεθα οὐχ ὡς μιαινόμενοι, ἀλλ' εἴ τι καὶ πρότερον ἔχομεν μύσος, τοῦτο ἀφαγνιού-
60 μενοι.[2] καὶ περὶ μὲν τῶν καθαρμῶν οὕτω μοι
61 δοκεῖ ἔχειν.

V. Τὸ δὲ νόσημα τοῦτο οὐδέν τί μοι δοκεῖ θειότερον εἶναι τῶν λοιπῶν, ἀλλὰ φύσιν ἔχει ἣν καὶ τὰ ἄλλα νοσήματα, καὶ πρόφασιν ὅθεν ἕκαστα γίνεται·[3] καὶ ἰητὸν εἶναι, καὶ οὐδὲν ἧσσον ἑτέρων, ὅ τι ἂν μὴ ἤδη ὑπὸ χρόνου πολλοῦ καταβεβιασμένον ᾖ, ὥστε ἤδη[4] ἰσχυρότερον εἶναι τῶν φαρμάκων τῶν προσφερομένων. ἄρχεται δὲ ὥσπερ καὶ τἄλλα νοσήματα κατὰ γένος· εἰ γὰρ ἐκ φλεγματώδεος φλεγματώδης, καὶ ἐκ χολώδεος
10 χολώδης γίνεται, καὶ ἐκ φθινώδεος φθινώδης, καὶ ἐκ σπληνώδεος σπληνώδης,[5] τί κωλύει ὅτῳ πατὴρ ἢ μήτηρ εἴχετο νοσήματι, τούτῳ[6] καὶ τῶν ἐκγόνων ἔχεσθαί τινα; ὡς ὁ γόνος ἔρχεται πάντοθεν τοῦ σώματος, ἀπό τε τῶν ὑγιηρῶν ὑγιηρός, καὶ ἀπὸ τῶν νοσερῶν νοσερός. ἕτερον δὲ

[1] ἀποδείκνυμεν Ermerins and Reinhold: ἀποδεικνύμενοι (δείκνυνται θι) MSS. Reinhold also reads οἵους for ὡς, an ingenious correction. In θ we have τεμεν and then a gap followed by δείκνυνται.

[2] From ἀλλ' to ἀφαγνιούμενοι is omitted by θ but is found in M. Probably the eye of the scribe of θ passed from the first -μενοι to the second.

[3] The MSS. (with slight variations) read μὲν after φύσιν, and after γίνεται have φύσιν δὲ τοῦτο καὶ πρόφασιν ἀπὸ ταὐτοῦ τὸ θεῖον γίνεσθαι ἀφ' ὅτου καὶ τἄλλα πάντα. There is obviously corruption here as in Chapter I, one passage having been compared by a scribe to the other. It is hard to mark off the two passages as they were written originally. Reinhold

boundaries to the sanctuaries and precincts of the gods, so that nobody may cross them unless he be pure; and when we enter we sprinkle ourselves, not as defiling ourselves thereby, but to wash away any pollution we may have already contracted. Such is my opinion about purifications.

V. But this disease is in my opinion no more divine than any other; it has the same nature as other diseases, and the cause that gives rise to individual diseases.[1] It is also curable, no less than other illnesses, unless by long lapse of time it be so ingrained as to be more powerful than the remedies that are applied. Its origin, like that of other diseases, lies in heredity. For if a phlegmatic parent has a phlegmatic child, a bilious parent a bilious child, a consumptive parent a consumptive child, and a splenetic parent a splenetic child, there is nothing to prevent some of the children suffering from this disease when one or the other of the parents suffered from it; for the seed comes from every part of the body, healthy seed from the healthy parts, diseased seed from the diseased parts.

[1] Possibly ὅθεν ἕκαστα γίνεται is also part of the gloss; in which case the translation will be, "it has the same nature and cause as other diseases."

---

emends Chapter I and reads here τῶν λοιπῶν, ἀλλ' ἀπὸ ταὐτοῦ γίγνεσθαι ἀφ' ὅτου καὶ τἄλλα πάντα, καὶ ἰητὸν εἶναι κ.τ.λ. I believe that not only has there been corruption due to comparison, but also glosses have crept in.

[4] θ has ὡς for ὥστε ἤδη.

[5] θ has σπληνίας.

[6] εἴχετο νοσήματι, τούτῳ Reinhold: εἴχετο τούτω τῶ νοσήματι θι.

μέγα τεκμήριον ὅτι οὐδὲν θειότερόν ἐστι τῶν λοιπῶν νοσημάτων· τοῖσι γὰρ[1] φλεγματώδεσι φύσει γίνεται· τοῖσι δὲ χολώδεσιν οὐ προσπίπτει· καίτοι εἰ θειότερόν ἐστι τῶν ἄλλων, τοῖσιν ἅπασιν
20 ὁμοίως ἔδει γίνεσθαι τὴν νοῦσον ταύτην, καὶ μὴ
21 διακρίνειν μήτε χολώδεα μήτε φλεγματώδεα.

VI. Ἀλλὰ γὰρ αἴτιος ὁ ἐγκέφαλος τούτου τοῦ πάθεος, ὥσπερ καὶ τῶν ἄλλων νοσημάτων τῶν μεγίστων· ὅτῳ δὲ τρόπῳ καὶ ἐξ οἵης προφάσιος γίνεται, ἐγὼ φράσω σάφα. ὁ ἐγκέφαλός ἐστι τοῦ ἀνθρώπου διπλόος ὥσπερ καὶ τοῖσιν ἄλλοισι ζῴοις ἅπασιν· τὸ δὲ μέσον αὐτοῦ διείργει μῆνιγξ λεπτή· διὸ οὐκ αἰεὶ κατὰ τωὐτὸ τῆς κεφαλῆς ἀλγεῖ, ἀλλ' ἐν μέρει ἑκάτερον, ὁτὲ δὲ ἅπασαν. καὶ φλέβες δ' ἐς αὐτὸν τείνουσιν ἐξ
10 ἅπαντος τοῦ σώματος, πολλαὶ καὶ λεπταί, δύο δὲ παχεῖαι, ἡ μὲν ἀπὸ τοῦ ἥπατος, ἡ δὲ ἀπὸ τοῦ σπληνός. καὶ ἡ μὲν ἀπὸ τοῦ ἥπατος ὧδ' ἔχει· τὸ μέν τι τῆς φλεβὸς[2] κάτω τείνει διὰ τῶν ἐπὶ δεξιὰ παρ' αὐτὸν τὸν νεφρὸν καὶ τὴν ψυὴν ἐς τὸ ἐντὸς τοῦ μηροῦ, καὶ καθήκει ἐς τὸν πόδα, καὶ καλεῖται κοίλη φλέψ· ἡ δὲ ἑτέρη ἄνω τείνει διὰ φρενῶν τῶν δεξιῶν[3] καὶ τοῦ πλεύμονος· ἀπέσχισται δὲ καὶ ἐς τὴν καρδίην καὶ ἐς τὸν βραχίονα τὸν δεξιόν· καὶ τὸ λοιπὸν ἄνω φέρει διὰ τῆς
20 κληῖδος ἐς τὰ δεξιὰ τοῦ αὐχένος, ἐς αὐτὸ τὸ δέρμα, ὥστε κατάδηλος εἶναι· παρ' αὐτὸ δὲ τὸ οὖς κρύπτεται καὶ ἐνταῦθα σχίζεται, καὶ τὸ μὲν παχύτατον καὶ μέγιστον καὶ κοιλότατον ἐς τὸν ἐγκέφαλον τελευτᾷ, τὸ δὲ ἐς τὸ οὖς τὸ δεξιόν, τὸ δὲ ἐς τὸν ὀφθαλμὸν τὸν δεξιόν, τὸ δὲ ἐς τὸν

Another strong proof that this disease is no more divine than any other is that it affects the naturally phlegmatic, but does not attack the bilious. Yet, if it were more divine than others, this disease ought to have attacked all equally, without making any difference between bilious and phlegmatic.

VI. The fact is that the cause of this affection, as of the more serious diseases generally, is the brain. The manner and the cause I will now set forth clearly. The brain of man, like that of all animals, is double, being parted down its centre by a thin membrane. For this reason pain is not always felt in the same part of the head, but sometimes on one side, sometimes on the other, and occasionally all over. Veins lead up to it from all the body, many of which are thin, while two are stout, one coming from the liver, the other from the spleen. The vein from the liver has the following character. One part of it stretches downwards on the right side, close by the kidney and the loin, to the inner part of the thigh, reaching down to the foot; it is called the hollow vein. The other part of it stretches upwards through the right diaphragm and lung. It branches away to the heart and the right arm. The rest leads upwards through the collar-bone to the right of the neck, to the very skin, so as to be visible. Right by the ear it hides itself, and here it branches, the thickest, largest and most capacious part ending in the brain, another in the right ear, another in the right eye, and the last in the nostril.

---

[1] *τοῖσι γὰρ* M : *τοῖσι δὲ λοιποῖσι θ.*

[2] *τῆς φλεβὸς* M : *τοῦ σπληνὸς θ.*

[3] M and *θ* read *τῶν φλεβῶν* and place *τῶν δεξιῶν* after *πλεύμονος.*

μυκτῆρα. ἀπὸ μὲν τοῦ ἥπατος οὕτως ἔχει τὰ τῶν φλεβῶν. διατέταται δὲ καὶ ἀπὸ τοῦ σπληνὸς φλὲψ ἐς τὰ ἀριστερὰ καὶ κάτω καὶ ἄνω, ὥσπερ καὶ ἡ ἀπὸ τοῦ ἥπατος, λεπτοτέρη δὲ καὶ
30 ἀσθενεστέρη.

VII. Κατὰ ταύτας δὲ τὰς φλέβας καὶ ἐπαγόμεθα τὸ πολὺ τοῦ πνεύματος· αὗται γὰρ ἡμῖν εἰσὶν ἀναπνοαὶ τοῦ σώματος τὸν ἠέρα ἐς σφᾶς ἕλκουσαι, καὶ ἐς τὸ σῶμα τὸ λοιπὸν ὀχετεύουσι κατὰ τὰ φλέβια, καὶ ἀναψύχουσι καὶ πάλιν ἀφιᾶσιν. οὐ γὰρ οἷόν τε τὸ πνεῦμα στῆναι, ἀλλὰ χωρεῖ ἄνω τε καὶ κάτω· ἢν γὰρ στῇ που καὶ ἀποληφθῇ, ἀκρατὲς γίνεται ἐκεῖνο τὸ μέρος ὅπου[1] ἂν στῇ· τεκμήριον δέ· ὅταν κατακειμένῳ
10 ἢ καθημένῳ φλέβια πιεσθῇ, ὥστε τὸ πνεῦμα[2] μὴ διεξιέναι διὰ τῆς φλεβός, εὐθὺς νάρκη ἔχει.
12 περὶ μὲν τῶν φλεβῶν οὕτως ἔχει.

VIII. Ἡ δὲ νοῦσος αὕτη γίνεται τοῖσι μὲν φλεγματίῃσι, τοῖσι δὲ χολώδεσιν οὔ. ἄρχεται δὲ φύεσθαι ἐπὶ τοῦ ἐμβρύου ἔτι ἐν τῇ μήτρῃ ἐόντος· καθαίρεται γὰρ καὶ ἀνθεῖ, ὥσπερ τἄλλα μέρεα, πρὶν γενέσθαι, καὶ ὁ ἐγκέφαλος. ἐν ταύτῃ δὲ τῇ καθάρσει ἢν μὲν καλῶς καὶ μετρίως καθαρθῇ καὶ μήτε πλέον μήτε ἔλασσον τοῦ δέοντος ἀπορρυῇ, οὕτως ὑγιηροτάτην τὴν κεφαλὴν ἔχει· ἢν δὲ πλέονα ῥυῇ ἀπὸ παντὸς τοῦ ἐγκεφάλου
10 καὶ ἀπότηξις πολλὴ γένηται, νοσώδεά τε τὴν κεφαλὴν ἕξει αὐξανόμενος καὶ ἤχου πλέην, καὶ οὔτε ἥλιον οὔτε ψῦχος ἀνέξεται· ἢν δὲ ἀπὸ ἑνός

[1] καθὸ M : καθότι θ.
[2] πνεῦμα most MSS. : αἷμα θ.

Such is the character of the veins from the liver. From the spleen too extends a vein downwards and upwards to the left; it is similar to the one from the liver, but thinner and weaker.

VII. By these veins we take in the greater part of our breath, for they are vents of our body, drawing the air to themselves, and they spread it over the body in general through the minor veins and cool it; then they breathe it out again. For the breath cannot rest, but moves up and down. If it is caught anywhere and rests, that part of the body where it rests becomes paralysed. A proof is that should minor veins be so compressed, when a man is lying or seated, that the breath cannot pass through the vein, a numbness immediately seizes him. Such is the character of the veins.[1]

VIII. This disease attacks the phlegmatic, but not the bilious. Its birth begins in the embryo while it is still in the womb, for like the other parts, the brain too is purged and has its impurities[2] expelled before birth. In this purging if the action be thorough and regulated, and if there flow away neither more nor less than is proper, the infant has a perfectly healthy head. But if the flux from all the brain be too abundant, and a great melting[3] take place, he will have as he grows a diseased head, and one full of noise, and he will not be able to endure either sun or cold. If an excessive flux come from one eye

[1] Compare with this the argument of the treatise *Breaths.*

[2] ἀνθεῖ is a difficult word. It seems to be equivalent to ἐξανθεῖ, but may be corrupt. The meaning, however, is plain. The old explanation was that ἀνθεῖ means "grows," but it surely is connected with ἐξανθεῖ lower down.

[3] "Deliquescence" would be the modern technical term.

τινος γένηται ἢ ὀφθαλμοῦ ἢ ὠτός, ἢ φλέψ τις συνισχνανθῇ, ἐκεῖνο κακοῦται τὸ μέρος, ὅπως ἂν καὶ τῆς ἀποτήξιος ἔχῃ· ἢν δὲ κάθαρσις μὴ ἐπιγένηται, ἀλλὰ συστραφῇ τῷ ἐγκεφάλῳ, οὕτως ἀνάγκη φλεγματώδεα εἶναι. καὶ οἷσι μὲν παιδίοις ἐοῦσιν ἐξανθεῖ ἕλκεα καὶ ἐς τὴν κεφαλὴν καὶ ἐς τὰ ὦτα καὶ ἐς τὸν χρῶτα, καὶ σιαλώδεα γίνεται
20 *καὶ μυξόρροα, ταῦτα μὲν ῥήϊστα διάγει προϊούσης τῆς ἡλικίης· ἐνταῦθα γὰρ ἀφίει καὶ ἐκκαθαίρεται τὸ φλέγμα, ὃ ἐχρῆν ἐν τῇ μήτρῃ καθαρθῆναι· καὶ τὰ οὕτω καθαρθέντα*[1] *οὐ γίνεται ἐπίληπτα τῇ νούσῳ ταύτῃ ἐπὶ τὸ πολύ. ὅσα δὲ καθαρά τέ ἐστι, καὶ μήθ' ἕλκος μηδὲν μήτε μύξα μήτε σίαλον αὐτοῖς προέρχεται, μήτε ἐν τῇσι μήτρῃσι πεποίηται τὴν κάθαρσιν, τούτοισι δὲ ἐπικίνδυνόν*
28 *ἐστιν ἁλίσκεσθαι ὑπὸ ταύτης τῆς νούσου.*

IX. *Ἢν δὲ ἐπὶ τὴν καρδίην ποιήσηται ὁ κατάρροος τὴν πορείην, παλμὸς ἐπιλαμβάνει καὶ ἆσθμα, καὶ τὰ στήθεα διαφθείρεται, ἔνιοι δὲ καὶ κυφοὶ γίνονται· ὅταν γὰρ ἐπικατέλθῃ τὸ φλέγμα*[2] *ψυχρὸν ἐπὶ τὸν πλεύμονα καὶ τὴν καρδίην, ἀποψύχεται τὸ αἷμα· αἱ δὲ φλέβες πρὸς βίην ψυχόμεναι πρὸς τῷ πλεύμονι καὶ τῇ καρδίῃ πηδῶσι, καὶ ἡ καρδίη πάλλεται, ὥστε ὑπὸ τῆς ἀνάγκης ταύτης τὸ ἆσθμα ἐπιπίπτειν καὶ τὴν*
10 *ὀρθοπνοίην. οὐ γὰρ δέχεται τὸ πνεῦμα ὅσον ἐθέλει, ἄχρι*[3] *κρατηθῇ τοῦ φλέγματος τὸ ἐπιρρυὲν καὶ διαθερμανθὲν διαχυθῇ ἐς τὰς φλέβας· ἔπειτα παύεται τοῦ παλμοῦ καὶ τοῦ ἄσθματος· παύεται*

[1] *καθαρθέντα*: four MSS. (including θι) have *παιδευθέντα.*
[2] *φλέγμα* M: *πνεῦμα* θ.
[3] *ἄχρι* θι: *μέχρις* M: *μέχρις ἂν* some MSS. and the editors.

or one ear, or if a vein be reduced in size, that part suffers a lesion in proportion to the melting. Should the purging not take place, but congestion occur in the brain, then the infants cannot fail to be phlegmatic. If while they are children sores break out on head, ears and skin, and if saliva and mucus be abundant, as age advances such enjoy very good health, for in this way the phlegm is discharged and purged away which should have been purged away in the womb. Those who have been so purged are in general not attacked by this disease. Those children, on the other hand, that are clean,[1] do not break out in sores, and discharge neither mucus nor saliva, run a risk of being attacked by this disease, if the purging has not taken place in the womb.

IX. Should the discharge make its way to the heart, palpitation and difficulty of breathing supervene, the chest becomes diseased, and a few even become hump-backed; for when the phlegm descends cold to the lungs and to the heart, the blood is chilled; and the veins, being forcibly chilled, beat against the lungs and the heart, and the heart palpitates, so that under this compulsion difficulty of breathing and orthopnoea result. For the patient does not get as much breath as he wants until the phlegm that has flowed in has been mastered, warmed and dispersed into the veins. Then the palpitation and difficulty of breathing cease. It ceases in pro-

[1] This use of *καθαρός* in the sense of "unpurged," "showing no discharge," is peculiar. It should mean "needing no purgation," not that the necessary purging does not take place. One suspects that the correct reading should be: *ὅσα δὲ μήτε καθαρά ἐστι.*

δὲ ὅπως ἂν καὶ τοῦ πλήθεος ἔχῃ· ἢν μὲν γὰρ πλέον ἐπικαταρρυῇ, σχολαίτερον,[1] ἢν δὲ ἔλασσον, θᾶσσον· καὶ ἢν πυκνότεροι ἔωσιν οἱ κατάρροοι, πυκνότερα ἐπίληπτος γίνεται. ταῦτα μὲν οὖν[2] πάσχει, ἢν ἐπὶ τὸν πλεύμονα καὶ τὴν καρδίην

19 ἴῃ·[3] ἢν δὲ ἐς τὴν κοιλίην, διάρροιαι λαμβάνουσιν.

X. Ἢν δὲ τούτων μὲν τῶν ὁδῶν ἀποκλεισθῇ, ἐς δὲ τὰς φλέβας, ἃς προείρηκα, τὸν κατάρροον ποιήσηται, ἄφωνος γίνεται καὶ πνίγεται, καὶ ἀφρὸς ἐκ τοῦ στόματος ἐκρεῖ,[4] καὶ οἱ ὀδόντες συνηρείκασι, καὶ αἱ χεῖρες συσπῶνται, καὶ τὰ ὄμματα διαστρέφονται, καὶ οὐδὲν φρονέουσιν, ἐνίοισι δὲ καὶ ὑποχωρεῖ ἡ κόπρος[5] κάτω.[6] ὅπως δὲ τούτων ἕκαστον πάσχει ἐγὼ φράσω· ἄφωνος μέν ἐστιν ὅταν ἐξαίφνης τὸ φλέγμα[7] ἐπικατελθὸν

10 ἐς[8] τὰς φλέβας ἀποκλείσῃ τὸν ἠέρα καὶ μὴ παραδέχηται μήτε ἐς τὸν ἐγκέφαλον μήτε ἐς τὰς φλέβας τὰς κοίλας μήτε ἐς τὰς κοιλίας,[9] ἀλλ' ἐπιλάβῃ τὴν ἀναπνοήν· ὅταν γὰρ λάβῃ ἄνθρωπος κατὰ τὸ στόμα καὶ τοὺς μυκτῆρας τὸ πνεῦμα, πρῶτον μὲν ἐς τὸν ἐγκέφαλον ἔρχεται, ἔπειτα δὲ ἐς τὴν κοιλίην τὸ πλεῖστον μέρος, τὸ δὲ ἐπὶ τὸν πλεύμονα, τὸ δὲ ἐπὶ τὰς φλέβας. ἐκ τούτων δὲ σκίδναται ἐς[10] τὰ λοιπὰ μέρεα κατὰ τὰς φλέβας· καὶ ὅσον μὲν ἐς τὴν κοιλίην ἔρχεται, τοῦτο μὲν

20 τὴν κοιλίην διαψύχει, καὶ ἄλλο οὐδὲν συμβάλλεται· ὁ δ' ἐς τὸν πλεύμονά τε καὶ τὰς φλέβας

[1] σχολαίτερον M: σχολέτερον θ.

[2] θ omits οὖν.

[3] ἴῃ θ: εἴη M.

[4] ἐκρεῖ M: ῥεῖ θ.

[5] ἡ κόπρος omitted by θ.

[6] After κάτω the MSS. have (with slight variations) καὶ ταῦτα γίνεται ἐνίοτε μὲν ἐς τὰ ἀριστερά, ὁτὲ δὲ ἐς τὰ δεξιά, ὁτὲ δὲ ἐς ἀμφότερα. It is surely a gloss.

portion to the quantity of the flux, that is, slower if the flux be great, quicker if it be less. And if the fluxes be frequent, the attacks are frequent. Such are the symptoms when the flux goes to the lungs and heart; when it goes to the bowels, the result is diarrhoea.

X. If the phlegm be cut off from these passages, but makes its descent into the veins I have mentioned above, the patient becomes speechless and chokes; froth flows from the mouth; he gnashes his teeth and twists[1] his hands; the eyes roll and intelligence fails, and in some cases excrement is discharged.[2] I will now explain how each symptom occurs. The sufferer is speechless when suddenly the phlegm descends into the veins and intercepts the air, not admitting it either into the brain, or into the hollow veins, or into the cavities, thus checking respiration. For when a man takes in breath by the mouth or nostrils, it first goes to the brain, then most of it goes to the belly, though some goes to the lungs and some to the veins. From these parts it disperses, by way of the veins, into the others. The portion that goes into the belly cools it, but has no further use; but the air that goes into the lungs and the veins is of use

[1] Possibly "clenches." The word can denote any sort of convulsion.

[2] The omitted words mean: "These symptoms manifest themselves sometimes on the left, sometimes on the right, sometimes on both sides."

---

[7] φλέγμα θ: πνεῦμα M. [8] ἐς M: ἐπὶ θ.

[9] μήτε ἐς τὰς κοιλίας is in M but is omitted by θ, perhaps rightly.

[10] ἐς M: ἐπὶ θ.

ἀὴρ συμβάλλεται ἐς τὰς κοιλίας ἐσιὼν καὶ ἐς τὸν ἐγκέφαλον,[1] καὶ οὕτω τὴν φρόνησιν καὶ τὴν κίνησιν τοῖσι μέλεσι παρέχει, ὥστε, ἐπειδὰν ἀποκλεισθῶσιν αἱ φλέβες τοῦ ἠέρος ὑπὸ τοῦ φλέγματος καὶ μὴ παραδέχωνται,[2] ἄφωνον καθιστᾶσι καὶ ἄφρονα τὸν ἄνθρωπον. αἱ δὲ χεῖρες ἀκρατεῖς γίνονται καὶ σπῶνται, τοῦ αἵματος ἀτρεμίσαντος καὶ οὐ διαχεομένου[3] ὥσπερ 30 εἰώθει. καὶ οἱ ὀφθαλμοὶ διαστρέφονται, τῶν φλεβίων ἀποκλειομένων τοῦ ἠέρος καὶ σφυζόντων. ἀφρὸς δὲ ἐκ τοῦ στόματος προέρχεται ἐκ τοῦ πλεύμονος· ὅταν γὰρ τὸ πνεῦμα μὴ ἐσίῃ ἐς αὐτόν, ἀφρεῖ καὶ ἀναβλύει ὥσπερ ἀποθνήσκων. ἡ δὲ κόπρος ὑπέρχεται ὑπὸ βίης πνιγομένου· πνίγεται δὲ τοῦ ἥπατος καὶ τῆς ἄνω κοιλίης πρὸς τὰς φρένας προσπεπτωκότων καὶ τοῦ στομάχου τῆς γαστρὸς ἀπειλημμένου·[4] προσπίπτει δ' ὅταν τὸ πνεῦμα μὴ ἐσίῃ ἐς τὸ στόμα[5] ὅσον εἰώθει. 40 λακτίζει δὲ τοῖσι ποσὶν ὅταν ὁ ἀὴρ ἀποκλεισθῇ ἐν τοῖσι μέλεσι καὶ μὴ οἷός τε ᾖ διεκδῦναι ἔξω ὑπὸ τοῦ φλέγματος· ἀΐσσων δὲ διὰ τοῦ αἵματος ἄνω καὶ κάτω σπασμὸν ἐμποιεῖ καὶ ὀδύνην, διὸ λακτίζει. ταῦτα δὲ πάσχει πάντα, ὁπόταν τὸ[6] φλέγμα παραρρυῇ ψυχρὸν ἐς τὸ αἷμα θερμὸν ἐόν· ἀποψύχει γὰρ καὶ ἵστησι τὸ αἷμα· καὶ ἢν μὲν πολὺ ᾖ τὸ ῥεῦμα καὶ παχύ, αὐτίκα ἀποκτείνει· κρατεῖ γὰρ τοῦ αἵματος τῷ ψύχει[7] καὶ πήγνυσιν· ἢν δὲ ἔλασσον ᾖ, τὸ μὲν παραυτίκα κρατεῖ 50 ἀποφράξαν τὴν ἀναπνοήν· ἔπειτα τῷ χρόνῳ

[1] Here θ has ἔρχεται.
[2] Both M and θ have παραδέχονται.

when it enters the cavities and the brain, thus causing intelligence and movement of the limbs, so that when the veins are cut off from the air by the phlegm and admit none of it, the patient is rendered speechless and senseless. The hands are paralysed and twisted when the blood is still, and is not distributed as usual. The eyes roll when the minor veins are shut off from the air and pulsate. The foaming at the mouth comes from the lungs; for when the breath fails to enter them they foam and boil as though death were near. Excrement is discharged when the patient is violently compressed, as happens when the liver and the upper bowel are forced against the diaphragm and the mouth of the stomach is intercepted; this takes place when the normal amount of breath does not enter the mouth.[1] The patient kicks when the air is shut off in the limbs, and cannot pass through to the outside because of the phlegm; rushing upwards and downwards through the blood it causes convulsions and pain; hence the kicking. The patient suffers all these things when the phlegm flows cold into the blood which is warm; for the blood is chilled and arrested. If the flow be copious and thick, death is immediate, for it masters the blood by its coldness and congeals it. If the flow be less, at the first it is master, having cut off respiration;

[1] With the reading of θ, "body." Perhaps this reading is correct.

---

[3] For διαχεομένου M and some other MSS. have διαδεχομένου.

[4] ἀπειλημμένου M : κατειλημμένων θ.

[5] στόμα M : σῶμα θ.

[6] ὁκόταν τὸ M : ὁποιαν θ.

[7] ψύχει M : ψυχρῷ θ.

ὁπόταν σκεδασθῇ κατὰ τὰς φλέβας καὶ μιγῇ τῷ αἵματι πολλῷ ἐόντι καὶ θερμῷ, ἢν κρατηθῇ οὕτως, ἐδέξαντο τὸν ἠέρα αἱ φλέβες, καὶ
54 *ἐφρόνησαν.*

XI. *Καὶ ὅσα μὲν σμικρὰ παιδία κατάληπτα γίνεται τῇ νούσῳ ταύτῃ, τὰ πολλὰ ἀποθνήσκει, ἢν πολὺ τὸ ῥεῦμα ἐπιγένηται καὶ νότιον· τὰ γὰρ φλέβια λεπτὰ ἐόντα οὐ δύναται ὑποδέχεσθαι τὸ φλέγμα ὑπὸ πάχεος καὶ πλήθεος, ἀλλ' ἀποψύχεται καὶ πήγνυται τὸ αἷμα, καὶ οὕτως ἀποθνήσκει. ἢν δὲ ὀλίγον ᾖ καὶ ἐς ἀμφοτέρας τὰς φλέβας τὸν κατάρροον ποιήσηται, ἢ ἐς τὰς ἐπὶ θάτερα, περιγίνεται ἐπίσημα ἐόντα· ἢ γὰρ στόμα*
10 *παρέσπασται ἢ ὀφθαλμὸς ἢ χεὶρ ἢ αὐχήν, ὁπόθεν ἂν τὸ φλέβιον πληρωθὲν τοῦ φλέγματος κρατηθῇ καὶ ἀπισχνανθῇ. τούτῳ οὖν τῷ φλεβίῳ ἀνάγκη ἀσθενέστερον εἶναι καὶ ἐνδεέστερον τοῦτο τοῦ σώματος τὸ βλαβέν· ἐς δὲ τὸν πλείω χρόνον ὠφελεῖ ὡς ἐπὶ τὸ πολύ· οὐ γὰρ ἔτι ἐπίληπτον γίνεται, ἢν ἅπαξ ἐπισημανθῇ, διὰ τόδε· ὑπὸ τῆς ἀνάγκης ταύτης αἱ φλέβες αἱ λοιπαὶ κακοῦνται καὶ μέρος τι συνισχναίνονται, ὡς*[1] *τὸν μὲν ἠέρα δέχεσθαι, τοῦ δὲ φλέγματος τὸν κατάρροον μηκέτι*
20 *ὁμοίως ἐπικαταρρεῖν· ἀσθενέστερα μέντοι*[2] *τὰ μέλεα εἰκὸς εἶναι, τῶν φλεβῶν κακωθεισῶν. οἷσι δ' ἂν βόρειόν τε καὶ πάνυ ὀλίγον παραρρυῇ καὶ ἐς τὰ δεξιά, ἀσήμως περιγίνονται· κίνδυνος δὲ συντραφῆναι καὶ συναυξηθῆναι, ἢν μὴ θεραπευθῶσι τοῖσιν ἐπιτηδείοισιν. τοῖσι μὲν οὖν*
26 *παιδίοισιν οὕτω γίνεται, ἢ ὅτι τούτων ἐγγυτάτω.*

[1] *ὡς* M: *ὥστε* θ.
[2] After *μέντοι* both M and θ have *ὁμοίως*. It is omitted in

but in course of time, when it is dispersed throughout the veins and mixed with the copious, warm blood, if in this way it be mastered, the veins admit the air and intelligence returns.

XI. Little children when attacked by this disease generally die, if the flow come on copious and with a south wind; for the minor veins being thin cannot admit the phlegm because of its thickness and abundance, but the blood is chilled and congeals, causing death. But if the flow be slight, and make its descent either into both veins or into one or the other, the child recovers but bears the marks of the disease—a distortion of mouth, eye, hand or neck, according to the part from which the minor vein, filled with phlegm, was mastered and reduced. So by reason of this minor vein this part of the body which has been injured must be weaker and more defective. But the injury generally proves beneficial in the long run, as a child is no longer subject to the malady if it be once marked, the reason being as follows. In sympathy with this lesion the other veins too suffer and are partially reduced, so that while they admit the air the flux of phlegm that flows down into them is lessened. The limbs, however, are naturally weaker, the veins having suffered injury. When the flux takes place with the wind in the north, and is very slight and to the right, the children recover without a mark. There is a risk however that the disease will be nourished and grow with the patient, unless appropriate remedies be used. Children, then, suffer in this way, or very nearly so.

---

several Paris MSS. It is probably a repetition of the preceding ὁμοίως.

XII. Τοὺς δὲ πρεσβυτέρους οὐκ ἀποκτείνει, ὅταν ἐπιγένηται, οὐδὲ διαστρέφει· αἵ τε γὰρ φλέβες εἰσὶ κοῖλαι καὶ αἵματος μεσταὶ θερμοῦ, διὸ οὐ[1] δύναται ἐπικρατῆσαι τὸ φλέγμα, οὐδ' ἀποψῦξαι τὸ αἷμα, ὥστε καὶ πῆξαι, ἀλλ' αὐτὸ κρατεῖται καὶ καταμείγνυται τῷ αἵματι ταχέως· καὶ οὕτω παραδέχονται αἱ φλέβες τὸν ἠέρα, καὶ τὸ φρόνημα ἐγγίνεται, τά τε σημεῖα τὰ προειρημένα ἧσσον ἐπιλαμβάνει διὰ τὴν ἰσχύν. τοῖσι δὲ 10 πρεσβυτάτοις ὅταν ἐπιγένηται τοῦτο τὸ νόσημα, διὰ τόδε ἀποκτείνει ἢ παράπληκτον ποιεῖ, ὅτι αἱ φλέβες κεκένωνται καὶ τὸ αἷμα ὀλίγον τέ ἐστι καὶ λεπτὸν καὶ ὑδαρές. ἢν μὲν οὖν πολὺ καταρρυῇ καὶ χειμῶνος, ἀποκτείνει· ἀπέφραξε[2] γὰρ τὰς ἀναπνοὰς καὶ ἀπέπηξε τὸ αἷμα, ἢν ἐπ' ἀμφότερα ὁ κατάρροος γένηται· ἢν δ' ἐπὶ θάτερα μοῦνον, παράπληκτον ποιεῖ· οὐ γὰρ δύναται τὸ αἷμα ἐπικρατῆσαι τοῦ φλέγματος λεπτὸν ἐὸν καὶ ψυχρὸν καὶ ὀλίγον, ἀλλ' αὐτὸ κρατηθὲν ἐπάγη, 20 ὥστε ἀκρατέα εἶναι ἐκεῖνα καθ' ἃ τὸ αἷμα 21 διεφθάρη.

XIII. Ἐς δὲ τὰ δεξιὰ μᾶλλον καταρρεῖ ἢ ἐς τὰ ἀριστερά, ὅτι αἱ φλέβες ἐπικοιλότεραί[3] εἰσι καὶ πλέονες ἢ ἐν τοῖς ἀριστεροῖς.[4] ἐπικαταρρεῖ δὲ καὶ ἀποτήκεται τοῖσι μὲν παιδίοισι μάλιστα, οἷς ἂν διαθερμανθῇ ἡ κεφαλὴ ἤν τε ὑπὸ ἡλίου, ἤν τε ὑπὸ πυρός, καὶ ἐξαπίνης[5] φρίξῃ ὁ ἐγκέ-

[1] διὸ οὐ Ermerins, Reinhold: ἃ οὐ θ: ὅτι οὐ M: ἃ οὐδὲ Littré.

[2] ἀπέφραξε θ: ἀπέπνιξε M.

[3] Before ἐπικοιλότεραι Ermerins adds ἐνταῦθα.

[4] After ἀριστέροις θ has ὅτι ἀπὸ τοῦ ἥπατος (αἵματος μ)

XII. Older people are not killed by an attack of the disease, nor are they distorted; for their veins are capacious and full of hot blood, so that the phlegm cannot gain the mastery, nor chill the blood so as to congeal it; but is itself quickly mastered by the blood and mixed with it. So the veins admit the air, intelligence is present, and the symptoms already mentioned attack less violently because the patient is strong. When this disease attacks very old people it kills or paralyses them, the reason being that their veins are emptied, and their blood is scanty, thin and watery. Now if the flux be copious and in winter, death results; for it chokes respiration and congeals the blood should the flux take place to both sides. If on the other hand the flux be to one side only it causes paralysis; for the thin, cold, scanty blood cannot master the phlegm, but is itself mastered and congealed, so that those parts are powerless where the blood has been corrupted.

XIII. The flux is to the right rather than to the left because the veins are more capacious and more in number than on the left. The flux and melting occur mostly in children when the head has been heated by sun or fire, and then suddenly the brain

---

τείνουσι καὶ ἀπὸ τοῦ σπληνός. M has *ἀπὸ γὰρ τοῦ ἥπατος τείνουσι καὶ ἀπὸ τοῦ σπληνός.* Ermerins (after Dietz) reads *ἀπὸ γὰρ τοῦ ἥπατος τείνουσι καὶ οὐκ ἀπὸ τοῦ σπληνός.* Reinhold rewrites this: *ὅτι αἱ φλέβες αἱ ἀπὸ τοῦ ἥπατος τείνουσαι ἐπικοιλότεραί εἰσι καὶ πλέονες ἢ ἐν τοῖσιν ἀριστεροῖσιν αἱ ἀπὸ τοῦ σπληνός.* I feel that the sentence is a note which has crept into the text.

[5] Before *καὶ ἐξαπίνης* the MSS. have *ἤν τε.* Littré, followed by Ermerins, deletes. Reinhold adds *ἐπειδὰν* before *διαθερμανθῇ* and reads *ἔπειτα* for *ἤν τε καὶ* before *ἐξαπίνης.*

φαλος· τότε γὰρ ἀποκρίνεται τὸ φλέγμα. ἀποτήκεται μὲν γὰρ ἀπὸ τῆς θερμασίης καὶ διαχύσιος τοῦ ἐγκεφάλου· ἐκκρίνεται δὲ ἀπὸ τῆς ψύξιός
10 τε καὶ συστάσιος, καὶ οὕτως ἐπικαταρρεῖ. τοῖσι μὲν αὕτη ἡ πρόφασις γίνεται, τοῖσι δὲ καὶ ἐπειδὰν ἐξαπίνης μετὰ βόρεια πνεύματα νότος μεταλάβῃ, συνεστηκότα τὸν ἐγκέφαλον καὶ εὐσθενέοντα[1] ἔλυσε καὶ ἐχάλασεν, ὥστε πλημμυρεῖν τὸ φλέγμα, καὶ οὕτω τὸν κατάρροον ποιεῖται. ἐπικαταρρεῖ δὲ καὶ ἐξ ἀδήλου[2] φόβου γινομένου, καὶ ἢν δείσῃ βοήσαντός τινος, ἢ μεταξὺ κλαίων μὴ οἷός τε ᾖ τὸ πνεῦμα ταχέως ἀναλαβεῖν, οἷα γίνεται παιδίοισι πολλάκις· ὅ τι δ' ἂν τούτων αὐτῷ
20 γένηται, εὐθὺς ἔφριξε τὸ σῶμα, καὶ ἄφωνος γενόμενος τὸ πνεῦμα οὐχ εἵλκυσεν, ἀλλὰ τὸ πνεῦμα ἠρέμησε, καὶ ὁ ἐγκέφαλος συνέστη, καὶ τὸ αἷμα ἐστάθη, καὶ οὕτως ἀπεκρίθη καὶ ἐπικατερρύη τὸ φλέγμα. τοῖσι μὲν παιδίοισιν αὗται αἱ προφάσιες τῆς ἐπιλήψιός εἰσι τὴν ἀρχήν. τοῖσι δὲ πρεσβύτῃσιν ὁ χειμὼν πολεμιώτατός ἐστιν· ὅταν γὰρ παρὰ πυρὶ πολλῷ διαθερμανθῇ τὴν κεφαλὴν καὶ τὸν ἐγκέφαλον, ἔπειτα ἐν ψύχει γένηται καὶ ῥιγώσῃ, ἢ καὶ ἐκ ψύχεος εἰς ἀλέην ἔλθῃ καὶ παρὰ
30 πῦρ πολύ, τὸ αὐτὸ τοῦτο πάσχει, καὶ οὕτως ἐπίληπτος γίνεται κατὰ τὰ προειρημένα. κίνδυνος δὲ πολὺς καὶ ἦρος παθεῖν τωὐτὸ τοῦτο, ἢν ἡλιωθῇ ἡ κεφαλή· τοῦ δὲ θέρεος[3] ἥκιστα, οὐ γὰρ γίνονται μεταβολαὶ ἐξαπιναῖοι. ὅταν δὲ εἴκοσιν

[1] εὐσθενέοντα Littré, with one MS.: ἀσθενέα ὄντα θ: ἀσθενέοντα M.

[2] I have adopted the readings of θμ in this sentence. The editors omit καὶ before ἢν and put a comma at ἀδήλου, as

has been chilled, for then it is that the phlegm separates off. It melts owing to the heat and diffusion of the brain; it separates owing to the chill and contraction, and so flows down. This is one cause. In other cases the cause is that the south wind, suddenly coming on after north winds, loosens and relaxes the brain when it is braced and strong, so that the phlegm overflows, and thus it produces the flux. It is also caused by fear of the mysterious, if the patient be afraid at a shout, or if while weeping he be unable quickly to recover his breath, things which often happen to children. Whichever of them occur, the body is immediately chilled, the patient loses the power of speech and does not breathe, the breath stops, the brain hardens, the blood stays, and so the phlegm separates off and flows down. Such among children are the causes of the seizure[1] to begin with. Of old patients the greatest enemy is winter. For when an old man has been heated in head and brain by a large fire, and then comes into the cold and is chilled, or if he leave the cold for warmth and a large fire, he experiences the same symptoms and has a seizure, according to what has been said already. There is a serious risk of the same thing happening in spring also, if the head be struck by the sun. In summer the risk is least, as there are no sudden

[1] ἐπίληψις occurs only here in this treatise.

---

though the meaning were, "obscure causes too produce it, for instance a shout, etc." The objection to this is that the examples given are certainly not ἄδηλα.

[3] τοῦ δὲ θέρεος M : τὸ δὲ θέρος θ.

ἔτεα παρέλθῃ, οὐκέτι ἡ νοῦσος αὕτη ἐπιλαμβάνει, ἢν μὴ ἐκ παιδίου σύντροφος ᾖ, ἀλλ' ἢ ὀλίγους ἢ οὐδένα· αἱ γὰρ φλέβες αἵματος μεσταὶ πολλοῦ εἰσίν, καὶ ὁ ἐγκέφαλος συνέστηκε καί ἐστι στρυφνός, ὥστε οὐκ ἐπικαταρρεῖ ἐπὶ τὰς φλέβας·
40 ἢν δ' ἐπικαταρρυῇ, τοῦ αἵματος οὐ κρατεῖ,[1]
41 πολλοῦ ἐόντος καὶ θερμοῦ.

XIV. Ὧι δὲ ἀπὸ παιδίου συνηύξηται καὶ συντέθραπται, ἔθος πεποίηται ἐν τῇσι μεταβολῇσι τῶν πνευμάτων τοῦτο πάσχειν, καὶ ἐπίληπτον γίνεται ὡς τὰ πολλά, καὶ μάλιστα ἐν τοῖσι νοτίοισιν· ἥ τε ἀπάλλαξις χαλεπὴ γίνεται· ὁ γὰρ ἐγκέφαλος ὑγρότερος γέγονε τῆς φύσιος καὶ πλημμυρεῖ ὑπὸ τοῦ φλέγματος, ὥστε τοὺς μὲν καταρρόους πυκνοτέρους γίνεσθαι, ἐκκριθῆναι δὲ μηκέτι οἷόν τε εἶναι τὸ φλέγμα, μηδὲ ἀναξηραν-
10 θῆναι τὸν ἐγκέφαλον, ἀλλὰ διαβεβρέχθαι καὶ εἶναι ὑγρόν. γνοίη δ' ἄν τις τόδε[2] μάλιστα τοῖσι προβάτοισι τοῖσι καταλήπτοισι γινομένοις ὑπὸ τῆς νούσου ταύτης καὶ μάλιστα τῇσιν αἰξίν· αὗται γὰρ πυκνότατα λαμβάνονται· ἢν διακόψῃς[3] τὴν κεφαλήν, εὑρήσεις τὸν ἐγκέφαλον ὑγρὸν ἐόντα καὶ ὕδρωπος περίπλεων καὶ κακὸν ὄζοντα, καὶ ἐν τούτῳ δηλονότι γνώσει ὅτι οὐχ ὁ θεὸς τὸ σῶμα λυμαίνεται, ἀλλ' ἡ νοῦσος. οὕτω δ' ἔχει καὶ τῷ ἀνθρώπῳ· ὁπόταν γὰρ ὁ χρόνος γένηται τῇ νούσῳ,
20 οὐκ ἔτι ἰήσιμος γίνεται· διεσθίεται γὰρ ὁ ἐγκέφαλος ὑπὸ τοῦ φλέγματος καὶ τήκεται, τὸ δὲ ἀποτηκόμενον ὕδωρ γίνεται, καὶ περιέχει τὸν ἐγκέφαλον ἐκτὸς καὶ περικλύζει· καὶ διὰ τοῦτο πυκνότερον ἐπίληπτοι γίνονται καὶ ῥᾷον. διὸ δὴ πολυχρόνιος ἡ νοῦσος, ὅτι τὸ περιρρέον λεπτόν ἐστιν ὑπὸ

changes. After the twentieth year this disease does not occur, or occurs but rarely, unless it has been present from infancy. For the veins are full of abundance of blood, and the brain is compact and hard, so that either there is no flux to the veins, or, if there be a flux, it does not master the blood, which is copious and hot.

XIV. But when the disease dates from infancy and has grown and been nourished with the body, the habit has been formed of the flux occurring at the changes of the winds, and the patient generally has an attack then, especially if the wind be in the south. Recovery, too, proves difficult; the brain is unnaturally moist, and flooded with phlegm, so that not only do fluxes occur more frequently but the phlegm can no longer separate, nor the brain be dried, being on the contrary soaked and moist. The truth of this is best shown by the cattle that are attacked by this disease, especially by the goats, which are the most common victims. If you cut open the head you will find the brain moist, very full of dropsy and of an evil odour, whereby you may learn that it is not a god but the disease which injures the body. So is it also with a man. In fact, when the disease has become chronic it then proves incurable, for the brain is corroded by phlegm and melts, and the part which melts becomes water, surrounding the brain outside and flooding it, for which reason such people are attacked more frequently and more readily. Wherefore the disease lasts a long time, because the surrounding fluid is thin

---

[1] *κρατέει* θ: *κατακρατέει* M.
[2] *τόδε* M: *τῷδε* θ.
[3] *διακόψῃς* M: *διακόψας οραις* (*sic*) θ.

πολυπληθείης, καὶ εὐθὺς κρατεῖται ὑπὸ τοῦ
27 αἵματος καὶ διαθερμαίνεται.

XV. Ὅσοι δὲ ἤδη ἐθάδες εἰσὶ τῇ νούσῳ, προγινώσκουσιν ὅταν μέλλωσι λήψεσθαι, καὶ φεύγουσιν ἐκ τῶν ἀνθρώπων, ἢν μὲν ἐγγὺς ᾖ αὐτῷ τὰ οἰκία,[1] οἴκαδε, ἢν δὲ μή, ἐς τὸ ἐρημότατον, ὅπη μέλλουσιν αὐτὸν ἐλάχιστοι ὄψεσθαι πεσόντα, εὐθύς τε ἐγκαλύπτεται· τοῦτο δὲ ποιεῖ ὑπ' αἰσχύνης τοῦ πάθεος καὶ οὐχ ὑπὸ φόβου, ὡς οἱ πολλοὶ νομίζουσι,[2] τοῦ δαιμονίου. τὰ δὲ παιδάρια τὸ μὲν πρῶτον πίπτουσιν ὅπη ἂν τύχωσιν ὑπὸ
10 ἀηθίης· ὅταν δὲ πολλάκις[3] καταληπτοι γένωνται, ἐπειδὰν προαίσθωνται, φεύγουσι παρὰ τὰς μητέρας ἢ παρὰ ἄλλον ὅντινα μάλιστα γινώσκουσιν, ὑπὸ δέους καὶ φόβου τῆς πάθης· τὸ γὰρ αἰσχύ-
14 νεσθαι[4] οὔπω γινώσκουσιν.

XVI. Ἐν δὲ τῇσι μεταβολῇσι τῶν πνευμάτων διὰ τάδε φημὶ ἐπιλήπτους γίνεσθαι, καὶ μάλιστα τοῖσι νοτίοισιν, ἔπειτα τοῖσι βορείοισιν, ἔπειτα τοῖσι λοιποῖσι πνεύμασι· ταῦτα γὰρ τῶν λοιπῶν πνευμάτων ἰσχυρότατά ἐστι καὶ ἀλλήλοις ἐναντιώτατα κατὰ τὴν στάσιν καὶ κατὰ τὴν δύναμιν. ὁ μὲν γὰρ βορέης συνίστησι τὸν ἠέρα καὶ τὸ θολερόν τε καὶ τὸ νοτῶδες ἐκκρίνει καὶ λαμπρόν τε καὶ διαφανέα ποιεῖ· κατὰ δὲ τὸν αὐτὸν τρόπον
10 καὶ τἄλλα πάντα ἐκ τῆς θαλάσσης ἀρξάμενα[5] καὶ τῶν ἄλλων ὑδάτων· ἐκκρίνει γὰρ ἐξ ἁπάντων τὴν νοτίδα καὶ τὸ δνοφερόν, καὶ ἐξ αὐτῶν τῶν ἀνθρώπων, διὸ καὶ ὑγιηρότατός ἐστι τῶν ἀνέμων.

[1] ᾖ αὐτῷ τὰ οἰκία θ: αὐτῶν ὁ οἶκος ἔῃ M.
[2] Here θ adds καί.

through its abundance, and is immediately mastered and warmed by the blood.

XV. Such as are habituated to their disease have a presentiment when an attack is imminent, and run away from men, home, if their house be near, if not, to the most deserted spot, where the fewest people will see the fall, and immediately hide their heads. This is the result of shame at their malady, and not, as the many hold, of fear of the divine. Young children at first fall anywhere, because they are unfamiliar with the disease; but when they have suffered several attacks, on having the presentiment they run to their mothers or to somebody they know very well, through fear and terror at what they are suffering, since they do not yet know what shame is.

XVI. At the changes of the winds for these reasons do I hold that patients are attacked, most often when the south wind blows, then the north wind, and then the others. In fact the north and south are stronger than any other winds, and the most opposite, not only in direction but in power. For the north wind contracts the air and separates from it what is turbid and damp, making it clear and transparent. It acts in the same way upon everything as well that rises from the sea or waters generally. For it separates the moist and the dull from everything, including men themselves, for which reason it is the most

---

[3] πολλάκις θ: πλεονάκις other MSS. and the editors.

[4] Before οὔπω the MSS. except M and θμ have παῖδες ὄντες. Littré retains, and so does Reinhold. I think it must be a gloss (we should expect ἐόντες) and so, I find, do Ermerins and Wilamowitz.

[5] ἀρξάμενα MSS.: ἐξάρμενα Mack's Codex Mediceus: ἀρξάμενος Ermerins: ἐξαερούμενα Reinhold.

ὁ δὲ νότος τἀναντία τούτῳ ἐργάζεται· πρῶτον μεν ἄρχεται τὸν ἠέρα συνεστηκότα κατατήκειν καὶ διαχεῖν, καθότι καὶ οὐκ εὐθὺς πνεῖ μέγας, ἀλλὰ γαληνίζει[1] πρῶτον, ὅτι οὐ δύναται ἐπικρατῆσαι τοῦ ἠέρος αὐτίκα, τοῦ πρόσθεν πυκνοῦ τε ἐόντος καὶ συνεστηκότος,[2] ἀλλὰ τῷ χρόνῳ διαλύει· τὸ δ' 20 αὐτὸ τοῦτο καὶ τὴν γῆν ἐργάζεται καὶ τὴν θάλασσαν καὶ ποταμοὺς καὶ κρήνας καὶ φρέατα καὶ ὅσα φύεται καὶ ἐν οἷς τι ὑγρόν ἐστιν· ἔστι δὲ ἐν παντί, ἐν τῷ μὲν πλέον, ἐν τῷ δὲ ἔλασσον· ἅπαντα δὲ ταῦτα αἰσθάνεται τοῦ πνεύματος τούτου, καὶ ἔκ τε λαμπρῶν δνοφώδεα γίνεται, καὶ ἐκ ψυχρῶν θερμά, καὶ ἐκ ξηρῶν νοτώδεα· ὅσα δ' ἐν οἰκήμασι κεράμια ἢ κατὰ γῆς ἐστι μεστὰ οἴνου ἢ ἄλλου τινὸς ὑγροῦ, πάντα ταῦτα αἰσθάνεται τοῦ νότου καὶ διαλλάσσει τὴν 30 μορφὴν ἐς ἕτερον εἶδος· τόν τε ἥλιον καὶ τὴν σελήνην καὶ τἄλλα ἄστρα πολὺ ἀμβλυωπότερα καθίστησι τῆς φύσιος. ὅτε οὖν καὶ τούτων οὕτω μεγάλων ἐόντων καὶ ἰσχυρῶν τοσοῦτον ἐπικρατεῖ καὶ τὸ σῶμα ποιεῖ αἰσθάνεσθαι καὶ μεταβάλλειν ἐν τῶν ἀνέμων τούτων τῇσι μεταλλαγῇσιν, ἀνάγκη τοῖσι μὲν νοτίοισι λύεσθαί τε καὶ φλυδᾶν τὸν ἐγκέφαλον καὶ τὰς φλέβας χαλαρωτέρας γίνεσθαι, τοῖσι δὲ βορείοισι συνίστασθαι τὸ ὑγιηρότατον τοῦ ἐγκεφάλου, τὸ δὲ νοσηλότατον 40 καὶ ὑγρότατον ἐκκρίνεσθαι καὶ περικλύζειν ἔξωθεν, καὶ οὕτω τοὺς καταρρόους ἐπιγίνεσθαι ἐν τῇσι μεταβολῇσι τούτων τῶν πνευμάτων. οὕτως αὕτη ἡ νοῦσος γίνεται καὶ θάλλει ἀπὸ τῶν προσιόντων τε καὶ ἀπιόντων, καὶ οὐδέν ἐστιν ἀπορωτέρη τῶν ἄλλων οὔτε ἰῆσθαι οὔτε γνῶναι, 46 οὐδὲ θειοτέρη ἢ αἱ ἄλλαι.

healthy of the winds. But the action of the south wind is the opposite. At first it begins to melt and diffuse the condensed air, inasmuch as it does not blow strong immediately, but is calm at first, because it cannot at once master the air, that before was thick and condensed, but requires time to dissolve it. In exactly the same way it acts upon earth, sea, rivers, springs, wells, and everything that grows in which there is moisture, and moisture is in everything, though more in some things than in others. All these things feel the effects of this wind, and become dull instead of bright, hot instead of cold, wet instead of dry. Vessels of pottery too kept in rooms or underground, which are full of wine or other liquid always feel the effects of the south wind and change their shape to a different form. The sun, moon and stars it makes much duller than they naturally are. Since then it so masters even things that are so big and strong, makes the body feel its effects and change with the changes of these winds, of necessity a south wind relaxes and moistens the brain and enlarges the veins, while north winds press together the healthiest part of the brain, separating the most diseased and moist, and washing it out; for which reason the fluxes occur at the changes of these winds. Thus this disease is born and grows from the things that come to the body and leave it, is no more troublesome to understand and cure than are others, and is no more divine than others are.

---

[1] γαληνίζει θμ : λαγανίζει M : λαγχρίζει Ermerins.
[2] So M: θ has αὐτίκα τοῦ πρόσθεν ἠέρος πυκνοῦ κ.τ.λ.

XVII. Εἰδέναι δὲ χρὴ τοὺς ἀνθρώπους, ὅτι ἐξ οὐδενὸς ἡμῖν αἱ ἡδοναὶ γίνονται καὶ[1] εὐφροσύναι καὶ γέλωτες καὶ παιδιαὶ ἢ ἐντεῦθεν,[2] καὶ λῦπαι καὶ ἀνίαι[3] καὶ δυσφροσύναι καὶ κλαυθμοί. καὶ τούτῳ φρονέομεν μάλιστα[4] καὶ βλέπομεν καὶ ἀκούομεν καὶ διαγινώσκομεν τά τε αἰσχρὰ καὶ καλὰ καὶ κακὰ καὶ ἀγαθὰ καὶ ἡδέα καὶ ἀηδέα, τὰ μὲν νόμῳ διακρίνοντες, τὰ δὲ τῷ συμφέροντι αἰσθανόμενοι.[5] τῷ δὲ αὐτῷ τούτῳ καὶ μαινόμεθα[6]
10 καὶ παραφρονέομεν, καὶ δείματα καὶ φόβοι παρίστανται ἡμῖν, τὰ μὲν νύκτωρ, τὰ δὲ καὶ μεθ' ἡμέρην, καὶ ἀγρυπνίαι καὶ πλάνοι ἄκαιροι, καὶ φροντίδες οὐχ ἱκνεύμεναι, καὶ ἀγνωσίαι τῶν καθεστώτων καὶ ἀηθίαι. καὶ ταῦτα πάσχομεν ἀπὸ τοῦ ἐγκεφάλου πάντα, ὅταν οὗτος μὴ ὑγιαίνῃ, ἀλλὰ θερμότερος τῆς φύσιος γένηται ἢ ψυχρότερος ἢ ὑγρότερος ἢ ξηρότερος, ἤ τι ἄλλο πεπόνθῃ πάθος παρὰ τὴν φύσιν ὃ μὴ ἐώθει. καὶ μαινόμεθα μὲν ὑπὸ ὑγρότητος· ὅταν γὰρ ὑγρότερος
20 τῆς φύσιος ᾖ, ἀνάγκη κινεῖσθαι, κινευμένου δὲ μήτε τὴν ὄψιν ἀτρεμίζειν μήτε τὴν ἀκοήν, ἀλλ' ἄλλοτε ἄλλα ὁρᾶν καὶ ἀκούειν, τήν τε γλῶσσαν τοιαῦτα διαλέγεσθαι οἷα ἂν βλέπῃ τε καὶ ἀκούῃ ἑκάστοτε· ὅσον δ' ἂν ἀτρεμήσῃ ὁ ἐγκέφαλος
25 χρόνον, τοσοῦτον καὶ φρονεῖ ὁ ἄνθρωπος.

XVIII. Γίνεται δὲ ἡ διαφθορὴ τοῦ ἐγκεφάλου ὑπὸ φλέγματος καὶ χολῆς· γνώσει δὲ ἑκάτερα

[1] Before εὐφροσύναι some MSS. have αἱ. It is omitted by θμ, and in M was first omitted and then restored.

[2] After ἐντεῦθεν θμ have ὅθεν, which is read by Wilamowitz

[3] ἀνίαι M : μανίαι θ.

[4] After μάλιστα the MSS. (except θ) and the editors have καὶ νοεῦμεν.

XVII. Men ought to know that from the brain, and from the brain only, arise our pleasures, joys, laughter and jests, as well as our sorrows, pains, griefs and tears. Through it, in particular, we think, see, hear, and distinguish the ugly from the beautiful, the bad from the good, the pleasant from the unpleasant, in some cases using custom as a test, in others perceiving them from their utility. It is the same thing which makes us mad or delirious, inspires us with dread and fear, whether by night or by day, brings sleeplessness, inopportune mistakes, aimless anxieties, absent-mindedness, and acts that are contrary to habit. These things that we suffer all come from the brain, when it is not healthy, but becomes abnormally hot, cold, moist, or dry, or suffers any other unnatural affection to which it was not accustomed. Madness comes from its moistness. When the brain is abnormally moist, of necessity it moves, and when it moves neither sight nor hearing are still, but we see or hear now one thing and now another, and the tongue speaks in accordance with the things seen and heard on any occasion. But all the time the brain is still a man is intelligent.

XVIII. The corruption of the brain is caused not only by phlegm but by bile. You may distinguish

---

[5] After *αἰσθανόμενοι* the MSS. have: *τῷ δὲ καὶ τὰς ἡδονὰς καὶ τὰς ἀηδίας τοῖς καιροῖς διαγινώσκοντες, καὶ οὐ* (*οὖ* without *καὶ* *θμ*) *ταὐτὰ ἀρέσκει ἡμῖν*. Reinhold reads *διαγιγνώκουσιν οὐ*. Littré and Ermerins retain. I reject the phrase, as being a gloss. Wilamowitz has *τῷ δὲ τὰς ἡδονὰς καὶ τὰς ἀηδίας τοῖσι καιροῖσι διαγιγνώσκοντες, οὐ ταὐτὰ ἀρέσκει ἡμῖν*. This restores the grammar to a simple anacoluthon, but in sense it is little more than a repetition of the preceding words.

[6] *θ* has *μαινομενόμεθα*.

ὧδε· οἱ μὲν ὑπὸ φλέγματος μαινόμενοι ἥσυχοί τέ εἰσι καὶ οὐ βοηταὶ οὐδὲ θορυβώδεες, οἱ δὲ ὑπὸ χολῆς κεκράκταί τε καὶ κακοῦργοι καὶ οὐκ ἀτρεμαῖοι, ἀλλ' αἰεί τι ἄκαιρον δρῶντες. ἢν μὲν οὖν συνεχῶς μαίνωνται, αὗται αἱ προφάσιές εἰσιν· ἢν δὲ δείματα καὶ φόβοι παριστῶνται, ὑπὸ μεταστάσιος τοῦ ἐγ- 10 *κεφάλου· μεθίσταται δὲ θερμαινόμενος· θερμαίνεται δὲ ὑπὸ τῆς χολῆς, ὅταν ὁρμήσῃ ἐπὶ τὸν ἐγκέφαλον κατὰ τὰς φλέβας τὰς αἱματίτιδας ἐκ τοῦ σώματος· καὶ ὁ φόβος παρέστηκε μέχρι ἀπέλθῃ πάλιν ἐς τὰς φλέβας καὶ τὸ σῶμα· ἔπειτα πέπαυται. ἀνιᾶται δὲ καὶ ἀσᾶται παρὰ καιρὸν ψυχομένου τοῦ ἐγκεφάλου καὶ συνισταμένου παρὰ τὸ ἔθος· τοῦτο δὲ ὑπὸ φλέγματος πάσχει· ὑπ' αὐτοῦ δὲ τοῦ πάθεος καὶ ἐπιλήθεται. ἐκ νυκτῶν δὲ βοᾷ καὶ κέκραγεν, ὅταν ἐξαπίνης ὁ* 20 *ἐγκέφαλος διαθερμαίνηται· τοῦτο δὲ πάσχουσιν οἱ χολώδεες, οἱ δὲ φλεγματώδεες οὔ·*[1] *διαθερμαίνεται δὲ καὶ ἐπὴν τὸ αἷμα ἐπέλθῃ ἐπὶ τὸν ἐγκέφαλον πολὺ καὶ ἐπιζέσῃ. ἔρχεται δὲ κατὰ τὰς φλέβας πολὺ τὰς προειρημένας, ὅταν τυγχάνῃ ὥνθρωπος ἐνύπνιον ὁρῶν φοβερὸν καὶ ἐν τῷ φόβῳ*[2] *ᾖ· ὥσπερ οὖν καὶ ἐγρηγορότι τότε μάλιστα τὸ πρόσωπον φλογιᾷ, καὶ οἱ ὀφθαλμοὶ ἐρεύθονται, ὅταν φοβῆται, καὶ ἡ γνώμη ἐπινοῇ τι κακὸν ἐργάσασθαι, οὕτω καὶ ἐν τῷ ὕπνῳ πάσχει.* 30 *ὅταν δὲ ἐπέγρηται καὶ καταφρονήσῃ καὶ τὸ αἷμα* 31 *πάλιν σκεδασθῇ ἐς τὰς φλέβας*[3] *πέπαυται.*

[1] M places the δὲ after φλεγματώδεες.
[2] τῷ φόβῳ M: πόνῳ θ.

them thus. Those who are mad through phlegm are quiet, and neither shout nor make a disturbance; those maddened through bile are noisy, evil-doers and restless, always doing something inopportune. These are the causes of continued madness. But if terrors and fears attack, they are due to a change in the brain. Now it changes when it is heated, and it is heated by bile which rushes to the brain from the rest of the body by way of the blood-veins. The fear besets the patient until the bile re-enters the veins and the body. Then it is allayed. The patient suffers from causeless distress and anguish when the brain is chilled and contracted contrary to custom. These effects are caused by phlegm, and it is these very effects that cause loss of memory. Shouts and cries at night are the result of the sudden heating of the brain, an affection from which the bilious suffer but not the phlegmatic. The brain is heated also when the blood rushes to it in abundance and boils. The blood comes in abundance by the veins mentioned above, when the patient happens to see a fearful dream and is in fear. Just as in the waking state the face is flushed, and the eyes are red, mostly when a man is afraid and his mind contemplates some evil act, even so the same phenomena are displayed in sleep. But they cease when the man wakes to consciousness[1] and the blood is dispersed again into the veins.

[1] Or, "and comes to his senses."

---

[3] Littré with some inferior MSS. inserts τὰς προειρημένας before πέπαυται· Reinhold reads τὰς κατὰ τὸ σῶμα.

XIX. Κατὰ ταῦτα νομίζω τὸν ἐγκέφαλον δυναμιν ἔχειν πλείστην ἐν τῷ ἀνθρώπῳ· οὗτος γὰρ ἡμῖν ἐστι τῶν ἀπὸ τοῦ ἠέρος γινομένων ἑρμηνεύς, ἢν ὑγιαίνων τυγχάνῃ· τὴν δὲ φρόνησιν ὁ ἀὴρ παρέχεται. οἱ δὲ ὀφθαλμοὶ καὶ τὰ ὦτα καὶ ἡ γλῶσσα καὶ αἱ χεῖρες καὶ οἱ πόδες οἷα ἂν ὁ ἐγκέφαλος γινώσκῃ, τοιαῦτα πρήσσουσι· † γίνεται γὰρ ἐν ἅπαντι τῷ σώματι τῆς φρονήσιος, ὡς[1] ἂν μετέχῃ τοῦ ἠέρος. † ἐς δὲ τὴν σύνεσιν ὁ ἐγκέφαλός
10 ἐστιν ὁ διαγγέλλων· ὅταν γὰρ σπάσῃ τὸ πνεῦμα ὥνθρωπος ἐς ἑωυτόν, ἐς τὸν ἐγκέφαλον πρῶτον ἀφικνεῖται, καὶ οὕτως ἐς τὸ λοιπὸν σῶμα σκίδναται ὁ ἀήρ, καταλελοιπὼς ἐν τῷ ἐγκεφάλῳ ἑωυτοῦ τὴν ἀκμὴν καὶ ὅ τι ἂν ᾖ φρόνιμόν τε καὶ γνώμην ἔχον· εἰ γὰρ ἐς τὸ σῶμα πρῶτον ἀφικνεῖτο καὶ ὕστερον ἐς τὸν ἐγκέφαλον, ἐν τῇσι σαρξὶ καὶ ἐν τῇσι φλεψὶ καταλελοιπὼς τὴν διάγνωσιν ἐς τὸν ἐγκέφαλον ἂν ἴοι[2] θερμὸς ἐὼν καὶ οὐκ ἀκραιφνής, ἀλλ' ἐπιμεμιγμένος τῇ ἰκμάδι τῇ ἀπό τε τῶν
20 σαρκῶν καὶ τοῦ αἵματος, ὥστε μηκέτι εἶναι
21 ἀκριβής.

XX. Διὸ φημὶ τὸν ἐγκέφαλον εἶναι τὸν ἑρμηνεύοντα τὴν σύνεσιν. αἱ δὲ φρένες ἄλλως ὄνομα ἔχουσι τῇ τύχῃ κεκτημένον καὶ τῷ νόμῳ, τῷ δ' ἐόντι οὔκ, οὐδὲ τῇ φύσει, οὐδὲ οἶδα ἔγωγε τίνα δύναμιν ἔχουσιν αἱ φρένες ὥστε νοεῖν τε καὶ φρονεῖν, πλὴν εἴ τι ὥνθρωπος ὑπερχαρείη ἐξ ἀδοκήτου ἢ ἀνιηθείη,[3] πηδῶσι καὶ ἅλσιν[4] παρέχουσιν ὑπὸ λεπτότητος καὶ ὅτι ἀνατέτανται μάλιστα

[1] τε ὡς θ M: ὡς Littré. But see *Postscript*.
[2] ἀνήει θ M: ἂν ἴοι Littré. Perhaps we should read ἂν ᾔει.

XIX. In these ways I hold that the brain is the most powerful organ of the human body, for when it is healthy it is an interpreter to us of the phenomena caused by the air, as it is the air that gives it intelligence. Eyes, ears, tongue, hands and feet act in accordance with the discernment of the brain; in fact the whole body participates in intelligence in proportion to its participation in air. To consciousness the brain is the messenger. For when a man draws breath into himself, the air first reaches the brain, and so is dispersed through the rest of the body, though it leaves in the brain its quintessence, and all that it has of intelligence and sense. If it reached the body first and the brain afterwards, it would leave discernment in the flesh and the veins, and reach the brain hot, and not pure but mixed with the humour from flesh and blood, so as to have lost its perfect nature.[1]

XX. Wherefore I assert that the brain is the interpreter of consciousness. The diaphragm has a name due merely to chance and custom, not to reality and nature, and I do not know what power the diaphragm has for thought and intelligence. It can only be said that, if a man be unexpectedly over-joyed or grieved, the diaphragm jumps and causes him to start. This is due, however, to its

[1] Modern psychology has no terms exactly corresponding to *σύνεσις*, *γνώμη*, *φρόνησις*, and *διάγνωσις* in this chapter. It is doubtful if the author distinguished them very clearly. Contrast with this Chapter *Breaths*, XIV.

---

[3] I follow Littré with much diffidence. M has *εἴ τι ὁ ἄνθρωπος ὑπερχαρῇ ἐξ ἀδοκήτου ἢ ἀνιαθείη*: θ has *ἤν τι ὥνθρωπος ὑπερχαρῇ ἐξ ἀπροσδοκήτου πάθους*. The sense is clear but the true reading seems lost.

[4] *ἄσηην* θ M: *ἄλσιν* Littré with several Paris MSS.

ἐν τῷ σώματι, καὶ κοιλίην οὐκ ἔχουσι ἐς ἥντινα
10 χρὴ δέξασθαι ἢ ἀγαθὸν ἢ κακὸν προσπῖπτον,
ἀλλ' ὑπ' ἀμφοτέρων τούτων τεθορύβηνται διὰ
τὴν ἀσθενείην τῆς φύσιος· ἐπεὶ αἰσθάνονταί γε
οὐδενὸς πρότερον τῶν ἐν τῷ σώματι ἐόντων, ἀλλὰ
μάτην τοῦτο τὸ ὄνομα ἔχουσι καὶ τὴν αἰτίην,
ὥσπερ τὰ[1] πρὸς τῇ καρδίῃ ὦτα καλεῖται,
οὐδὲν ἐς τὴν ἀκοὴν συμβαλλόμενα. λέγουσι δέ
τινες ὡς καὶ φρονέομεν τῇ καρδίῃ καὶ τὸ
ἀνιώμενον τοῦτό ἐστι καὶ τὸ φροντίζον· τὸ δὲ
οὐχ οὕτως ἔχει, ἀλλὰ σπᾶται μὲν ὥσπερ αἱ
20 φρένες καὶ μᾶλλον διὰ ταύτας τὰς αἰτίας·
ἐξ ἅπαντος τοῦ σώματος φλέβες ἐς αὐτὴν
τείνουσι, καὶ συγκλείσασα[2] ἔχει ὥστε αἰσθάνεσθαι,
ἤν τις πόνος ἢ τάσις γίνηται τῷ
ἀνθρώπῳ· ἀνάγκη δὲ καὶ ἀνιώμενον φρίσσειν τε
τὸ σῶμα καὶ συντείνεσθαι, καὶ ὑπερχαίροντα τὸ
αὐτὸ τοῦτο πάσχειν· ὅτι ἡ καρδίη αἰσθάνεταί
τε μάλιστα καὶ αἱ φρένες. τῆς μέντοι φρονήσιος
οὐδετέρῳ μέτεστιν, ἀλλὰ πάντων τούτων
αἴτιος ὁ ἐγκέφαλός ἐστιν· ὡς οὖν καὶ τῆς
30 φρονήσιος[3] τοῦ ἠέρος πρῶτος αἰσθάνεται τῶν ἐν
τῷ σώματι ἐόντων, οὕτω καὶ ἤν τις μεταβολὴ
ἰσχυρὴ γένηται ἐν τῷ ἠέρι ὑπὸ τῶν ὡρέων, καὶ
αὐτὸς ἑωυτοῦ διάφορος γίνεται[4] ὁ ἐγκέφαλος.
διὸ καὶ τὰ νοσήματα ἐς αὐτὸν ἐμπίπτειν φημὶ
ὀξύτατα καὶ μέγιστα καὶ θανατωδέστατα καὶ
36 δυσκριτώτατα τοῖς ἀπείροισιν.

[1] ὥσπερ τὰ M: ὥστε θ.
[2] συγκλείσασα θμ: ξυγκλύσιας Reinhold.
[3] φρονήσιος MSS.: ἐφορμήσιος Reinhold.
[4] After γίνεται (for which θ reads γένηται) the MSS. have

being thin, and having a wider extent than any other organ; it has no cavity where it can receive any accident, good or bad, but it is disturbed by both owing to the weakness of its nature. Since it perceives nothing before the other parts do, but is idly named as though it were the cause of perception; just like the parts by the heart called "ears,"[1] though they contribute nothing to hearing. Some people say that the heart is the organ with which we think, and that it feels pain and anxiety. But it is not so; it merely is convulsed, as is the diaphragm, only more so for the following reasons. From all the body veins extend to it, and it so encloses them that it feels any pain or tension that comes upon a man. The body must, too, when in pain, shiver and be strained, and the same effects are produced by excess of joy, because the heart and the diaphragm are best endowed with feeling. Neither, however, has any share of intelligence, but it is the brain which is the cause of all the things I have mentioned.[2] As therefore it is the first of the bodily organs to perceive the intelligence coming from the air, so too if any violent change has occurred in the air owing to the seasons, the brain also becomes different from what it was. Therefore I assert that the diseases too that attack it are the most acute, most serious, most fatal, and the hardest for the inexperienced to judge of.

[1] Our "auricles." The Greek word *φρένες* can mean either "sense" or "diaphragm."

[2] The author can distinguish between *αἴσθησις* and *φρόνησις*.

ἐν τῷ ἠέρι, and, after *ἐγκέφαλος*, *πρῶτος αἰσθάνεται*. Both appear to be repetitions of phrases which have just occurred.

XXI. Αὕτη δὲ ἡ νοῦσος ἡ ἱερὴ καλεομένη ἀπο τῶν αὐτῶν προφασίων γίνεται ἀφ' ὧν[1] καὶ αἱ λοιπαὶ ἀπὸ τῶν προσιόντων καὶ ἀπιόντων, καὶ ψύχεος καὶ ἡλίου καὶ πνευμάτων μεταβαλλομένων τε καὶ οὐδέποτε ἀτρεμιζόντων. ταῦτα δ' ἐστὶ θεῖα, ὥστε μηδὲν δεῖ[2] ἀποκρίνοντα τὸ νόσημα θειότερον τῶν λοιπῶν νομίσαι, ἀλλὰ πάντα θεῖα καὶ πάντα ἀνθρώπινα· φύσιν δὲ ἕκαστον ἔχει καὶ δύναμιν ἐφ' ἑωυτοῦ, καὶ οὐδὲν
10 ἄπορόν[3] ἐστιν οὐδὲ ἀμήχανον· ἀκεστά τε τὰ πλεῖστά ἐστι τοῖς αὐτοῖσι τούτοισιν ἀφ' ὧν καὶ γίνεται. ἕτερον γὰρ ἑτέρῳ τροφή ἐστι, τοτὲ δὲ καὶ κάκωσις. τοῦτο οὖν δεῖ τὸν ἰητρὸν ἐπίστασθαι, ὅπως τὸν καιρὸν διαγινώσκων ἑκάστου τῷ μὲν ἀποδώσει τὴν τροφὴν καὶ αὐξήσει, τῷ δὲ ἀφαιρήσει καὶ κακώσει. χρὴ γὰρ καὶ ἐν ταύτῃ τῇ νούσῳ καὶ ἐν τῇσιν ἄλλῃσιν ἁπάσῃσι μὴ αὔξειν τὰ νοσήματα, ἀλλὰ τρύχειν προσφέροντα τῇ νούσῳ τὸ πολεμιώτατον ἑκάστῃ καὶ μὴ
20 τὸ σύνηθες· ὑπὸ μὲν γὰρ τῆς συνηθείης θάλλει καὶ αὔξεται, ὑπὸ δὲ τοῦ πολεμίου φθίνει τε καὶ ἀμαυροῦται. ὅστις δὲ ἐπίσταται ἐν ἀνθρώποισι ξηρὸν καὶ ὑγρὸν ποιεῖν, καὶ ψυχρόν καὶ θερμόν, ὑπὸ διαίτης, οὗτος καὶ ταύτην τὴν νοῦσον ἰῷτο ἄν, εἰ τοὺς καιροὺς διαγινώσκοι τῶν συμφερόντων,
26 ἄνευ καθαρμῶν καὶ μαγείης.[4]

[1] θ omits ἀφ' ὧν, perhaps rightly.

[2] δεῖ is not in the MSS. It was added by Ermerins, who reads μὴ δεῖ; Reinhold has μηδὲν δεῖ ἀποκρίνοντα. In θ the phrase appears as μηδένη (*sic*) ἀποκρίνοντα. M has διακρίνοντα.

[3] ἄπορόν M: ἄπειρόν θ.

[4] The last sentence in nearly all the MSS. contains many

XXI. This disease styled sacred comes from the same causes as others, from the things that come to and go from the body, from cold, sun, and from the changing restlessness of winds. These things are divine. So that there is no need to put the disease in a special class and to consider it more divine than the others; they are all divine and all human. Each has a nature and power of its own; none is hopeless or incapable of treatment. Most are cured by the same things as caused them. One thing is food for one thing, and another for another, though occasionally each actually does harm. So the physician must know how, by distinguishing the seasons for individual things, he may assign to one thing nutriment and growth, and to another diminution and harm. For in this disease as in all others it is necessary, not to increase the illness, but to wear it down by applying to each what is most hostile to it, not that to which it is conformable. For what is conformity gives vigour and increase; what is hostile causes weakness and decay. Whoever knows how to cause in men by regimen moist or dry, hot or cold, he can cure this disease also, if he distinguish the seasons for useful treatment, without having recourse to purifications and magic.

---

glosses; *τὴν τοιαύτην μεταβολὴν καὶ δύναμιν* after *ἀνθρώποισι*, after *διαίτης* the words *τὸν ἄνθρωπον*, and for *μαγείης* the phrase *μαγευμάτων καὶ πάσης ἄλλης βαναυσίης τοιαύτης*. I have kept the readings of *θ*, merely changing the *ποιέει* of this MS. to *ποιεῖν*. The reading of M is *ὑγρὸν καὶ ξηρὸν ποιέειν, καὶ θερμὸν καὶ ψυχρὸν ὑπὸ διαίτης, οὗτος καὶ ταύτην τὴν νοῦσον ἰῷτο ἄν, εἰ τοὺς καιροὺς διαγινώσκοι τῶν ξυμφερόντων, ἄνευ καθαρμῶν καὶ μαγευμάτων καὶ πάσης τῆς τοιαύτης βαναυσίης.*

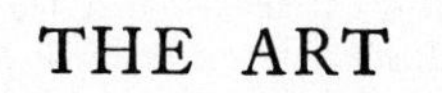

THE ART

INTRODUCTION

The little treatise called *The Art* has as its object to prove that there is such a thing as an art of medicine. After a few preliminary remarks, in which the writer attacks the unreasonableness of denying the reality of a thing which is seen to exist, the art of medicine is defined as the relief of suffering caused by disease, and the refusal to treat incurable disorders. Then four objections are dealt with in some detail. Detractors are said to urge:—

(1) That cures are due to luck;
(2) That patients often recover without medical help;
(3) That some patients die although treated by a physician;
(4) That physicians refuse to treat some diseases, knowing that they are powerless.

After meeting these objections the writer goes on to divide diseases into two main classes, external and internal. The former are said to be easy to cure, the latter difficult. These difficulties are then discussed at some length, and the failures that occur are attributed to circumstances, not to medicine itself.

It is quite plain from even a cursory reading of the treatise that its author was not a physician. His interest lies in subtle reasonings and in literary style, not in science. Besides this, in the last chapter he speaks of "those who are skilled in the art" as giving a proof of the existence of medicine based on works, and not, like the proofs given in the present book, on words. He evidently distinguishes himself from medical men.

The two most striking characteristics of *The Art* are an attenuated logic and a fondness for sophistic rhetoric. The rhetorical character of the whole book is so striking that without doubt it must be attributed to a sophist. The elaborate parallels, verbal antitheses, and balancing of phrase with phrase, can have no other explanation.

When, however, we attempt to advance further than this we are met by serious difficulties. Gomperz, relying among other things upon the second chapter, declares that the author must have been Protagoras. Professor Taylor,[1] relying on the same chapter, calls him an adherent of the Eleatic doctrine of being. The key-sentence to this chapter, "Things that exist are seen and are known; things that do not exist are neither seen nor known," does not seem strikingly either Protagorean or Eleatic; indeed in its context it does not seem to have any metaphysical reference at all, but merely points out the absurdity of denying the obvious.

The writer of *Epidemics III.* was known in Abdera, the native town of Protagoras, and the two men may well have met. There is, on the whole, some evidence that Gomperz is right. On the other hand, almost as good a case could be made out for considering the author to be Hippias. In the *Protagoras* Plato represents him as making a speech[2]

[1] *Varia Socratica*, p. 225. [2] 337 C–338 B.

full of sophistic rhetoric, and insisting on the contrast between φύσις and νόμος,[1] besides containing the word εἶδος, which occurs so frequently in *The Art* as to be almost a peculiarity. In the same dialogue Protagoras slyly criticizes[2] Hippias for making "the arts" instruments of education, the implication being that they were considered of great importance by Hippias but were slighted by Protagoras. The first sentence of *The Art* refers to those who "make an art of vilifying the arts in order to show off their learning." We should not be surprised to find that it was the famous polymath who took up the cudgels in defence of medicine, but the evidence is much too slight to warrant any conclusion being drawn. It is nevertheless curious, to say the least, to find that Gomperz notices a magisterial complacency and pedagogic self-confidence in *The Art*, which are the very traits we observe[3] in the Platonic Hippias. The irony of Gomperz's position is all the greater in that he attributes to the author of *The Art* "encyclopaedic learning," to which Protagoras could lay no claim, though it is a commonplace to attribute it to Hippias. Here the matter must be left, in that tantalizing uncertainty which darkens so many of the questions springing out of the study of the Hippocratic collection. We may, however, with some confidence put the date of *The Art* in the great sophistic period, namely the end of the fifth century B.C. It is in Erotian's list.

[1] See *The Art*, Chapter II (end). [2] 318 E.

[3] See especially *Protagoras* 315 C, where Hippias is described as sitting on a magisterial seat giving answers on abstruse points to his questioners.

INTRODUCTION

Manuscripts and Editions

The chief manuscripts are A and M, and the book is also included in many of the inferior manuscripts. It has been edited with great learning and enthusiasm by Gomperz.[1] Many interesting remarks will also be found in the first volume of the same author's *Greek Thinkers.* I have not thought it necessary, after the labours of Gomperz, to record all the readings of A and M, and a similar remark applies to *Breaths,* which has been ably edited by Nelson.

[1] *Die Apologie der Heilkunst* von Theodor Gomperz. Zweite durchgesehene Auflage, Leipzig, 1910.

ΠΕΡΙ ΤΕΧΝΗΣ

I. Εἰσί τινες οἳ τέχνην πεποίηνται τὸ τὰς τέχνας αἰσχροεπεῖν, ὡς μὲν οἴονται οὐ τοῦτο διαπρησσόμενοι ὃ ἐγὼ λέγω,[1] ἀλλ' ἱστορίης οἰκείης ἐπίδειξιν ποιεύμενοι. ἐμοὶ δὲ τὸ μέν τι τῶν μὴ εὑρημένων ἐξευρίσκειν, ὅ τι καὶ εὑρεθὲν κρέσσον ᾖ ἢ ἀνεξεύρετον, συνέσιος δοκεῖ ἐπιθύμημά τε καὶ ἔργον εἶναι, καὶ τὸ τὰ ἡμίεργα ἐς τέλος ἐξεργάζεσθαι ὡσαύτως· τὸ δὲ λόγων οὐ καλῶν τέχνῃ τὰ τοῖς ἄλλοις εὑρημένα αἰσχύνειν 10 προθυμεῖσθαι, ἐπανορθοῦντα μὲν μηδέν, διαβάλλοντα δὲ τὰ τῶν εἰδότων πρὸς τοὺς μὴ εἰδότας ἐξευρήματα, οὐκέτι συνέσιος δοκεῖ ἐπιθύμημά τε καὶ ἔργον εἶναι, ἀλλὰ κακαγγελίη μᾶλλον φύσιος ἢ ἀτεχνίη· μούνοισι γὰρ δὴ τοῖσιν ἀτέχνοισιν ἡ ἐργασίη αὕτη ἁρμόζει, φιλοτιμεομένων μέν, οὐδαμὰ δὲ δυναμένων κακίῃ ὑπουργεῖν ἐς τὸ τὰ τῶν πέλας ἔργα ἢ ὀρθὰ ἐόντα διαβάλλειν, ἢ οὐκ ὀρθὰ μωμεῖσθαι. τοὺς μὲν οὖν ἐς τὰς ἄλλας τέχνας τούτῳ τῷ τρόπῳ ἐμπίπτοντας, οἷσι μέλει 20 τε, καὶ ὧν μέλει, οἱ δυνάμενοι κωλυόντων· ὁ δὲ παρεὼν λόγος τοῖσιν ἐς ἰητρικὴν οὕτως ἐμπορευομένοις ἐναντιώσεται, θρασυνόμενος μὲν διὰ τούτους οὓς ψέγει, εὐπορέων δὲ διὰ τὴν τέχνην 24 ᾗ βοηθεῖ, δυνάμενος δὲ διὰ σοφίην ᾗ πεπαίδευται.

II. Δοκεῖ δή μοι τὸ μὲν σύμπαν τέχνη εἶναι

[1] οὐ τοῦτο διαπρησσόμενοι ὃ ἐγὼ λέγω so Gomperz: οἱ τοῦτο διαπρησσόμενοι οὐχ ὃ ἐγὼ λέγω Littré with some Paris MSS.

THE ART

I. Some there are who have made an art of vilifying the arts, though they consider, not that they are accomplishing the object I mention, but that they are making a display of their own knowledge. In my opinion, however, to discover that was unknown before, when the discovery of it is better than a state of ignorance, is the ambition and task of intelligence, and so is to bring to completion what was already accomplished in part. On the other hand, to be eager to bring shame through the art of abuse upon the discoveries of others, improving nothing, but disparaging before those who do not know the discoveries of those who do, seems to me to be not the ambition and work of intelligence, but the sign of a nasty nature, or of want of art. Indeed it becomes only those who are without art to act in this manner, with the ambition, though not the power, to indulge their malevolence by disparaging what is right in their neighbours' works and by cavilling at what is amiss. Now as for the attacks of this kind that are made on the other arts, let them be repelled by those who care to do so and can, and with regard to those points about which they care; the present discussion will oppose those who thus invade the art of medicine, and it is emboldened by the nature of those it blames, well equipped through the art it defends, and powerful through the wisdom in which it has been educated.

II. Now it seems to me that generally speaking

οὐδεμία οὐκ ἐοῦσα· καὶ γὰρ ἄλογον τῶν ἐόντων τι ἡγεῖσθαι μὴ ἐόν· ἐπεὶ τῶν γε μὴ ἐόντων τίνα ἂν τίς οὐσίην θεησάμενος ἀπαγγείλειεν ὡς ἔστιν; εἰ γὰρ δὴ ἔστι γ' ἰδεῖν τὰ μὴ ἐόντα, ὥσπερ τὰ ἐόντα, οὐκ οἶδ' ὅπως ἄν τις αὐτὰ νομίσειε μὴ ἐόντα, ἅ γε εἴη καὶ ὀφθαλμοῖσιν ἰδεῖν καὶ γνώμῃ νοῆσαι ὡς ἔστιν· ἀλλ' ὅπως μὴ οὐκ ᾖ τοῦτο τοιοῦτον· ἀλλὰ τὰ μὲν ἐόντα αἰεὶ ὁρᾶταί τε καὶ
10 *γινώσκεται, τὰ δὲ μὴ ἐόντα οὔτε ὁρᾶται οὔτε γινώσκεται. γινώσκεται τοίνυν δεδειγμένων ἤδη*[1] *τῶν τεχνέων, καὶ οὐδεμία ἐστὶν ἥ γε ἔκ τινος εἴδεος οὐχ ὁρᾶται. οἶμαι δ' ἔγωγε καὶ τὰ ὀνόματα αὐτὰς διὰ τὰ εἴδεα λαβεῖν· ἄλογον γὰρ ἀπὸ τῶν ὀνομάτων ἡγεῖσθαι τὰ εἴδεα βλαστάνειν, καὶ ἀδύνατον· τὰ μὲν γὰρ ὀνόματα νομοθετήματά ἐστι, τὰ δὲ εἴδεα οὐ νομοθετήματα, ἀλλὰ βλαστή-*
18 *ματα φύσιος.*[2]

III. *Περὶ μὲν οὖν τούτων εἴ γέ τις μὴ ἱκανῶς ἐκ τῶν εἰρημένων συνίησιν, ἐν ἄλλοισιν ἂν λόγοισιν σαφέστερον διδαχθείη. περὶ δὲ ἰητρικῆς, ἐς ταύτην γὰρ ὁ λόγος, ταύτης οὖν τὴν ἀπόδειξιν ποιήσομαι, καὶ πρῶτόν γε διοριεῦμαι ὃ νομίζω ἰητρικὴν εἶναι· τὸ δὴ πάμπαν ἀπαλλάσσειν τῶν νοσεόντων τοὺς καμάτους καὶ τῶν νοσημάτων τὰς σφοδρότητας ἀμβλύνειν, καὶ τὸ μὴ ἐγχειρεῖν τοῖσι κεκρατημένοις ὑπὸ τῶν νοσημάτων, εἰδότας ὅτι*
10 *ταῦτα οὐ δύναται ἰητρική. ὡς οὖν ποιεῖ τε*

[1] After *ἤδη* Gomperz would add *εἴδεα*.

[2] In the MSS. *φύσιος* occurs after *ὀνόματα*; it was transposed by Gomperz. Possibly the transposition is not necessary, as *φύσιος* is easily understood after *βλαστήματα*.

there is no art which does not exist; in fact it is absurd to regard as non-existent one of the things that exist. Since what substance could there be of non-existents, and who could behold them and declare that[1] they exist? For if really it be possible to see the non-existent, as it is to see the existent, I do not know how a man could regard as non-existent what he can both see with his eyes and with his mind think that[1] it exists. Nay, it cannot be so; but the existent is always seen and known, and the non-existent is neither seen nor known. Now reality is known when the arts have been already revealed, and there is no art which is not seen as the result of[2] some real essence.[3] I for my part think that the names also of the arts have been given them because of their real essences; for it is absurd—nay impossible—to hold that real essences spring from names. For names are conventions, but real essences are not conventions but the offspring of nature.

III. As to this subject in general, if it is not sufficiently understood from what I have said, other treatises will give clearer instruction. I will now turn to medicine, the subject of the present treatise, and set forth the exposition of it. First I will define what I conceive medicine to be. In general terms, it is to do away with the sufferings of the sick, to lessen the violence of their diseases, and to refuse to treat those who are overmastered by their diseases, realizing that in such cases medicine is powerless. That medicine fulfils these conditions,

[1] Or, "how." [2] Or "springing from."

[3] εἶδος is often used with this meaning in the present treatise.

ταῦτα, καὶ οἵη τέ ἐστι διὰ παντὸς ποιεῖν, περὶ τούτου μοι ὁ λοιπὸς λόγος ἤδη ἔσται. ἐν δὲ τῇ τῆς τέχνης ἀποδείξει ἅμα καὶ τοὺς λόγους τῶν αἰσχύνειν αὐτὴν οἰομένων ἀναιρήσω, ᾗ ἂν ἕκαστος 15 *αὐτῶν πρήσσειν τι οἰόμενος τυγχάνῃ.*

IV. *Ἔστι μὲν οὖν μοι ἀρχὴ τοῦ λόγου ἣ καὶ ὁμολογηθήσεται παρὰ πᾶσιν· ὅτι μὲν ἔνιοι ἐξυγιαίνονται τῶν θεραπευομένων ὑπὸ ἰητρικῆς ὁμολογεῖται· ὅτι δὲ οὐ πάντες, ἐν τούτῳ ἤδη ψέγεται ἡ τέχνη, καὶ φασὶν οἱ τὰ χείρω λέγοντες διὰ τοὺς ἁλισκομένους ὑπὸ τῶν νοσημάτων τοὺς ἀποφεύγοντας αὐτὰ τύχῃ ἀποφεύγειν καὶ οὐ διὰ τὴν τέχνην. ἐγὼ δὲ οὐκ ἀποστερέω μὲν οὐδ' αὐτὸς τὴν τύχην ἔργου οὐδενός, ἡγεῦμαι δὲ τοῖσι* 10 *μὲν κακῶς θεραπευομένοισι νοσήμασι τὰ πολλὰ τὴν ἀτυχίην ἕπεσθαι, τοῖσι δὲ εὖ τὴν εὐτυχίην. ἔπειτα δὲ καὶ πῶς οἷόν τ' ἐστὶ τοῖς ὑγιασθεῖσιν ἄλλο τι αἰτιήσασθαι ἢ τὴν τέχνην, εἴπερ χρώμενοι αὐτῇ καὶ ὑπουργέοντες ὑγιάσθησαν; τὸ μὲν γὰρ τῆς τύχης εἶδος ψιλὸν οὐκ ἐβουλήθησαν θεήσασθαι, ἐν ᾧ τῇ τέχνῃ ἐπέτρεψαν σφέας αὐτούς, ὥστε τῆς μὲν ἐς τὴν τύχην ἀναφορῆς ἀπηλλαγμένοι εἰσί, τῆς μέντοι ἐς τὴν τέχνην οὐκ ἀπηλλαγμένοι· ἐν ᾧ γὰρ ἐπέτρεψαν αὐτῇ σφέας* 20 *καὶ ἐπίστευσαν, ἐν τούτῳ αὐτῆς καὶ τὸ εἶδος ἐσκέψαντο καὶ τὴν δύναμιν περανθέντος τοῦ* 22 *ἔργου ἔγνωσαν.*

---

[1] Literally, "effects," "works."

[2] That is, they refused to see nothing but luck in the sphere of medicine and therapeutics. It is impossible to bring out in a translation all the associations of the words used in this passage. Is εἶδος "form," "face," as is sug-

and is able constantly to fulfil them, will be the subject of my treatise from this point. In the exposition of the art I shall at the same time refute the arguments of those who think to shame it, and I shall do so just in those points where severally they believe they achieve some success.

IV. The beginning of my discourse is a point which will be conceded by all. It is conceded that of those treated by medicine some are healed. But because not all are healed the art is blamed, and those who malign it, because there are some who succumb to diseases, assert that those who escape do so through luck and not through the art. Now I, too, do not rob luck of any of its prerogatives,[1] but I am nevertheless of opinion that when diseases are badly treated ill-luck generally follows, and good luck when they are treated well. Again, how is it possible for patients to attribute their recoveries to anything else except the art, seeing that it was by using it and serving it that they recovered? For in that they committed themselves to the art they showed their unwillingness to behold nothing but the reality of luck,[2] so that while freed from dependence upon luck they are not freed from dependence upon the art. For in that they committed themselves with confidence to the art, they thereby acknowledged also its reality, and when its work was accomplished they recognized its power.

gested by θεήσασθαι? So Gomperz, who translates "das nackte Antlitz des Zufalls wollten sie nicht erschauen." Or is it "essence," as A. E. Taylor thinks (*Varia Socratica*, p. 226, where τὸ τῆς τύχης εἶδος is equated with ἡ τύχη). Though I translate εἶδος by "reality" I think that the meaning "form," "face" is not excluded.

V. Ἐρεῖ δὴ ὁ τἀναντία λέγων, ὅτι πολλοὶ ἤδη καὶ οὐ χρησάμενοι ἰητρῷ νοσέοντες ὑγιάσθησαν, καὶ ἐγὼ τῷ λόγῳ οὐκ ἀπιστέω· δοκεῖ δέ μοι οἷόν τε εἶναι καὶ ἰητρῷ μὴ χρωμένους ἰητρικῇ περιτυχεῖν, οὐ μὴν ὥστε εἰδέναι ὅ τι ὀρθὸν ἐν αὐτῇ ἔνι ἢ[1] ὅ τι μὴ ὀρθόν, ἀλλ' ὥσει ἐπιτύχοιεν[2] τοιαῦτα θεραπεύσαντες ἑωυτούς, ὁποῖά περ ἂν ἐθεραπεύθησαν εἰ καὶ ἰητροῖσιν ἐχρῶντο. καὶ τοῦτό γε τεκμήριον μέγα τῇ οὐσίῃ τῆς τέχνης, ὅτι 10 ἐοῦσά τέ ἐστι καὶ μεγάλη, ὅπου γε φαίνονται καὶ οἱ μὴ νομίζοντες αὐτὴν εἶναι σῳζόμενοι δι' αὐτήν· πολλὴ γὰρ ἀνάγκη καὶ τοὺς μὴ χρωμένους ἰητροῖσι νοσήσαντας δὲ καὶ ὑγιασθέντας εἰδέναι, ὅτι ἢ δρῶντές τι ἢ μὴ δρῶντες ὑγιάσθησαν· ἢ γὰρ ἀσιτίῃ ἢ πολυφαγίῃ, ἢ ποτῷ πλέονι ἢ δίψῃ, ἢ λουτροῖς, ἢ ἀλουσίῃ, ἢ πόνοισιν ἢ ἡσυχίῃ, ἢ ὕπνοισιν ἢ ἀγρυπνίῃ, ἢ τῇ[3] ἁπάντων τούτων ταραχῇ χρώμενοι ὑγιάσθησαν. καὶ τῷ ὠφελῆσθαι πολλὴ ἀνάγκη αὐτοῖς ἐστὶν ἐγνωκέναι ὅ 20 τι ἦν τὸ ὠφελῆσαν, καὶ ὅτε ἐβλάβησαν τῷ βλαβῆναι ὅ τι ἦν[4] τὸ βλάψαν. τὰ γὰρ τῷ

[1] ἔνι ἢ Gomperz, from the ἐνῆ ἢ of A.

[2] ἀλλ' ὥσει my emendation: ἀλως τε εἰ A: ἀλλως τε M: ἄλλ' ὥστ' ἂν Littré: ἀλλ' ὥστε Gomperz: perhaps ἀλλ' ὥστε ἐπιτυχεῖν (with θεραπεύσαντας).

[3] ἢ τῇ M: ἢ τὶ A: ἤ τινι Gomperz.

[4] With some misgiving I omit the τι after ἦν, which A has in the second clause and Gomperz adds in the first. Gomperz reads ὅτι not ὅ τι.

---

[1] The sense is clear but the reading is uncertain. No scholar will accept that of Gomperz or that of Littré, as both are impossible Greek. Perhaps the optative was the result

V. Now my opponent will object that in the past many, even without calling in a physician, have been cured of their sickness, and I agree that he is right. But I hold that it is possible to profit by the art of medicine even without calling in a physician, not indeed so as to know what is correct medical treatment and what is incorrect, but so as by chance[1] to employ in self-treatment the same means as would have been employed had a physician actually been called in. And it is surely strong proof of the existence of the art, that it both exists and is powerful, if it is obvious that even those who do not believe in it recover through it. For even those who, without calling in a physician, recovered from a sickness must perforce know that their recovery was due to doing something or to not doing something; it was caused in fact by fasting or by abundant diet, by excess of drink or by abstinence therefrom, by bathing or by refraining therefrom, by violent exercise or by rest, by sleep or by keeping awake, or by using a combination of all these things. And they must perforce have learnt, by having been benefited, what it was that benefited them, just as when they were harmed they must have learnt, by having been harmed, what it was that harmed them.[2]

of ἐπι- being read as εἰ (which A has), and ἐπιτυχεῖν was the original reading.

[2] I cannot think that Gomperz's reading, with ὅτι for ὅ τι, is correct. It would surely make the sentence a flat repetition of the preceding one. I take the sequence of thought to be this. Cures apparently spontaneous are not really so. The cure has its cause, *e. g.* a bath or a sleep, and the fact that the cure followed the bath or sleep proves that the latter was the cause. To distinguish the beneficial in this way is not guesswork, but implies the existence of an art.

ὠφελῆσθαι καὶ τὰ τῷ βεβλάφθαι ὡρισμένα οὐ πᾶς ἱκανὸς γνῶναι· εἰ τοίνυν ἐπιστήσεται ἢ ἐπαινεῖν ἢ ψέγειν ὁ νοσήσας τῶν διαιτημάτων τι οἷσιν ὑγιάσθη, πάντα ταῦτα τῆς ἰητρικῆς.[1] *καὶ ἔστιν οὐδὲν ἧσσον τὰ ἁμαρτηθέντα τῶν ὠφελησάντων μαρτύρια τῇ τέχνῃ ἐς τὸ εἶναι· τὰ μὲν γὰρ ὠφελήσαντα τῷ ὀρθῶς προσενεχθῆναι ὠφέλησαν, τὰ δὲ βλάψαντα τῷ μηκέτι ὀρθῶς*
30 *προσενεχθῆναι ἔβλαψαν. καίτοι ὅπου τό τε ὀρθὸν καὶ τὸ μὴ ὀρθὸν ὅρον ἔχει ἑκάτερον, πῶς τοῦτο οὐκ ἂν τέχνη εἴη; τοῦτο γὰρ ἔγωγέ φημι ἀτεχνίην εἶναι, ὅπου μήτε ὀρθὸν ἔνι μηδὲν μήτε οὐκ ὀρθόν· ὅπου δὲ τούτων ἔνεστιν ἑκάτερον,*
35 *οὐκέτι ἂν τοῦτο ἔργον ἀτεχνίης εἴη.*

VI. *Ἔτι τοίνυν εἰ μὲν ὑπὸ φαρμάκων τῶν τε καθαιρόντων καὶ τῶν ἱστάντων ἡ ἴησις τῇ τε ἰητρικῇ καὶ τοῖσιν ἰητροῖσι μοῦνον ἐγίνετο, ἀσθενὴς ἦν ἂν ὁ ἐμὸς λόγος· νῦν δὲ φαίνονται τῶν ἰητρῶν οἱ μάλιστα ἐπαινεόμενοι καὶ διαιτήμασιν ἰώμενοι καὶ ἄλλοισί γε εἴδεσιν, ἃ οὐκ ἄν τις φαίη, μὴ ὅτι ἰητρός, ἀλλ᾽ οὐδὲ ἰδιώτης ἀνεπιστήμων ἀκούσας, μὴ οὐ τῆς τέχνης εἶναι. ὅπου οὖν οὐδὲν οὔτ᾽ ἐν τοῖς ἀγαθοῖσι τῶν ἰητρῶν οὔτ᾽*
10 *ἐν τῇ ἰητρικῇ αὐτῇ ἀχρεῖόν ἐστιν, ἀλλ᾽ ἐν τοῖσι πλείστοισι τῶν τε φυομένων καὶ τῶν ποιευμένων ἔνεστι τὰ εἴδεα τῶν θεραπειῶν καὶ τῶν φαρμάκων, οὐκ ἔστιν ἔτι οὐδενὶ τῶν ἄνευ ἰητροῦ ὑγιαζομένων τὸ αὐτόματον αἰτιήσασθαι ὀρθῷ λόγῳ· τὸ μὲν γὰρ αὐτόματον οὐδὲν φαίνεται ἐὸν ἐλεγχόμενον· πᾶν γὰρ τὸ γινόμενον διά τι εὑρίσκοιτ᾽ ἂν γινό-*

[1] After *ἰητρικῆς* in many MSS. occur the words *ὄντα εὑρήσει*. M has *ἰητρικῆς ἔστι καὶ ἔστιν οὐδέν*. A has *ἰητρικῆς*

For it is not everybody who is capable of discerning things distinguished by benefit and things distinguished by harm. If therefore the patient will know how to praise or to blame what composed the regimen under which he recovered, all these things belong to the art of medicine. Again, mistakes, no less than benefits, witness to the existence of the art; for what benefited did so because correctly administered, and what harmed did so because incorrectly administered. Now where correctness and incorrectness each have a defined limit, surely there must be an art. For absence of art I take to be absence of correctness and of incorrectness; but where both are present art cannot be absent.

VI. Moreover, if the medical art and medical men brought about a cure only by means of medicines, purgative or astringent, my argument would be weak. As it is, the physicians of greatest repute obviously cure by regimen and by other substances, which nobody—not only a physician but also an unlearned layman, if he heard of them—would say do not belong to the art. Seeing then that there is nothing that cannot be put to use by good physicians and by the art of medicine itself, but in most things that grow or are made are present the essential substances of cures and of drugs, no patient who recovers without a physician can logically attribute the recovery to spontaneity. Indeed, under a close examination spontaneity disappears; for everything that occurs will be found to do so through

---

ἔστιν οὐδέν. Gomperz reads *ἰητρικῆς εὑρήσει ὡς ἔστιν. καὶ ἔστιν οὐδὲν κ.τ.λ.* Littré follows the majority of the late MSS. (*ὄντα εὑρήσει*).

μενον, καὶ ἐν τῷ διά τι τὸ αὐτόματον οὐ φαίνεται οὐσίην ἔχον οὐδεμίην ἀλλ' ἢ ὄνομα· ἡ δὲ ἰητρικὴ καὶ ἐν τοῖσι διά τι καὶ ἐν τοῖσι προνοουμένοισι
20 φαίνεταί τε καὶ φανεῖται αἰεὶ οὐσίην ἔχουσα.

VII. Τοῖσι μὲν οὖν τῇ τύχῃ τὴν ὑγιείην προστιθεῖσι τὴν δὲ τέχνην ἀφαιρέουσι τοιαῦτ' ἄν τις λέγοι· τοὺς δ' ἐν τῇσι τῶν ἀποθνῃσκόντων συμφορῇσι τὴν τέχνην ἀφανίζοντας θαυμάζω, ὅτεῳ ἐπαιρόμενοι ἀξιοχρέῳ λόγῳ τὴν μὲν τῶν ἀποθνῃσκόντων ἀτυχίην ἀναιτίην[1] καθιστᾶσι, τὴν δὲ τῶν τὴν ἰητρικὴν μελετησάντων σύνεσιν αἰτίην· ὡς τοῖσι μὲν ἰητροῖς ἔνεστι τὰ μὴ δέοντα ἐπιτάξαι, τοῖσι δὲ νοσέουσιν οὐκ ἔνεστι τὰ προσταχθέντα
10 παραβῆναι. καὶ μὴν πολύ γε εὐλογώτερον τοῖσι κάμνουσιν ἀδυνατεῖν τὰ προστασσόμενα ὑπουργεῖν, ἢ τοῖς ἰητροῖσι τὰ μὴ δέοντα ἐπιτάσσειν. οἱ μὲν γὰρ ὑγιαινούσῃ γνώμῃ μεθ' ὑγιαίνοντος σώματος ἐγχειρέουσι, λογισάμενοι τά τε παρεόντα, τῶν τε παροιχομένων τὰ ὁμοίως διατεθέντα τοῖσι παρεοῦσιν, ὥστε ποτὲ θεραπευθέντα[2] εἰπεῖν ὡς ἀπήλλαξαν· οἱ δὲ οὔτε ἃ κάμνουσιν οὔτε δι' ἃ κάμνουσιν εἰδότες, οὐδ' ὅ τι ἐκ τῶν παρεόντων ἔσται, οὐδ' ὅ τι ἐκ τῶν τούτοισιν ὁμοίων γίνεται,
20 ἐπιτάσσονται, ἀλγέοντες μὲν ἐν τῷ παρεόντι, φοβεύμενοι δὲ τὸ μέλλον, καὶ πλήρεις μὲν τῆς νούσου, κενεοὶ δὲ σιτίων, θέλοντες τὰ πρὸς τὴν νοῦσον ἤδη μᾶλλον ἢ τὰ πρὸς τὴν ὑγιείην προσδέχεσθαι, οὐκ ἀποθανεῖν ἐρῶντες ἀλλὰ καρτερεῖν ἀδυνατέοντες. οὕτως δὲ διακειμένους πότερον

[1] ἀτυχίην ἀναιτίαν A: ἀκρασίην αἰτίην M: ἀκρασίην ἀναιτίην Gomperz: ἀκρησίην οὐκ αἰτίην Littré with several MSS
[2] θεραπευθέντας Gomperz.

something, and this "through something" shows that spontaneity is a mere name, and has no reality. Medicine, however, because it acts "through something," and because its results may be forecasted, has reality, as is manifest now and will be manifest for ever.

VII. Such then might be the answer to those who attribute recovery to chance and deny the existence of the art. As to those who would demolish the art by fatal cases of sickness, I wonder what adequate reason induces them to hold innocent the ill-luck[1] of the victims, and to put all the blame upon the intelligence of those who practised the art of medicine. It amounts to this: while physicians may give wrong instructions, patients can never disobey orders. And yet it is much more likely that the sick cannot follow out the orders than that the physicians give wrong instructions. The physician sets about his task with healthy mind and healthy body, having considered the case and past cases of like characteristics to the present, so as to say how they were treated and cured. The patient knows neither what he is suffering from, nor the cause thereof; neither what will be the outcome of his present state, nor the usual results of like conditions. In this state he receives orders, suffering in the present and fearful of the future; full of the disease, and empty of food; wishful of treatment rather to enjoy immediate alleviation of his sickness than to recover his health; not in love with death, but powerless to endure. Which is the more likely:

[1] With the reading of Gomperz, "weakness." I follow A here, but it is one of the few cases where the other tradition has the more vigorous reading, which may be correct.

εἰκὸς τούτους τὰ ὑπὸ τῶν ἰητρῶν ἐπιτασσόμενα ποιεῖν ἢ ἄλλα ποιεῖν ἢ ἃ ἐπετάχθησαν,[1] ἢ τοὺς ἰητροὺς τοὺς ἐκείνως διακειμένους ὡς ὁ πρόσθεν λόγος ἡρμήνευσεν ἐπιτάσσειν τὰ μὴ δέοντα; ἆρ'
30 οὐ πολὺ μᾶλλον, τοὺς μὲν δεόντως ἐπιτάσσειν τοὺς δὲ εἰκότως ἀδυνατεῖν πείθεσθαι, μὴ πειθομένους δὲ περιπίπτειν τοῖσι θανάτοις, ὧν οἱ μὴ ὀρθῶς λογιζόμενοι τὰς αἰτίας τοῖς οὐδὲν αἰτίοις
34 ἀνατιθέασι, τοὺς αἰτίους ἐλευθεροῦντες;

VIII. Εἰσὶ δέ τινες οἳ καὶ διὰ τοὺς μὴ θέλοντας ἐγχειρεῖν τοῖσι κεκρατημένοις ὑπὸ τῶν νοσημάτων μέμφονται τὴν ἰητρικήν, λέγοντες ὡς ταῦτα μὲν καὶ αὐτὰ ὑφ' ἑωυτῶν ἂν ἐξυγιάζοιτο ἃ ἐγχειρέουσιν ἰῆσθαι, ἃ δ' ἐπικουρίης δεῖται μεγάλης οὐχ ἅπτονται, δεῖν δέ, εἴπερ ἦν ἡ τέχνη, πάνθ' ὁμοίως ἰῆσθαι. οἱ μὲν οὖν ταῦτα λέγοντες, εἰ ἐμέμφοντο τοῖς ἰητροῖς, ὅτι αὐτῶν τοιαῦτα λεγόντων οὐκ ἐπιμέλονται ὡς παραφρονεύντων, εἰκότως ἂν ἐμεμφοντο
10 μᾶλλον ἢ ἐκεῖνα μεμφόμενοι. εἰ γάρ τις ἢ τέχνην ἐς ἃ μὴ τέχνη, ἢ φύσιν ἐς ἃ μὴ φύσις πέφυκεν, ἀξιώσειε δύνασθαι, ἀγνοεῖ ἄγνοιαν ἁρμόζουσαν μανίῃ μᾶλλον ἢ ἀμαθίῃ. ὧν γάρ ἔστιν ἡμῖν τοῖσί τε τῶν φυσίων τοῖσι τε τῶν τεχνέων ὀργάνοις ἐπικρατεῖν, τούτων ἔστιν ἡμῖν δημιουργοῖς εἶναι, ἄλλων δὲ οὐκ ἔστιν. ὅταν οὖν τι πάθῃ ὤνθρωπος κακὸν ὃ κρέσσον ἐστὶ τῶν ἐν ἰητρικῇ ὀργάνων,

[1] Littré (Ermerins, Reinhold) reads ἐπιτασσόμενα <μὴ> ποιέειν, ἢ ἀλλὰ ποιέειν, ἃ οὐκ ἐπετάχθησαν. He inserts μὴ on his own authority and reads οὐκ with many MSS. A reads ἡ α, M has ἢ ἃ.

---

[1] The word φύσις (and φυσίων below) is difficult to translate. It refers to the natural powers of the human constitu-

that men in this condition obey, instead of varying, the physician's orders, or that the physician, in the condition that my account has explained above, gives improper orders? Surely it is much more likely that the physician gives proper orders, which the patient not unnaturally is unable to follow; and not following them he meets with death, the cause of which illogical reasoners attribute to the innocent, allowing the guilty to go free.

VIII. Some too there are who blame medicine because of those who refuse to undertake desperate cases, and say that while physicians undertake cases which would cure themselves, they do not touch those where great help is necessary; whereas, if the art existed, it ought to cure all alike. Now if those who make such statements charged physicians with neglecting them, the makers of the statements, on the ground that they are delirious, they would bring a more plausible charge than the one they do bring. For if a man demand from an art a power over what does not belong to the art, or from nature[1] a power over what does not belong to nature, his ignorance is more allied to madness than to lack of knowledge. For in cases where we may have the mastery through the means afforded by a natural constitution or by an art, there we may be craftsmen, but nowhere else. Whenever therefore a man suffers from an ill which is too strong for the means at the disposal

tion, which may be too weak to resist the attack of a severe disease. Its *ὄργανα* are the means whereby we can influence the *φύσις*, the various bodily "organs" which can be affected by medicine and treatment generally. Gomperz translates *φύσις* by "Natur," and *τοῖσι τῶν φυσίων ὀργάνοις* by "durch die Kräfte der Körper.'

οὐδὲ προσδοκᾶσθαι τοῦτό που δεῖ ὑπὸ ἰητρικῆς κρατηθῆναι ἄν· αὐτίκα γὰρ τῶν ἐν ἰητρικῇ καιόντων
20 *τὸ πῦρ*[1] *ἐσχάτως καίει, τούτου δὲ ἧσσον καὶ*[2] *ἄλλα πολλα· τῶν μὲν οὖν ἡσσόνων τὰ κρέσσω οὔπω δηλονότι ἀνίητα· τῶν δὲ κρατίστων τὰ κρέσσω πῶς οὐ δηλονότι ἀνίητα; ἃ γὰρ πῦρ δημιουργεῖ, πῶς οὐ τὰ τούτῳ*[3] *μὴ ἁλισκόμενα δηλοῖ ὅτι ἄλλης τέχνης δεῖται, καὶ οὐ ταύτης, ἐν ᾗ τὸ πῦρ ὄργανον; ὡὑτὸς δέ μοι λόγος καὶ ὑπὲρ τῶν ἄλλων ὅσα τῇ ἰητρικῇ συνεργεῖ, ὧν ἁπάντων φημὶ δεῖν ἑκάστου μὴ*[4] *κατατυχόντα τὸν ἰητρὸν τὴν δύναμιν αἰτιᾶσθαι τοῦ πάθεος, μὴ τὴν τέχνην. οἱ μὲν οὖν*
30 *μεμφόμενοι τοὺς τοῖσι κεκρατημένοισι μὴ ἐγχειρέοντας παρακελεύονται καὶ ὧν μὴ προσήκει ἅπτεσθαι οὐδὲν ἧσσον ἢ ὧν προσήκει· παρακελευόμενοι δὲ ταῦτα ὑπὸ μὲν τῶν ὀνόματι ἰητρῶν θαυμάζονται, ὑπὸ δὲ τῶν καὶ τέχνῃ καταγελῶνται. οὐ μὴν οὕτως ἀφρόνων οἱ ταύτης τῆς δημιουργίης ἔμπειροι οὔτε μωμητέων οὔτ' ἐπαινετέων δέονται, ἀλλὰ λελογισμένων πρὸς ὅ τι αἱ ἐργασίαι τῶν δημιουργῶν τελευτώμεναι πλήρεις εἰσί, καὶ ὅτευ ὑπολειπόμεναι ἐνδεεῖς, ἔτι τῶν ἐνδειῶν, ἅς τε τοῖς*
40 *δημιουργεῦσιν ἀναθετέον ἅς τε τοῖς δημιουργεο-*
41 *μένοισιν.*

IX. Τὰ *μὲν οὖν κατὰ τὰς ἄλλας τέχνας ἄλλος χρόνος μετ' ἄλλου λόγου δείξει· τὰ δὲ κατὰ τὴν ἰητρικὴν οἷά τέ ἐστιν ὥς τε κριτέα, τὰ μὲν ὁ παροιχόμενος τὰ δὲ ὁ παρεὼν διδάξει λόγος. ἔστι*

[1] τὸ πῦρ A: πῦρ M. Either dittography in A or τὸ has fallen out after -των in M.
[2] Gomperz reads ἡσσόνως for ἧσσον καὶ.
[3] Gomperz reads τούτων τὰ τούτῳ for τὰ τούτῳ.

of medicine, he surely must not even expect that it can be overcome by medicine. For example, of the caustics employed in medicine fire is the most powerful, though there are many others less powerful than it. Now affections that are too strong for the less powerful caustics plainly are not for this reason incurable; but those which are too strong for the most powerful plainly are incurable. For when fire operates, surely affections not overcome thereby show that they need another art, and not that wherein fire is the means. I apply the same argument to the other agents employed in medicine; when any one of them plays the physician false, the blame should be laid on the power of the affection, and not on the art. Now those who blame physicians who do not undertake desperate cases, urge them to take in hand unsuitable patients just as much as suitable ones. When they urge this, while they are admired by physicians in name, they are a laughing-stock of really scientific physicians. Those experienced in this craft have no need either of such foolish blame or of such foolish praise; they need praise only from those who have considered where the operations of craftsmen reach their end and are complete, and likewise where they fall short; and have considered moreover which of the failures should be attributed to the craftsmen, and which to the objects on which they practise their craft.

IX. The scope of the other arts shall be discussed at another time and in another discourse; the scope of medicine, the nature of things medical and how they are to be judged, my discourse has or will set

[4] Neither A nor M have a negative after ἑκάστου: Littré reads μὴ with a few MSS.: Gomperz inserts οὐ.

γὰρ τοῖσι ταύτην τὴν τέχνην ἱκανῶς εἰδόσι τὰ μὲν τῶν νοσημάτων οὐκ ἐν δυσόπτῳ κείμενα καὶ οὐ πολλά, τὰ δὲ οὐκ ἐν εὐδήλῳ καὶ πολλά. ἔστι δὲ τὰ μὲν ἐξανθεῦντα ἐς τὴν χροιὴν ἢ χροιῇ ἢ οἰδήμασιν ἐν εὐδήλῳ· παρέχει γὰρ ἑωυτῶν τῇ τε ὄψει
10 τῷ τε ψαῦσαι τὴν στερεότητα καὶ τὴν ὑγρότητα αἰσθάνεσθαι, καὶ ἅ τε αὐτῶν θερμὰ ἅ τε ψυχρά, ὧν τε ἑκάστου ἢ παρουσίῃ ἢ ἀπουσίῃ τοιαῦτά ἐστιν. τῶν μὲν οὖν τοιούτων πάντων ἐν πᾶσι τὰς ἀκέσιας ἀναμαρτήτους δεῖ εἶναι, οὐχ ὡς ῥηϊδίας, ἀλλ' ὅτι ἐξεύρηνται· ἐξεύρηνταί γε μὴν οὐ τοῖσι βουληθεῖσιν, ἀλλὰ τούτων τοῖσι δυνηθεῖσι· δύνανται δὲ οἷσι τά τε τῆς παιδείης μὴ ἐκποδὼν τά
18 τε τῆς φύσιος μὴ ταλαίπωρα.[1]

X. Πρὸς μὲν οὖν τὰ φανερὰ τῶν νοσημάτων οὕτω δεῖ εὐπορεῖν τὴν τέχνην· δεῖ γε μὴν αὐτὴν μηδὲ[2] πρὸς τὰ ἧσσον φανερὰ ἀπορεῖν· ἔστι δὲ ταῦτα ἃ πρός τε τὰ ὀστέα τέτραπται καὶ τὴν νηδύν· ἔχει δὲ τὸ σῶμα οὐ μίαν, ἀλλὰ πλείους· δύο μὲν γὰρ αἱ τὸ σιτίον δεχόμεναί τε καὶ ἀφιεῖσαι, ἄλλαι δὲ τούτων πλείους, ἃς ἴσασιν οἷσι τούτων ἐμέλησεν· ὅσα γὰρ τῶν μελέων ἔχει σάρκα περιφερέα, ἣν μῦν καλέουσι, πάντα νηδὺν ἔχει. πᾶν
10 γὰρ τὸ ἀσύμφυτον, ἤν τε δέρματι, ἤν τε σαρκὶ καλύπτηται, κοῖλόν ἐστιν· πληροῦταί τε ὑγιαῖνον μὲν πνεύματος ἀσθενῆσαν δὲ ἰχῶρος· ἔχουσι μὲν

[1] ταλαίπωρα M : ἀταλαίπωρα A.
[2] μηδὲ many MSS. : οὐδὲ AM.

---

[1] The word νηδύς is here used in a rather strange sense, and in particular the singular is peculiar. It must be either collective, "whatever is hollow," or generic with the article.

forth. Men with an adequate knowledge of this art realize that some, but only a few, diseases have their seat where they can be seen; others, and they are many, have a seat where they cannot be perceived. Those that can be perceived produce eruptions on the skin, or manifest themselves by colour or swelling; for they allow us to perceive by sight or touch their hardness, moistness, heat or cold, and what are the conditions which, by their presence or absence in each case, cause the diseases to be of the nature they are. Of all such diseases in all cases the cures should be infallible, not because they are easy, but because they have been discovered. However, they have not been discovered for those who have desire only, but for those of them who have power; this power belongs to those whose education has been adequate, and whose natural ability is not wretched.

X. Now such being its nature the art must be a match for the open diseases; it ought however not to be helpless before diseases that are more hidden. These are those which are determined to the bones or to the cavities.[1] The body has of these not one but several. There are two that take in food and discharge it, with several others besides these, known to men who are interested in these things; all limbs, in fact, have cavities that are surrounded by the flesh that is called muscle. Everything in fact not a continuous growth, whether it be skin or flesh that covers it, is hollow, and in health is filled with air, in disease with juice.[2]

[2] Apparently "pus," a sense which ἰχώρ has in *Wounds in the Head.*

τοίνυν οἱ βραχίονες σάρκα τοιαύτην· ἔχουσι δ' οἱ μηροί· ἔχουσι δ' αἱ κνῆμαι. ἔτι δὲ καὶ ἐν τοῖσιν ἀσάρκοισι τοιαύτη ἔνεστιν οἵη καὶ ἐν τοῖσιν εὐσάρκοισιν εἶναι δέδεικται· ὅ τε γὰρ θώρηξ καλεόμενος, ἐν ᾧ τὸ ἧπαρ στεγάζεται, ὅ τε τῆς κεφαλῆς κύκλος, ἐν ᾧ ὁ ἐγκέφαλος, τό τε νῶτον, πρὸς ᾧ ὁ πλεύμων, τούτων οὐδὲν ὅ τι οὐ καὶ αὐτὸ
20 κενόν ἐστι,[1] πολλῶν διαφυσίων μεστόν, ᾗσιν οὐδὲν ἀπέχει πολλῶν ἀγγεῖα εἶναι, τῶν μέν τι βλαπτόντων τὸν κεκτημένον, τῶν δὲ καὶ ὠφελεύντων. ἔτι δὲ καὶ πρὸς τούτοισι φλέβες πολλαὶ καὶ νεῦρα οὐκ ἐν τῇ σαρκὶ μετέωρα, ἀλλὰ πρὸς τοῖς ὀστέοισι προστεταμένα, σύνδεσμος ἔς τι τῶν ἄρθρων, καὶ αὐτὰ τὰ ἄρθρα, ἐν οἷσιν αἱ συμβολαὶ τῶν κινεομένων ὀστέων ἐγκυκλέονται· καὶ τούτων οὐδὲν ὅ τι οὐχ ὑπόφορόν[2] ἐστι καὶ ἔχον περὶ αὐτὸ θαλάμας, ἃς καταγγέλλει ἰχώρ, ὃς ἐκ διοιγομένων
30 αὐτέων πολύς τε καὶ πολλὰ λυπήσας ἐξέρχεται.

XI. Οὐ γὰρ δὴ ὀφθαλμοῖσί γε ἰδόντι τούτων τῶν εἰρημένων οὐδενὶ οὐδὲν ἔστιν εἰδέναι· διὸ καὶ ἄδηλα ἐμοί τε ὠνόμασται καὶ τῇ τέχνῃ κέκριται εἶναι. οὐ μὴν ὅτι ἄδηλα κεκράτηκεν, ἀλλ' ᾗ δυνατὸν κεκράτηται· δυνατὸν δέ, ὅσον αἵ τε[3] τῶν νοσεόντων φύσιες ἐς[4] τὸ σκεφθῆναι παρέχουσιν, αἵ τε τῶν ἐρευνησόντων ἐς τὴν ἔρευναν πεφύκασιν. μετὰ πλείονος μὲν γὰρ πόνου καὶ οὐ μετ' ἐλάσσονος χρόνου ἢ εἰ τοῖσιν ὀφθαλμοῖσιν ἑώρατο γινώσκε-
10 ται· ὅσα γὰρ τὴν τῶν ὀμμάτων ὄψιν ἐκφεύγει, ταῦτα τῇ τῆς γνώμης ὄψει κεκράτηται· καὶ ὅσα

[1] ὅ τι οὐ καὶ αὐτὸ κενόν ἐστι Littré with M. Gomperz reads ὅ τι τούτων οὐ κενεών ἐστιν.

Such flesh then the arms have, and so have the thighs and the legs. Moreover, in the fleshless parts also there are cavities like those we have shown to be in the fleshy parts. For the trunk, as it is called, in which the liver is covered, the sphere of the head, in which is the brain, the back, by which are the lungs—all these are themselves hollow, being full of interstices, which do not at all fail to be vessels to contain many things, some of which do harm to the possessor and some do good. Moreover, in addition to these there are many veins, and sinews that are not near the surface of the flesh but stretched along the bones, binding the joints to a certain point, and the joints themselves, at which the movable bones meet and turn round. Of these none is not porous; all have cells about them, which are made known by juice, which, when the cells are opened, comes out in great quantity, causing many pains.

XI. Without doubt no man who sees only with his eyes can know anything of what has been here described. It is for this reason that I have called them obscure, even as they have been judged to be by the art. Their obscurity, however, does not mean that they are our masters, but as far as is possible they have been mastered, a possibility limited only by the capacity of the sick to be examined and of researchers to conduct research. More pains, in fact, and quite as much time, are required to know them as if they were seen with the eyes; for what

---

[2] *ὑπόφορόν* Littré and Gomperz after Zwinger: *ὕπαφρον* MSS.: *ὕποφρον* Erotian.

[3] *ὅσον αἵ τε* Littré: *δὲ ὥσαι τε* A: *δὲ ὅσαι τε* M′: *δ' ἕως αἵ τε* Gomperz. [4] Gomperz deletes *ἐs* after *φύσιεs*.

δὲ ἐν τῷ μὴ ταχὺ ὀφθῆναι οἱ νοσέοντες πάσχουσιν, οὐχ οἱ θεραπεύοντες αὐτοὺς αἴτιοι, ἀλλ' ἡ φύσις ἥ τε τοῦ νοσέοντος ἥ τε τοῦ νοσήματος· ὁ μὲν γάρ, ἐπεὶ οὐκ ἦν αὐτῷ ὄψει ἰδεῖν τὸ μοχθέον οὐδ' ἀκοῇ πυθέσθαι, λογισμῷ μετῄει. καὶ γὰρ δὴ καὶ ἃ πειρῶνται οἱ τὰ ἀφανέα νοσέοντες ἀπαγγέλλειν περὶ τῶν νοσημάτων τοῖσι θεραπεύουσιν, δοξάζοντες μᾶλλον ἢ εἰδότες ἀπαγγέλλουσιν· εἰ γὰρ
20 ἠπίσταντο, οὐκ ἂν περιέπιπτον αὐτοῖσιν· τῆς γὰρ αὐτῆς συνέσιός ἐστιν ἧσπερ τὸ εἰδέναι τῶν νούσων τὰ αἴτια καὶ τὸ θεραπεύειν αὐτὰς ἐπίστασθαι πάσῃσι τῇσι θεραπείῃσιν αἳ κωλύουσι τὰ νοσήματα μεγαλύνεσθαι. ὅτε οὖν οὐδὲ ἐκ τῶν ἀπαγγελλομένων ἔστι τὴν ἀναμάρτητον σαφήνειαν ἀκοῦσαι, προσοπτέον τι καὶ ἄλλο τῷ θεραπεύοντι· ταύτης οὖν τῆς βραδυτῆτος οὐχ ἡ τέχνη, ἀλλ' ἡ φύσις αἰτίη τῶν σωμάτων· ἡ μὲν γὰρ αἰσθανομένη ἀξιοῖ θεραπεύειν, σκοπεῦσα ὅπως μὴ τόλμῃ
30 μᾶλλον ἢ γνώμῃ, καὶ ῥᾳστώνῃ μᾶλλον ἢ βίῃ θεραπεύῃ· ἡ δ' ἢν μὲν διεξαρκέσῃ ἐς τὸ ὀφθῆναι, ἐξαρκέσει καὶ ἐς τὸ ὑγιανθῆναι· ἢν δ' ἐν ᾧ τοῦτο ὁρᾶται κρατηθῇ διὰ τὸ βραδέως αὐτὸν ἐπὶ τὸν θεραπεύσοντα ἐλθεῖν ἢ διὰ τὸ τοῦ νοσήματος τάχος, οἰχήσεται. ἐξ ἴσου μὲν γὰρ ὁρμώμενον τῇ θεραπείῃ οὐκ ἔστι θᾶσσον, προλαβὸν δὲ θᾶσσον· προλαμβάνει δὲ διά τε τὴν τῶν σωμάτων στεγνότητα, ἐν ᾗ οὐκ ἐν εὐόπτῳ οἰκέουσιν αἱ νοῦσοι, διά τε τὴν τῶν καμνόντων ὀλιγωρίην·

escapes the eyesight is mastered by the eye of the mind, and the sufferings of patients due to their not being quickly observed are the fault, not of the medical attendants, but of the nature of the patient and of the disease. The attendant in fact, as he could neither see the trouble with his eyes nor learn it with his ears, tried to track it by reasoning. Indeed, even the attempted reports of their illnesses made to their attendants by sufferers from obscure diseases are the result of opinion, rather than of knowledge. If indeed they understood their diseases they would never have fallen into them, for the same intelligence is required to know the causes of diseases as to understand how to treat them with all the treatment that prevents illnesses from growing worse. Now when not even the reports afford perfectly reliable information, the attendant must look out for fresh light. For the delay thus caused not the art is to blame, but the constitution of human bodies. For it is only when the art sees its way that it thinks it right to give treatment, considering how it may give it, not by daring but by judgment, not by violence but by gentleness. As to our human constitution, if it admits of being seen, it will also admit of being healed. But if, while the sight is being won, the body is mastered by slowness in calling in the attendant or by the rapidity of the disease, the patient will pass away. For if disease and treatment start together, the disease will not win the race, but it will if it start with an advantage, which advantage is due to the density of our bodies, in which diseases lurk unseen, and to the careless neglect of patients. This advantage is not to be wondered at, as it is

40 ἐπεὶ ἔοικε·[1] οὐ λαμβανόμενοι γάρ, ἀλλ' εἰλημμένοι
41 ὑπὸ τῶν νοσημάτων θέλουσι θεραπεύεσθαι.

XII. Ἐπεὶ τῆς[2] τέχνης τὴν δύναμιν ὁπόταν τινὰ τῶν τὰ ἄδηλα νοσεύντων ἀναστήσῃ, θαυμάζειν ἀξιώτερον, ἢ ὁπόταν ἐγχειρήσῃ τοῖς ἀδυνάτοις.[3] . . . οὔκουν ἐν ἄλλῃ γε δημιουργίῃ τῶν εὑρημένων οὐδεμιῇ ἔνεστιν οὐδὲν τοιοῦτον, ἀλλ' αὐτῶν ὅσαι πυρὶ δημιουργεῦνται, τούτου μὴ παρεόντος ἀεργοί εἰσι, μετὰ δὲ τοῦ ἀφθῆναι ἐνεργοί. καὶ ὅσαι τοι ἐν[4] εὐεπανορθώτοισι σώμασι δημιουργεῦνται, αἱ μὲν μετὰ ξύλων, αἱ δὲ μετὰ σκυτέων,
10 αἱ δὲ γραφῇ χαλκῷ τε καὶ σιδήρῳ καὶ τοῖσι τούτων ὁμοίοισιν αἱ πλεῖσται,[5] ἐόντα δὲ[6] τὰ ἐκ τουτέων καὶ μετὰ τούτων δημιουργεύμενα εὐεπανόρθωτα, ὅμως οὐ τῷ τάχει μᾶλλον, ἢ τῷ ὡς δεῖ δημιουργεῖται· οὐδ' ὑπερβατῶς, ἀλλ' ἢν ἀπῇ τι τῶν ὀργάνων, ἐλινύει· καίτοι κἀκείνῃσι τὸ βραδὺ πρὸς τὸ λυσιτελεῦν ἀσύμφορον, ἀλλ' ὅμως προτι-
17 μᾶται.

XIII. Ἰητρικὴ δὲ τοῦτο μὲν τῶν ἐμπύων τοῦτο

[1] ἐπεὶ ἔοικε Littré: ἐπεὶ τί θῶμα Gomperz: ἐπιτίθεται or ἐπιτίθενται MSS.

[2] ἐπεὶ τῆς γε Littré with some MSS.: ἔτι τῆς Gomperz: ἐπὶ τῆς A: ἐπὶ τῆς γε M.

[3] Gomperz marks an hiatus after ἀδυνάτοις.

[4] τοι ἐν Gomperz: καίτοι ἐν A: καὶ τοῖσιν M.

[5] ὁμοίοισιν αἱ πλεῖσται M: ὁμοιοις (?) χυμασιαι πλειται first hand in A, altered in various ways by later hands.

[6] Gomperz brackets δὲ after ἐόντα.

---

[1] The whole of this chapter, except the first sentence, arouses suspicion. A new subject is introduced. We may get over this difficulty by postulating a hiatus after ἀδυνάτοις, and supposing that it contained an objection to medicine

only when diseases have established themselves, not while they are doing so, that patients are ready to submit to treatment.

XII. Now the power of the art, when it raises a patient suffering from an obscure disease, is more surprising than its failure when it attempts to treat incurables. . . . So in the case of no other craft that has been discovered are such extravagant demands made; those that depend on fire are inoperative when fire is not present, but operative when one has been lighted. And the arts that are worked in materials easy to shape aright, using in some cases wood, in others leather, in others—these form the great majority—paint, bronze, iron and similar substances—the articles wrought, I say, through these arts and with these substances are easily shaped aright, and yet are wrought not so much with a view to speed as to correctness. Nor are they wrought in a casual manner, but functioning ceases if any instrument be lacking. Yet in these arts too slowness is contrary to their interests; but in spite of this it is preferred.[1]

XIII. Now medicine, being prevented, in cases of

based on the slowness of its cures. But there are other difficulties. The grammar is broken, while in the rest of the work it is very regular. The diction is curious; why, for instance, *μετὰ ξύλων*, *μετὰ σκυτέων*, but *γραφῇ*, *χαλκῷ* and *σιδήρῳ*? Why *ἐκ τουτέων* (sc. *τεχνέων*) but *μετὰ τούτων* (sc. *σωμάτων*)? Again, should not the active and not the middle (*δημιουργεῦνται*) be used with *τέχναι* as subject? Finally, the MSS. are more corrupt than usual, with readings that imply deep-seated corruption. The *ὁμοίοις σχημασίαι πλεῖσται*(?) of A (for *ὁμοίοισιν αἱ πλεῖσται*) seems to show that the text is mutilated. Perhaps the last pages of an early ancestor of our MSS. were lost, to be afterwards added from a corrupt and mutilated MS.

δὲ τῶν τὸ ἧπαρ ἢ τοὺς νεφροὺς τοῦτο δὲ τῶν συμπάντων τῶν ἐν τῇ νηδύι νοσεύντων ἀπεστερημένη τι ἰδεῖν ὄψει, ᾗ τὰ πάντα πάντες ἱκανωτάτως ὁρῶσι, ὅμως ἄλλας εὐπορίας συνεργοὺς εὗρε. φωνῆς τε γὰρ λαμπρότητι καὶ τρηχύτητι, καὶ πνεύματος ταχυτῆτι καὶ βραδυτῆτι, καὶ ῥευμάτων ἃ διαρρεῖν εἴωθεν ἑκάστοισι δι' ὧν ἔξοδοι δέδονται, ὧν[1] τὰ μὲν ὀδμῇσι τὰ δὲ χροιῇσι τὰ δὲ λεπτότητι καὶ παχύ-
10 τητι διασταθμωμένη τεκμαίρεται, ὧν τε σημεῖα ταῦτα, ἅ τε πεπονθότων ἅ τε παθεῖν δυναμένων. ὅταν δὲ ταῦτα μὴ[2] μηνύωνται, μηδ' αὐτὴ ἡ φύσις ἑκοῦσα ἀφίῃ, ἀνάγκας εὕρηκεν, ᾗσιν ἡ φύσις ἀζήμιος βιασθεῖσα μεθίησιν· μεθεῖσα[3] δὲ δηλοῖ τοῖσι τὰ τῆς τέχνης εἰδόσιν ἃ ποιητέα. βιάζεται δὲ τοῦτο μὲν φλέγμα[4] διαχεῖν σιτίων δριμύτητι καὶ πωμάτων, ὅπως τεκμήρηταί τι ὀφθὲν περὶ ἐκείνων ὧν αὐτῇ ἐν ἀμηχάνῳ τὸ ὀφθῆναι ἦν· τοῦτο δ' αὖ πνεῦμα ὧν κατήγορον ὁδοῖσί τε
20 προσάντεσι καὶ δρόμοις ἐκβιᾶται κατηγορεῖν· ἱδρῶτάς τε τούτοισι τοῖσι προειρημένοις ἄγουσα, ὑδάτων θερμῶν ἀποπνοίῃσι πυρὶ ὅσα τεκμαίρονται, τεκμαίρεται. ἔστι δὲ ἃ καὶ διὰ τῆς κύστιος διελθόντα ἱκανώτερα δηλῶσαι τὴν νοῦσόν

[1] Gomperz brackets ὧν.
[2] μὴ added by Littré (followed by Gomperz).
[3] μεθεῖσα Reinhold and Gomperz: ἀνεθῆσα or ἀνεθεῖσα MSS.
[4] Before φλέγμα A has πυου (another hand πoουσιν) τὸ σύντροφον, the other MSS. πῦρ τὸ σύντροφον.

---

[1] The natural subject of βιάζεται is ἡ τέχνη, and the natural object φύσις. The various readings seem to imply that either (*a*) the true reading is lost, or (*b*) a corrupt gloss has crept

empyema, and of diseased liver, kidneys, and the cavities generally, from seeing with the sight with which all men see everything most perfectly, has nevertheless discovered other means to help it. There is clearness or roughness of the voice, rapidity or slowness of respiration, and for the customary discharges the ways through which they severally pass, sometimes smell, sometimes colour, sometimes thinness or thickness furnishing medicine with the means of inferring, what condition these symptoms indicate, what symptoms mean that a part is already affected and what that a part may hereafter be affected. When this information is not afforded, and nature herself will yield nothing of her own accord, medicine has found means of compulsion, whereby nature is constrained, without being harmed, to give up her secrets; when these are given up she makes clear, to those who know about the art, what course ought to be pursued. The art, for example, forces ⟨nature⟩ [1] to disperse phlegm by acrid foods and drinks, so that it may form a conclusion by vision concerning those things which before were invisible. Again, when respiration is symptomatic, by uphill roads and by running [2] it compels nature to reveal symptoms. It brings on sweats by the means already stated, and forms the conclusions that are formed through fire when it makes hot water give out steam. There are also certain excretions through the bladder which indicate the disease better than those which

into the text. I adopt the second alternative because the agent dispersing the phlegm is δριμύτης, not πῦρ, whether with Littré πῦρ τὸ σύντροφον means "innate heat," or with Gomperz we render τὸ σύντροφον φλέγμα "thickened phlegm."

[2] Perhaps a hendiadys: "making patients run uphill."

ἐστιν ἢ διὰ τῆς σαρκὸς ἐξιόντα. ἐξεύρηκεν οὖν καὶ τοιαῦτα πώματα καὶ βρώματα, ἃ τῶν θερμαινόντων θερμότερα γινόμενα τήκει τε ἐκεῖνα καὶ διαρρεῖν ποιεῖ, ἃ οὐκ ἂν διερρύη μὴ τοῦτο παθόντα. ἕτερα μὲν οὖν πρὸς ἑτέρων, καὶ ἄλλα δι'
30 *ἄλλων ἐστὶ τά τε διιόντα τά τ' ἐξαγγέλλοντα, ὥστ' οὐ θαυμάσιον αὐτῶν τάς τε ἀπιστίας χρονιωτέρας γίνεσθαι τάς τ' ἐγχειρήσιας βραχυτέρας, οὕτω δι' ἀλλοτρίων ἑρμηνειῶν πρὸς τὴν*
34 *θεραπεύουσαν σύνεσιν ἑρμηνευομένων.*

XIV. *Ὅτι μὲν οὖν καὶ λόγους ἐν ἑωυτῇ εὐπόρους ἐς τὰς ἐπικουρίας ἔχει ἡ ἰητρική, καὶ οὐκ εὐδιορθώτοισι δικαίως οὐκ ἂν ἐγχειροίη τῇσι νούσοισιν, ἢ ἐγχειρευμένας ἀναμαρτήτους ἂν παρέχοι, οἵ τε νῦν λεγόμενοι λόγοι δηλοῦσιν αἵ τε τῶν εἰδότων τὴν τέχνην ἐπιδείξιες, ἃς ἐκ τῶν ἔργων*[1] *ἐπιδεικνύουσιν, οὐ τὸ λέγειν καταμελετήσαντες, ἀλλὰ τὴν πίστιν τῷ πλήθει ἐξ ὧν ἂν ἴδωσιν οἰκειοτέρην ἡγεύμενοι ἢ ἐξ ὧν ἂν*
10 *ἀκούσωσιν.*

[1] After *ἔργων* the MSS. except A add *ἥδιον ἢ ἐκ τῶν λόγων.*

---

[1] I do not see that there is any real difference between *ἕτερα πρὸς ἑτέρων* and *ἄλλα δι' ἄλλων.* The whole phrase is a piece of "legal tautology," bringing out the variability of the relation

come out through the flesh. So medicine has also discovered drinks and foods of such a kind that, becoming warmer than the natural heat, melt the matters I spoke of, and make them flow away, which they never would have done without this treatment. Now as the relation between excretions and the information they give is variable, and depends upon a variety of conditions,[1] it is accordingly not surprising that disbelief in this information is prolonged, but treatment is curtailed, for extraneous factors must be used in interpreting the information before it can be utilized by medical intelligence.

XIV. Now that medicine has plentiful reasoning in itself to justify its treatment, and that it would rightly refuse to undertake obstinate cases, or undertaking them would do so without making a mistake, is shown both by the present essay and by the expositions of those versed in the art, expositions set forth in acts, not by attention to words, under the conviction that the multitude find it more natural to believe what they have seen than what they have heard.

between the phenomena of the excretions and what may be diagnosed from them. A number of "extraneous factors," *e. g.* age of the patient, character of the disease, etc., have to be taken into account before the information has any real medical value.

BREATHS

INTRODUCTION

This work, like *The Art*, is a sophistic essay, probably written to be delivered to an audience.[1] The two books are similar in style,[2] and on this ground alone we might conjecture that they are not widely separated in date. The subject matter too points to the end of the fifth century B.C. as the time when *Breaths* was written. Diogenes of Apollonia, whose date indeed is very uncertain, though he probably flourished about 430 B.C., had revived the doctrine that air is the primal element from which all things are derived. The writer of *Breaths* would prove that air, powerful in nature generally, is also the prime factor in causing diseases. He is a rhetorical sophist who, either in earnest or perhaps merely to show his skill in supporting a

[1] See *e. g.* Chapter XIV (beginning) *τοὺς ἀκούοντας πείθειν πειρήσομαι.* These *ἐπιδείξεις* must have been to the average Athenian what our "reviews" are to the average Englishman.

[2] *Breaths* shows the tendency to similes and highly metaphorical language which Plato attributes (*Protagoras* 337 C–338 A) to Hippias. See *e. g.* Chapter III, air is *πάντων δυνάστης*: and (*sub finem*) *γῆ τούτου βάθρον, οὗτός τε γῆς ὄχημα*: VIII *ἐξεμόχλευσε τὸ στόμα*: X *χαραδρωθέωσιν οἱ πόροι*: XIV *τῆς νούσου καὶ τοῦ παρέοντος χειμῶνος*, and (*sub finem*) *γαλήνης ἐν τῷ σώματι γενομένης.* I do not suggest that Hippias was the author, but I do hold that the book must have been written at a time when the sophistry he represented was a living force.

ὑπόθεσις, adopted the fundamental tenet of a rather belated Ionian monist.[1]

The author shows no genuine interest in medicine, nor do his contentions manifest any serious study of physiology or pathology. Any impartial reader will detect in Chapter XIV (the discussion of epilepsy) just the illogical but confident dogmatism that is associated with half-educated, would-be scientists. The account of dropsy in Chapter XII is not only illogical but ludicrously absurd.

The work is a striking example of the necessity of experiment before accepting a hypothesis. The writer makes with a gay assurance a string of positive statements, unsupported by any evidence worth speaking of. It is easy enough to defend a hypothesis if you deal with an unexplored subject, pick out the phenomena which seem to support your view, ignore everything which tells against it, and never make an experiment to verify or condemn your generalization.

Nearly all Greek speculation in biology and physiology is open to this criticism. In no department of science is experiment more necessary, and in no department did the Greeks experiment to less purpose. Dissection of human bodies, too, and constant use of the microscope, together with an exact knowledge of chemistry, are all necessary before

[1] In Chapter III (*sub finem*) we have in τούτου βάθρον and γῆς ὄχημα phrases which cannot be independent of the famous lines in Euripides' *Troades* 884 foll. (ὦ γῆς ὄχημα κἀπὶ γῆς ἔχων ἕδραν κ.τ.λ.). If the author was not imitating Euripides they were both probably copying some famous philosophic *dictum*, as it is most unlikely that Euripides copied the Hippocratic writer, whose intellect is distinctly of an inferior type.

any substantial progress can be made in this direction.

But here, as elsewhere, the modern stands amazed before the intellectual activity of the Greek. His imagination, although unchecked and ill-disciplined, was alive and active. He loathed mystery; his curiosity remained unsatisfied until he had discovered a rational cause, even though that cause was grounded on insecure foundations. His confidence that the human intelligence was great enough to solve all problems often led him into the fallacy of imagining that it had already discovered what was still dark; his delight in a simple solution that satisfied his aesthetic sense often blinded him to its intellectual absurdities. The Greek lacked self-criticism; it was perhaps the greatest defect in his mental equipment. The astounding genius of Socrates is shown nowhere so plainly as in his constant insistence on the need of self-examination. We may laugh at the crudities of περὶ φυσῶν, which is "as windy in its rhetoric as in its subject matter";[1] but we must respect its inquiring spirit and its restless curiosity.

The theme of the writer takes us back to the speculations of Anaximenes, and even earlier still, for in the very infancy of thought man must have noticed that air is an essential condition of life. For centuries the conviction that air, or some essential principle behind the manifestations of air as wind, breath and vapour, was primal and elemental, kept arising in one form or in another. On its physical side the quest came to an end in the

[1] Sir Clifford Allbutt, *Greek Medicine in Rome*, p. 243.

discovery of oxygen;[1] on its spiritual side it has given us the fine ideas we associate with the word "spirit," which has come down to us through the Latin from the Greek *πνεῦμα*. The instinct of the Greeks in this matter was right, however pathetic their efforts may have been to satisfy it.

The writer of *περὶ φυσῶν* uses three words to describe air—*φῦσα*, *πνεῦμα* and *ἀήρ*. Though he defines *φῦσα* as *πνεῦμα* in the body and *ἀήρ* as *πνεῦμα* out of it, he is not careful in his use of these words, and to translate them is a matter of great difficulty. The natural renderings would be to translate *φῦσα* "air" and *πνεῦμα* "breath"; but what is one to do with *ἀήρ*? So I have throughout (except in one passage referring to respiration) equated *φῦσα* and "breath," *πνεῦμα* and "wind," *ἀήρ* and "air." I fully realize the objections to this course, but they are much less than those attaching to the plan of picking and choosing a translation to suit the context in each case. Such a plan would certainly give a faulty translation, with incongruous or wrong associations; it is surely better to use "breath," "wind," and "air," in technical senses for the purpose of translating this particular treatise.

It is at first sight surprising that a book of the character of *περὶ φυσῶν* should find its way into the Hippocratic collection. It is probable, however, that this collection represents, not works written by the Coan school, but works preserved in the library of the medical school at Cos. Knowing the vanity of

[1] See Sir Clifford Allbutt, *op. cit.* p. 224. Chapter X of this book contains the best account of pneumatism that I have seen. See also M. Wellmann, *Die Pneumatische Schule bis auf Archigenes*. Berlin, 1895.

the sophists[1] we ought not to be surprised that they sent "presentation copies" of their works on medical subjects to the chief centres where medicine was studied. Perhaps in this way were preserved both περὶ φυσῶν and περὶ τέχνης.[2] At quite an early date it became known as an Hippocratic work. It is referred to in Menon's *Iatrica* (Chapter V), and it is in the list of Erotian.

MSS. and Editions

Περὶ φυσῶν is found in many Paris manuscripts, including A, and in M. On these two MSS. the text is constructed, with occasional help from variants noted in the old editions, and from the Renaissance translations into Latin of Francesco Filelfo and Janus Lascaris. The manuscript A shows its usual superiority to M, but on one occasion at least M appears to preserve the original reading. There are also some extracts from περὶ φυσῶν in a Milan MS, which Nelson calls "a."

There is a modern edition of περὶ φυσῶν by Dr. Axel Nelson,[3] in which every scrap of information about the work has been carefully collected. The reader feels, however, that much of his time is taken up with insignificant points, and that the learned author might have omitted these to make room for a fuller account of the position of περὶ φυσῶν in the development of philosophic thought.

[1] See *e.g.* Plato, *Protagoras* 347 B, where Hippias in his vanity offers to deliver an ἐπίδειξις at a most inopportune moment.

[2] Perhaps too περὶ φύσιος ἀνθρώπων.

[3] *Die hippokratische Schrift* περὶ φυσῶν, *Text und Studien von* Axel Nelson. Uppsala 1909.

ΠΕΡΙ ΦΥΣΩΝ

I. Εἰσί τινες τῶν τεχνέων, αἳ τοῖσι μὲν κεκτημένοις εἰσὶν ἐπίπονοι, τοῖσι δὲ χρεωμένοις ὀνήισται,[1] καὶ τοῖσι μὲν δημότῃσι κοινὸν ἀγαθόν, τοῖσι δὲ μεταχειριζομένοις σφας λυπηραί. τῶν δὴ τοιούτων ἐστὶ τεχνέων ἣν οἱ Ἕλληνες καλέουσιν ἰητρικήν· ὁ μὲν γὰρ ἰητρὸς ὁρῇ τε δεινά, θιγγάνει τε ἀηδέων, ἐπ' ἀλλοτρίῃσί τε συμφορῇσιν ἰδίας καρποῦται λύπας· οἱ δὲ νοσέοντες ἀποτρέπονται διὰ τὴν τέχνην τῶν μεγίστων κακῶν, νούσων, 10 λύπης, πόνων, θανάτου· πᾶσι γὰρ τούτοις ἄντικρυς ἰητρικὴ εὑρίσκεται ἀκεστορίς.[2] ταύτης δὲ τῆς τέχνης τὰ μὲν φλαῦρα χαλεπὸν γνῶναι, τὰ δὲ σπουδαῖα ῥηίδιον· καὶ τὰ μὲν φλαῦρα τοῖσιν ἰητροῖσι μούνοις ἔστιν εἰδέναι, καὶ οὐ τοῖσι δημότῃσιν· οὐ γὰρ σώματος, ἀλλὰ γνώμης ἐστὶν ἔργα. ὅσα μὲν γὰρ χειρουργῆσαι χρή, συνεθισθῆναι δεῖ· τὸ γὰρ ἔθος τῇσι χερσὶ κάλλιστον διδασκάλιον γίνεται· περὶ δὲ τῶν ἀφανεστάτων καὶ χαλεπωτάτων νοσημάτων δόξῃ μᾶλλον ἢ 20 τέχνῃ κρίνεται· διαφέρει δὲ ἐν αὐτοῖσι πλεῖστον ἡ πείρη τῆς ἀπειρίης. ἓν δὲ δή τι τῶν τοιούτων ἐστὶ τόδε, τί ποτε τὸ αἴτιόν ἐστι τῶν νούσων, καὶ τίς ἀρχὴ καὶ πηγὴ γίνεται τῶν ἐν τῷ σώματι

[1] ὀνήισται Nelson : ὠφέλιμοι A : ὀνήιστοι or ὀνηισταὶ other MSS.

[2] ἄντικρυς . . . ἀκεστορίς most MSS. : ἀνθέστηκεν ἡ' ητρική Nelson.

BREATHS[1]

I. THERE are some arts which to those that possess them are painful, but to those that use them are helpful, a common good to laymen, but to those that practise them grievous. Of such arts there is one which the Greeks call medicine. For the medical man sees terrible sights, touches unpleasant things, and the misfortunes of others bring a harvest of sorrows that are peculiarly his; but the sick by means of the art rid themselves of the worst of evils, disease, suffering, pain and death. For medicine proves for all these evils a manifest cure. And of this art the weak points are difficult to apprehend, while the strong points are more easy; the weak points laymen cannot know, but only those skilled in medicine, as they are matters of the understanding and not of the body. For whenever surgical treatment is called for, training by habituation is necessary, for habit proves the best teacher of the hands; but to judge of the most obscure and difficult diseases is more a matter of opinion than of art, and therein there is the greatest possible difference between experience and inexperience. Now of these obscure matters one is the cause of diseases, what the beginning and source is whence come

[1] This word is a very inadequate rendering of φῦσα, which means, according to the definition in Chapter III, air in the body, as opposed to air outside it.

παθῶν; εἰ γάρ τις εἰδείη τὴν αἰτίην τοῦ νοσήματος, οἷός τ' ἂν εἴη τὰ συμφέροντα προσφέρειν τῷ σώματι·[1] αὕτη γὰρ ἡ ἰητρικὴ μάλιστα κατὰ φύσιν ἐστίν. αὐτίκα γὰρ λιμὸς νοῦσός ἐστιν· ὃ γὰρ ἂν λυπῇ τὸν ἄνθρωπον, τοῦτο καλεῖται νοῦσος· τί οὖν λιμοῦ φάρμακον; ὃ παύει λιμόν·
30 τοῦτο δ' ἐστὶ βρῶσις· τούτῳ ἄρα ἐκεῖνο ἰητέον. αὖτις αὖ δίψαν ἔπαυσε πόσις· πάλιν αὖ πλησμονὴν ἰῆται κένωσις· κένωσιν δὲ πλησμονή· πόνον δὲ ἀπονίη.[2] ἑνὶ δὲ συντόμῳ λόγῳ, τὰ ἐναντία τῶν ἐναντίων ἐστὶν ἰήματα· ἰητρικὴ γάρ ἐστιν ἀφαίρεσις καὶ πρόσθεσις, ἀφαίρεσις μὲν τῶν πλεοναζόντων, πρόσθεσις δὲ τῶν ἐλλειπόντων· ὁ δὲ τοῦτ' ἄριστα ποιέων ἄριστος ἰητρός· ὁ δὲ τούτου πλεῖστον ἀπολειφθεὶς πλεῖστον ἀπελείφθη καὶ τῆς τέχνης. ταῦτα μὲν οὖν ἐν
40 παρέργῳ τοῦ λόγου τοῦ μέλλοντος εἴρηται.

II. Τῶν δὲ δὴ νούσων ἁπασέων ὁ μὲν τρόπος ὡυτός, ὁ δὲ τόπος διαφέρει· δοκεῖ μὲν οὖν οὐδὲν ἐοικέναι τὰ νοσήματα ἀλλήλοισι διὰ τὴν ἀλλοιότητα[3] τῶν τόπων, ἔστι δὲ μία πασέων νούσων καὶ ἰδέη καὶ αἰτίη. ταύτην δὲ ἥτις ἐστὶ διὰ τοῦ
6 μέλλοντος λόγου φράσαι πειρήσομαι.

III. Τὰ σώματα καὶ τὰ τῶν ἄλλων ζῴων καὶ τὰ τῶν ἀνθρώπων ὑπὸ τρισσῶν τροφέων τρέφεται· τῇσι δὲ τροφῇσι τάδε ὀνόματά ἐστιν, σιτία ποτά, πνεῦμα. πνεῦμα δὲ τὸ μὲν ἐν τοῖσι σώμασι φῦσα καλεῖται, τὸ δὲ ἔξω τῶν σωμάτων ὁ ἀήρ.

[1] After σώματι most MSS. have ἐκ τῶν ἐναντίων ἐπιστάμενος τὰ νουσήματα, M however reading τῷ νουσήματι—a reading adopted by Nelson. ἱστάμενος τῷ νοσήματι is the reading of a. Littré changes νουσήματα to βοηθήματα. I believe the phrase to be a gloss. It is omitted by A.

affections of the body. For knowledge of the cause of a disease will enable one to administer to the body what things are advantageous. Indeed this sort of medicine is quite natural. For example, hunger is a disease, as everything is called a disease which makes a man suffer. What then is the remedy for hunger? That which makes hunger to cease. This is eating; so that by eating must hunger be cured. Again, drink stays thirst; and again repletion is cured by depletion, depletion by repletion, fatigue by rest. To sum up in a single sentence, opposites are cures for opposites. Medicine in fact is substraction and addition, substraction of what is in excess, addition of what is wanting. He who performs these acts best is the best physician; he who is farthest removed therefrom is also farthest removed from the art. These remarks I have made incidentally in passing to the discourse that is to come.

II. Now of all diseases the fashion is the same, but the seat varies. So while diseases are thought to be entirely unlike one another, owing to the difference in their seat, in reality all have one essence[1] and cause. What this cause is I shall try to declare in the discourse that follows.

III. Now bodies, of men and of animals generally, are nourished by three kinds of nourishment, and the names thereof are solid food, drink, and wind. Wind in bodies is called breath, outside bodies it is

[1] ἰδέη has the meaning of οὐσία here, as εἶδος has in περὶ τέχνης. See the discussion in Taylor's *Varia Socratica.*

---

[2] After ἀπονίῃ M has ἀπονίην δὲ πόνος.
[3] After ἀλλοιότητα many MSS. have καὶ ἀνομοιότητα.

οὗτος δὲ μέγιστος ἐν τοῖσι πᾶσι τῶν πάντων δυνάστης ἐστίν· ἄξιον δὲ αὐτοῦ θεήσασθαι τὴν δύναμιν. ἄνεμος γάρ ἐστιν ἠέρος ῥεῦμα καὶ χεῦμα· ὅταν οὖν πολὺς ἀὴρ ἰσχυρὸν ῥεῦμα ποιήσῃ, τά τε
10 δένδρα ἀνασπαστὰ πρόρριζα γίνεται διὰ τὴν βίην τοῦ πνεύματος, τό τε πέλαγος κυμαίνεται, ὁλκάδες τε ἄπειροι τῷ[1] μεγέθει διαρριπτεῦνται. τοιαύτην μὲν οὖν ἐν τούτοις ἔχει δύναμιν· ἀλλὰ μήν ἐστί γε τῇ μὲν ὄψει ἀφανής, τῷ δὲ λογισμῷ φανερός· τί γὰρ ἄνευ τούτου γένοιτ' ἄν; ἢ τίνος οὗτος ἄπεστιν; ἢ τίνι οὐ συμπάρεστιν; ἅπαν γὰρ τὸ μεταξὺ γῆς τε καὶ οὐρανοῦ πνεύματος ἔμπλεόν ἐστιν. τοῦτο καὶ χειμῶνος καὶ θέρεος αἴτιον, ἐν μὲν τῷ χειμῶνι πυκνὸν καὶ ψυχρὸν
20 γινόμενον, ἐν δὲ τῷ θέρει πρηῢ καὶ γαληνόν. ἀλλὰ μὴν ἡλίου τε καὶ σελήνης καὶ ἄστρων ὁδὸς διὰ τοῦ πνεύματός ἐστιν· τῷ γὰρ πυρὶ τὸ πνεῦμα τροφή· πῦρ δὲ ἠέρος στερηθὲν οὐκ ἂν δύναιτο ζῆν· ὥστε καὶ τὸν τοῦ ἡλίου βίον ἀέναον ὁ ἀὴρ λεπτὸς ἐὼν παρέχεται. ἀλλὰ μὴν ὅτι καὶ τὸ πέλαγος μετέχει πνεύματος, φανερόν· οὐ γὰρ ἄν ποτε τὰ πλωτὰ ζῷα ζῆν ἠδύνατο, μὴ μετέχοντα πνεύματος· μετέχοι δ' ἂν πῶς ἂν ἄλλως ἀλλ' ἢ τοῦ ὕδατος ἕλκοντα τὸν ἠέρα; ἀλλὰ μὴν καὶ ἡ
30 γῆ τούτου βάθρον, οὗτός τε γῆς ὄχημα, κενεόν τε
31 οὐδέν ἐστιν τούτου.

IV. Διότι μὲν οὖν ἐν τοῖς ὅλοις[2] ὁ ἀὴρ ἔρρωται, εἴρηται· τοῖς δ' αὖ θνητοῖσιν οὗτος αἴτιος τοῦ τε βίου, καὶ τῶν νούσων τοῖσι νοσέουσι· τοσαύτη δὲ

[1] ἄπειροι τῷ M: ἀπείρατοι A: ἀπείραντοι Diels: ἄπλετοι Nelson after Danielsson.

called air. It is the most powerful of all and in all, and it is worth while examining its power. A breeze is a flowing and a current of air. When therefore much air flows violently, trees are torn up by the roots through the force of the wind, the sea swells into waves, and vessels of vast bulk are tossed about. Such then is the power that it has in these things, but it is invisible to sight, though visible to reason. For what can take place without it? In what is it not present? What does it not accompany? For everything between earth and heaven is full of wind. Wind is the cause of both winter and summer, becoming in winter thick and cold, and in summer gentle and calm. Nay, the progress of sun, moon, and stars is because of wind; for wind is food for fire, and without air fire could not live. Wherefore, too, air being thin causes the life of the sun to be eternal. Nay, it is clear that the sea, too, partakes of wind, for swimming creatures would not be able to live did they not partake of wind.[1] Now how could they partake except by inhaling the air of the water? In fact the earth too is a base for air, and air is a vehicle of the earth,[2] and there is nothing that is empty of air.

IV. How air, then, is strong in the case of wholes[3] has been said; and for mortals too this is the cause of life, and the cause of disease in the sick. So

[1] This is one of the ancient guesses that modern science has shown to be correct.

[2] Cf. Euripides *Troades* 884: ὦ γῆς ὄχημα κἀπὶ γῆς ἔχων ἕδραν.

[3] *I.e.*, in the case of the sea and of the earth, etc., as wholes.

---

[2] *τοῖς ὅλοις* Nelson (after Schneider) : *τοῖς ὁδοῖς* A : *τοῖσιν ἄλλοισιν* M (so Littré).

τυγχάνει ἡ χρείη πᾶσι τοῖς σώμασι τοῦ πνεύματος ἐοῦσα, ὥστε τῶν μὲν ἄλλων ἁπάντων ἀποσχόμενος ὥνθρωπος καὶ σιτίων καὶ ποτῶν δύναιτ' ἂν ἡμέρας καὶ δύο καὶ τρεῖς καὶ πλέονας διάγειν· εἰ δέ τις ἀπολάβοι τὰς τοῦ πνεύματος ἐς τὸ σῶμα ἐσόδους,[1] ἐν βραχεῖ μέρει ἡμέρης
10 ἀπόλοιτ' ἄν, ὡς μεγίστης τῆς χρείης ἐούσης τῷ σώματι τοῦ πνεύματος. ἔτι τοίνυν τὰ μὲν ἄλλα πάντα διαλείπουσιν οἱ ἄνθρωποι πρήσσοντες. ὁ γὰρ βίος μεταβολέων πλέως· τοῦτο δὲ μοῦνον ἀεὶ διατελέουσιν ἅπαντα τὰ θνητὰ ζῷα πρήσ-
15 σοντα, τοτὲ μὲν ἐκπνέοντα, τοτὲ δὲ ἀναπνέοντα.[2]

V. Ὅτι μὲν οὖν μεγάλη κοινωνίη ἅπασι τοῖσι ζῴοισι τοῦ ἠέρος ἐστίν, εἴρηται· μετὰ τοῦτο τοίνυν ῥητέον, ὡς οὐκ ἄλλοθέν ποθεν εἰκός ἐστι γίνεσθαι τὰς ἀρρωστίας ἢ ἐντεῦθεν.[3] περὶ μὲν οὖν ὅλου τοῦ πρήγματος ἀρκεῖ μοι ταῦτα· μετὰ δὲ ταῦτα πρὸς αὐτὰ τὰ ἔργα τῷ αὐτῷ λόγῳ πορευθεὶς ἐπιδείξω τὰ νοσήματα τούτου[4] ἔκγονα
8 πάντα ἐόντα.

VI. Πρῶτον δὲ ἀπὸ τοῦ κοινοτάτου νοσήματος ἄρξομαι, πυρετοῦ· τοῦτο γὰρ τὸ νόσημα πᾶσιν ἐφεδρεύει τοῖσιν ἄλλοισιν νοσήμασι.[5] ἔστι δὲ δισσὰ ἔθνεα πυρετῶν, ὡς ταύτῃ διελθεῖν· ὁ μὲν κοινὸς ἅπασι καλεόμενος λοιμός· ὁ δὲ διὰ πονηρὴν δίαιταν[6] ἰδίῃ τοῖσι πονηρῶς διαιτεομένοισι γινόμενος· ἀμφοτέρων δὲ τούτων ὁ ἀὴρ αἴτιος. ὁ μὲν

[1] ἐσόδους Nelson : ἐξόδους A : διεξόδους most MSS.

[2] ἐκπνέοντα καὶ ἀναπνέοντα Nelson : ἐνπνέοντα καὶ ἀναπνέοντα A : ἐμπνέοντα καὶ ἐκπνέοντα M.

[3] After ἐντεῦθεν the MSS. have (with unimportant variations) ὅταν τοῦτο πλέον ἢ ἔλασσον ἢ ἀθροώτερον γένηται ἢ μεμιασμένον νοσηροῖσι μιάσμασιν ἐς τὸ σῶμα ἐσέλθῃ. Nelson brackets πλέον ἢ ἔλασσον and γένηται.

great is the need of wind for all bodies that while a man can be deprived of everything else, both food and drink, for two, three, or more days, and live, yet if the wind passages into the body be cut off he will die in a brief part of a day, showing that the greatest need for a body is wind. Moreover, all other activities of a man are intermittent, for life is full of changes; but breathing is continuous for all mortal creatures, inspiration and expiration being alternate.

V. Now I have said that all animals participate largely in air. So after this I must say that it is likely that maladies occur from this source and from no other. On the subject as a whole I have said sufficient; after this I will by the same reasoning proceed to facts and show that diseases are all the offspring of air.

VI. I will begin in the first place with the most common disease, fever, for this disease is associated with all other diseases. To proceed on these lines,[1] there are two kinds of fevers; one is epidemic, called pestilence, the other is sporadic, attacking those who follow a bad regimen. Both of these fevers, however, are caused by air. Now epidemic fever

[1] It is uncertain whether ταύτῃ refers to the first sentence or to the one to which ὡς ταύτῃ διελθεῖν is appended. The translation implies the first interpretation; if the other be correct the whole sentence will be: "There are two kinds of fevers, if I may be allowed to classify them thus."

---

[4] After τούτου M has ἀπόγονά τε καί.

[5] After νοσήμασι the MSS. except A have μάλιστα δὲ φλεγμονῇ· δηλοῖ δὲ τὰ γινόμενα προσκόμματα· ἅμα γὰρ τῇ φλεγμονῇ εὐθὺς βουβὼν καὶ πυρετὸς ἕπεται. The Paris MS. K omits δηλοῖ to φλεγμονῇ.

[6] Nelson deletes διὰ πονηρὴν δίαιταν, perhaps rightly.

οὖν κοινὸς πυρετὸς διὰ τοῦτο τοιοῦτός[1] ἐστιν, ὅτι τὸ πνεῦμα τωὐτὸ πάντες ἕλκουσιν· ὁμοίου 10 δὲ ὁμοίως τοῦ πνεύματος τῷ σώματι μιχθέντος, ὅμοιοι καὶ οἱ πυρετοὶ γίνονται. ἀλλ' ἴσως φήσει τις· τί οὖν οὐχ ἅπασι τοῖσι ζῴοισι, ἀλλ' ἔθνει τινὶ αὐτῶν ἐπιπίπτουσιν αἱ τοιαῦται νοῦσοι; ὅτι διαφέρει, φαίην ἄν, καὶ σῶμα σώματος, καὶ ἀὴρ ἠέρος, καὶ φύσις φύσιος, καὶ τροφὴ τροφῆς· οὐ γὰρ πᾶσι τοῖσιν ἔθνεσι τῶν ζῴων ταὐτὰ οὔτ' εὐάρμοστα οὔτ' ἀνάρμοστά ἐστιν, ἀλλ' ἕτερα ἑτέροισι σύμφορα, καὶ ἕτερα ἑτέροις ἀσύμφορα· ὅταν μὲν οὖν ὁ ἀὴρ τοιούτοισι χρωσθῇ μιάσμασιν, 20 ἃ τῇ ἀνθρωπείῃ φύσει πολέμιά ἐστιν, ἄνθρωποι τότε νοσέουσιν· ὅταν δὲ ἑτέρῳ τινὶ ἔθνει ζῴων 22 ἀνάρμοστος ὁ ἀὴρ γένηται, κεῖνα τότε νοσέουσιν.

VII. Αἱ μέν νυν δημόσιαι τῶν νούσων εἴρηνται,[2] καὶ οἷσι καὶ ἀφ' ὅτευ γίνονται· τὸν δὲ δὴ διὰ πονηρὴν δίαιταν γινόμενον πυρετὸν διέξειμι. πονηρὴ δέ ἐστιν ἡ τοιήδε δίαιτα, τοῦτο μὲν ὅταν τις πλέονας τροφὰς ὑγρὰς ἢ ξηρὰς διδῷ τῷ σώματι ἢ τὸ σῶμα δύναται φέρειν, καὶ πόνον μηδένα τῷ πλήθει τῶν τροφέων ἀντιτιθῇ, τοῦτο δ' ὅταν ποικίλας καὶ ἀνομοίους ἀλλήλῃσιν ἐσπέμπῃ τροφάς· τὰ γὰρ ἀνόμοια στασιάζει, καὶ τὰ μὲν 10 θᾶσσον, τὰ δὲ σχολαίτερον πέσσεται. μετὰ δὲ πολλῶν σιτίων ἀνάγκη καὶ πολὺ πνεῦμα ἐσιέναι· μετὰ πάντων γὰρ τῶν ἐσθιομένων τε καὶ πινομένων ἀπέρχεται πνεῦμα ἐς τὸ σῶμα ἢ πλέον ἢ ἔλασσον. φανερὸν δ' ἐστὶν τῷδε· ἐρυγαὶ γὰρ

[1] τοιοῦτός MSS. : δὴ ὡυτός Nelson.
[2] After εἴρηνται the MSS. except A have καὶ ὅτι καὶ ὅκως.

has this characteristic because all men inhale the same wind; when a similar wind has mingled with all bodies in a similar way, the fevers too prove similar. But perhaps someone will say, "Why then do such diseases attack, not all animals, but only one species of them?" I would reply that it is because one body differs from another, one air from another, one nature from another and one nutriment from another. For all species of animals do not find the same things either well or ill-adapted to themselves, but some things are beneficial to some things and other things to others, and the same is true of things harmful. So whenever the air has been infected with such pollutions as are hostile to the human race, then men fall sick, but when the air has become ill-adapted to some other species of animals, then these fall sick.

VII. Of epidemic diseases I have already spoken, as well as of the victims and of the cause thereof; I must now go on to describe the fever caused by bad regimen. By bad regimen I mean, firstly, the giving of more food, moist or dry, to the body than the body can bear, without counteracting the bulky food by exercise; and, secondly, the taking of foods that are varied and dissimilar. For dissimilar foods disagree,[1] and some are digested quickly and some more slowly. Now along with much food much wind too must enter, for everything that is eaten or drunk is accompanied into the body by wind, either in greater quantity or in less. This is shown by the following fact. After food and drink most

[1] The meaning of στάσις in the medical writers is generally "stagnation," "stopping," and στασιάζει possibly means here "stagnate," "do not digest."

γίνονται μετὰ τὰ σιτία καὶ τὰ ποτὰ τοῖσι πλείστοισιν· ἀνατρέχει γὰρ ὁ κατακλεισθεὶς ἀήρ, ὁπόταν ἀναρρήξῃ τὰς πομφόλυγας, ἐν ᾗσι κρύπτεται. ὅταν οὖν τὸ σῶμα πληρωθὲν τροφῆς[1] πλησθῇ, καὶ πνεύματος ἐπίπλεον γίνεται, τῶν
20 σιτίων χρονιζομένων· χρονίζεται δὲ τὰ σιτία διὰ τὸ πλῆθος οὐ δυνάμενα διελθεῖν· ἐμφραχθείσης δὲ τῆς κάτω κοιλίης, ἐς ὅλον τὸ σῶμα διέδραμον αἱ φῦσαι· προσπεσοῦσαι δὲ πρὸς τὰ ἐναιμότατα τοῦ σώματος ἔψυξαν· τούτων δὲ τῶν τόπων ψυχθέντων, ὅπου αἱ ῥίζαι καὶ αἱ πηγαὶ τοῦ αἵματός εἰσι, διὰ παντὸς τοῦ σώματος[2] φρίκη διῆλθεν· ἅπαντος δὲ τοῦ αἵματος[3] ψυχθέντος,
28 ἅπαν τὸ σῶμα φρίσσει.

VIII. Διὰ τοῦτο μέν νυν αἱ φρῖκαι γίνονται πρὸ τῶν πυρετῶν· ὅπως δ' ἂν ὁρμήσωσιν αἱ φῦσαι πλήθει καὶ ψυχρότητι, τοιοῦτον γίνεται τὸ ῥῖγος, ἀπὸ μὲν πλεόνων καὶ ψυχροτέρων ἰσχυρότερον, ἀπὸ δὲ ἐλασσόνων καὶ ἧσσον ψυχρῶν ἧσσον ἰσχυρόν. ἐν δὲ τῇσι φρίκῃσι καὶ οἱ τρόμοι τοῦ σώματος κατὰ τόδε γίνονται. τὸ αἷμα φοβεύμενον τὴν παρεοῦσαν φρίκην συντρέχει καὶ διαΐσσει διὰ παντὸς τοῦ σώματος ἐς τὰ
10 θερμότατα[4] αὐτοῦ. καθαλλομένου δὲ τοῦ αἵματος ἐκ των ἀκρωτηρίων τοῦ σώματος ἐς τὰ σπλάγχνα, τρέμουσιν· τὰ μὲν γὰρ τοῦ σώματος γίνεται πολύαιμα, τὰ δ' ἄναιμα· τὰ μὲν οὖν ἄναιμα διὰ τὴν ψύξιν οὐκ ἀτρεμέουσιν, ἀλλὰ σφάλλονται, τὸ γὰρ θερμὸν ἐξ αὐτῶν ἐκλέλοιπε· τὰ δὲ

[1] πληρωθὲν τροφῆς A : σιτίων M, followed by Nelson.
[2] σώματος AM : αἵματος Nelson, from one of Foes' variants.
[3] αἵματος A : σώματος M.

people suffer from belching, because the enclosed air rushes upwards when it has broken the bubbles in which it is concealed. When therefore the body is filled full of food, it becomes full of wind too, if the foods remain a long time; and they do remain a long time because owing to their bulk they cannot pass on. The lower belly being thus obstructed, the breaths spread through all the body, and striking the parts of the body that are most full of blood they chill them. These parts being chilled, where are the roots and springs of the blood, a shiver passes through all the body,[1] for when all the blood has been chilled all the body shivers.

VIII. Now this is the reason why shivering occurs before fevers. The character, however, of the rigor depends upon the volume and coldness of the breaths that burst out; from copious and colder breaths come more violent rigor, from less copious and less cold, less violent rigor. The tremors of the body in shivers are caused as follows. The blood, through fear of the shivers that are present, runs together and dashes throughout the body to the warmest parts of it. As the blood leaps from the extremities of the body to the viscera, the sick man shakes. The reason is that some parts of the body become over-full, but others depleted, of blood. Now the depleted parts cannot be still, but shake, because of their being chilled; for the heat has left them. But the over-filled parts tremble

[1] If we give δὲ the not uncommon sense of "for" we can keep the reading of the MSS. Otherwise we must with Nelson read αἵματος for σώματος.

[4] After θερμότατα most MSS. have αὗται μὲν οὖν ἄλλαι (αἱ ἄλαι H). Reinhold conjectured αὐτοῦ μὲν οὖν ἑάλη.

πολύαιμα διὰ τὸ πλῆθος τοῦ αἵματος τρεμουσιν· οὐ γὰρ δύναται πολὺ γενόμενον ἀτρεμίζειν. χασμῶνται δὲ πρὸ τῶν πυρετῶν, ὅτι πολὺς ἀὴρ ἀθροισθείς, ἀθρόως ἄνω διεξιών, ἐξεμόχλευσε καὶ
20 διέστησε τὸ στόμα· ταύτῃ γὰρ εὐδιέξοδός ἐστιν· ὥσπερ γὰρ ἀπὸ τῶν λεβήτων ἀτμὸς ἀνέρχεται πολὺς ἑψομένου τοῦ ὕδατος, οὕτω καὶ τοῦ σώματος θερμαινομένου διαΐσσει διὰ τοῦ στόματος ὁ ἀὴρ συνεστραμμένος καὶ βίῃ φερόμενος. τά τε ἄρθρα διαλύεται πρὸ τῶν πυρετῶν· χλιαινόμενα γὰρ τὰ νεῦρα διίσταται.[1] ὅταν δὲ δὴ συναλισθῇ τὸ πλεῖστον τοῦ αἵματος, ἀναθερμαίνεται πάλιν ὁ ἀὴρ ὁ ψύξας τὸ αἷμα, κρατηθεὶς ὑπὸ τῆς θέρμης· διάπυρος δὲ καὶ ἄνυδρος[2] γενόμενος
30 ὅλῳ τῷ σώματι τὴν θερμασίην ἐνειργάσατο. συνεργὸν δὲ αὐτῷ τὸ αἷμά ἐστιν· τήκεται γὰρ χλιαινόμενον[3] καὶ γίνεται ἐξ αὐτοῦ πνεῦμα· τοῦ δὲ πνεύματος προσπίπτοντος πρὸς τοὺς πόρους τοῦ σώματος ἱδρὼς γίνεται· τὸ γὰρ πνεῦμα συνιστάμενον ὕδωρ χεῖται, καὶ διὰ τῶν πόρων διελθὸν ἔξω περαιοῦται[4] τὸν αὐτὸν τρόπον ὅνπερ ἀπὸ τῶν ἑψομένων ὑδάτων ἀτμὸς ἐπανιών, ἢν ἔχῃ στερέωμα πρὸς ὅ τι χρὴ προσπίπτειν, παχύνεται καὶ πυκνοῦται, καὶ σταγόνες ἀπο-
40 πίπτουσιν ἀπὸ τῶν πωμάτων,[5] οἷς ἂν ὁ ἀτμὸς

[1] Nelson brackets *τά τε ἄρθρα . . . διίσταται.*

[2] *ἀμυδρὸς* AM : *μύδρος* many MSS. : *ἀλυκρὸς* Nelson : *ἄνυδρος* my conjecture.

[3] *χλιαινόμενον* A : *πυρούμενον* M.

[4] *ἔξω περαιοῦται* MSS. : *ἐξυδαροῦται* Nelson.

[5] *πομάτων* A corrected to *πωμάτων* : *σωμάτων* M.

because of the quantity of blood; having become great it cannot keep still. Gapes precede fevers because much air gathers together, and, passing upwards in a mass, unbolts the mouth and forces it open, as through it there is an easy passage. For just as copious steam rises from pots when the water boils, even so, as the body grows hot, the air rushes through the mouth compressed and violently carried along. The joints too relax before fevers, because the sinews stretch when they grow warm. But when the greater part of the blood has been massed together, the air that cooled the blood becomes warm again, being overcome by the heat; and when it has become fiery and waterless,[1] it imparts its heat to the whole body. Herein it is aided by the blood, which melts[2] as it grows warm, and wind arises out of it; as the wind strikes the channels of the body, sweat is formed. For the wind when it condenses flows as water, and going through the channels passes on to the surface, just as steam rising from boiling water, should it meet a solid object that it must strike, thickens and condenses, and drops fall away from the lids on

[1] The text is most uncertain. Neither *ἀμυδρὸς* ("faint") nor *μύδρος* ("mass of molten metal") gives a possible sense, and Nelson's *ἁλυκρὸς* is only a weak repetition of *διάπυρος*. If *ἄνυδρος* be the original reading (cold air becomes misty, see below), it would easily turn into *ἀμυδρός*, which would in its turn become *μύδρος*, a scribe perceiving that *ἀμυδρὸς* makes no sense, and knowing that *διάπυρος* and *μύδρος* often occur together.

[2] I am uncertain whether *τήκεται* means "evaporates" or "becomes thinner."

προσπίπτῃ. πόνοι δὲ κεφαλῆς ἅμα τῷ πυρετῷ γίνονται διὰ τόδε· στενοχωρίη τῇσι διεξόδοισιν ἐν τῇ κεφαλῇ γίνεται τοῦ αἵματος· πέπληνται γὰρ αἱ φλέβες ἠέρος, πλησθεῖσαι δὲ καὶ πρησθεῖσαι τὸν πόνον ποιέουσιν τῇ κεφαλῇ· βίῃ γὰρ τὸ αἷμα βιαζόμενον διὰ στενῶν ὁδῶν θερμὸν ἐὸν οὐ δύναται περαιοῦσθαι ταχέως· πολλὰ γὰρ ἐμποδὼν αὐτῷ κωλύματα καὶ ἐμφράγματα· διὸ δὴ καὶ 49 οἱ σφυγμοὶ γίνονται περὶ τοὺς κροτάφους.

IX. Οἱ μὲν οὖν πυρετοὶ διὰ ταῦτα γίνονται καὶ τὰ μετὰ τῶν πυρετῶν ἀλγήματα καὶ νοσήματα· τῶν δὲ ἄλλων ἀρρωστημάτων, ὅσοι μὲν εἰλεοί, ἢ ἀνειλήματα, ὅτι ἀποστηρίγματα φυσέων ἐστί, πᾶσιν ἡγεῦμαι φανερὸν εἶναι.[1] πάντων γὰρ τῶν τοιούτων ἰητρικὴ τοῦ πνεύματος ἀπαρύσαι. τοῦτο γὰρ ὅταν προσπέσῃ πρὸς τόπους ἀπαθέας[2] καὶ ἀήθεας,[3] ὥσπερ τόξευμα ἐγκείμενον διαδύνει διὰ τῆς σαρκός· προσπίπτει 10 δὲ τοτὲ μὲν πρὸς τὰ ὑποχόνδρια, τοτὲ δὲ πρὸς τὰς λαπάρας, τοτὲ δὲ ἐς ἀμφότερα· διὸ δὴ καὶ θερμαίνοντες ἔξωθεν πυριήμασι πειρέονται μαλθάσσειν τὸν πόνον· ἀραιούμενον γὰρ ὑπὸ τῆς θερμασίης τοῦ πυριήματος διέρχεται τὸ πνεῦμα τοῦ σώματος, ὥστε παῦλάν τινα γενέσθαι 16 τῶν πόνων.

X. Ἴσως ἄν τις εἴποι· πῶς οὖν καὶ τὰ ῥεύματα γίνεται διὰ τὰς φύσας; ἢ τίνα τρόπον τῶν

[1] ὅτι ἀποστηρίγματα φυσέων ἐστί, πᾶσιν ἡγεῦμαι φανερὸν εἶναι. So Nelson, slightly changing the reading of **A**, which has ἢ before, and ὅτι after, ἀποστηρίγματα.

[2] ἀπαθέας A: ἀπαλοὺς M.

[3] After ἀήθεας many MSS. read καὶ ἀθίκτους.

which the steam strikes. Headache with fever arises in the following manner. The blood passages in the head become narrowed. The veins in fact are filled with air, and when full and inflated cause the headache; for the hot blood, forcibly forced through the narrow passages, cannot traverse them quickly because of the many hindrances and barriers in the way. This too is the reason why pulsations occur about the temples.

IX. This then is the way fevers are caused, and the pains and illnesses that accompany fever. As to other maladies, ileus and tormina for example, it is obvious, I think, to everybody that they are settlements of breaths, for the medical treatment for such disorders is to draw off some of the wind. For when it strikes against places that are not usually attacked by it, it pierces the flesh like an arrow forcing its way. Sometimes it strikes against the hypochondria, sometimes against the flanks, sometimes against both. It is for this reason that attendants try to soothe the pain by applying hot fomentations to the skin. For by the heat of the fomentation the wind is rarefied and passes through the body, thus affording some relief of the pains.[1]

X. Perhaps it may be objected: "How then do breaths cause fluxes, and in what way is wind the

[1] The first part of this chapter presents a mass of variant readings in the MSS. See Littré VI. 104, and Nelson, p. 20. It seems impossible to fix the text with any certainty, the variants indicating that the true reading has been lost, and that its place has been taken by glosses and guesses. For example, where A has *τοιούτων ἰητρικὴ τοῦ πνεύματος ἀπαρύσαι* (surely an impossible use of *ἰητρική*), M has *τοιούτων μία ἰητρικὴ τοῦ πνεύματος ἡ διόδευσις*, and other MSS. *τοιούτων αἰτίη τοῦ πνεύματος ἡ διόδευσις*

αἱμορραγιῶν τῶν περὶ τὰ στέρνα τοῦτ' αἴτιόν ἐστιν; οἶμαι δὲ καὶ ταῦτα δηλώσειν διὰ τοῦτο γινόμενα. ὅταν αἱ περὶ τὴν κεφαλὴν φλέβες γεμισθῶσιν ἠέρος, πρῶτον μὲν ἡ κεφαλὴ βαρύνεται τῶν φυσέων ἐγκειμένων· ἔπειτα εἰλεῖται τὸ αἷμα, οὐ διαχεῖν δυναμένων[1] διὰ τὴν στενότητα τῶν ὁδῶν· τὸ δὲ λεπτότατον τοῦ
10 αἵματος διὰ τῶν φλεβῶν ἐκθλίβεται· τοῦτο δὴ τὸ ὑγρὸν ὅταν ἀθροισθῇ πολύ, ῥεῖ δι' ἄλλων πόρων· ὅπη δ' ἂν ἀθρόον ἀφίκηται τοῦ σώματος, ἐνταῦθα συνίσταται νοῦσος· ἢν μὲν οὖν ἐπὶ τὴν ὄψιν ἔλθῃ, ταύτῃ ὁ πόνος· ἢν δὲ ἐς τὰς ἀκοάς, ἐνταῦθ' ἡ νοῦσος·[2] ἢν δὲ ἐς τὰ στέρνα, βράγχος καλεῖται. τὸ γὰρ φλέγμα δριμέσι χυμοῖσι μεμιγμένον, ὅπη ἂν προσπέσῃ ἐς ἀήθεας τόπους, ἑλκοῖ· τῇ δὲ φάρυγγι ἁπαλῇ ἐούσῃ ῥεῦμα προσπῖπτον τρηχύτητας ἐμποιεῖ· τὸ γὰρ πνεῦμα
20 τὸ διὰ τῆς φάρυγγος διαπνεόμενον ἐς τὰ στέρνα βαδίζει,[3] καὶ πάλιν ἔξεισι διὰ τῆς ὁδοῦ ταύτης· ὅταν οὖν ἀπαντήσῃ τῷ ῥεύματι τὸ πνεῦμα[4] κάτωθεν ἰὸν κάτω ἰόντι, βὴξ ἐπιγίνεται, καὶ ἀναρρίπτεται ἄνω τὸ φλέγμα· τούτων δὲ τοιούτων ἐόντων φάρυγξ ἑλκοῦται καὶ τρηχύνεται καὶ θερμαίνεται καὶ ἕλκει τὸ ἐκ τῆς κεφαλῆς ὑγρὸν θερμὴ ἐοῦσα· ἡ δὲ κεφαλὴ παρὰ τοῦ ἄλλου σώματος λαμβάνουσα τῇ φάρυγγι διδοῖ. ὅταν

[1] Nelson reads οὐ διαχωρεῖν δυνάμενον, perhaps rightly.

[2] After νοῦσος most MSS read ἢν δὲ ἐς τὰς ῥῖνας, κόρυζα γίνεται.

[3] βαδίζει M : πορεύεται A.

[4] The reading in the text is that of Littré. A has ὅταν οὖν ἀπαντήσῃ τὸ ῥεῦμα τῷ πνεύματι κ.τ.λ. M has ὅταν δὲ

cause of chest hemorrhages?" I think I can show that these too are caused by this agent. When the veins about the head are loaded with air, at first the head becomes heavy through the breaths that press against it. Then the blood is compressed, the passages being unable, on account of their narrowness, to pour it through.[1] The thinnest part of the blood is pressed out through the veins, and when a great accumulation of this liquid has been formed, it flows through other channels. Any part of the body it reaches in a mass becomes the seat of a disease. If it go to the eyes, the pain is there; if it be to the ears, the disease is there. If it go to the chest, it is called sore throat; for phlegm, mixed with acrid humours, produces sores wherever it strikes an unusual spot, and the throat, being soft, is roughened when a flux strikes it. For the wind that is breathed in through the throat passes[2] into the chest, and comes out again through this passage. So when the ascending wind meets the descending flux, a cough comes on, and the phlegm is thrown upwards. This being so the throat becomes sore, rough and hot, and being hot draws the moisture from the head, which passes on to the throat the moisture it receives from the rest of the body.

[1] I keep the text of A, but with no great confidence. As it stands, ὁδῶν must be taken with δυναμένων, though this gives a strange sense to διαχεῖν. Can it be said that αἱ ὁδοὶ διαχέουσι τὸ αἷμα? Nelson's emendation (οὐ διαχωρεῖν δυνάμενον) is possibly right. I had myself thought of οὐ διαρρεῖν δυνάμενον.

[2] I have kept the reading of M, because *Breaths* is full of startling metaphors.

ξυμβάλῃ τῷ πνεύματι τὸ ῥεῦμα κάτωθεν τῷ κατιόντι. Other MSS. read κάτωθεν τῷ ἀνιόντι.

οὖν ἐθισθῇ τὸ ῥεῦμα ταύτῃ ῥεῖν καὶ χαραδρωθέω-
30 σιν οἱ πόροι, διαδιδοῖ ἤδη καὶ ἐς τὰ στέρνα·
δριμὺ δὲ ἐὸν τὸ φλέγμα προσπῖπτόν τε τῇ σαρκὶ
ἑλκοῖ καὶ ἀναρρηγνύει τὰς φλέβας. ὅταν δὲ
ἐκχυθῇ τὸ αἷμα, χρονιζόμενον καὶ σηπόμενον
γίνεται πῦον· οὔτε γὰρ ἄνω δύναται ἀνελθεῖν
οὔτε κάτω ὑπελθεῖν· ἄνω μὲν γὰρ οὐκ εὔπορος ἡ
πορείη πρὸς ἄναντες ὑγρῷ χρήματι πορεύεσθαι,
κάτω δὲ κωλύει ὁ φραγμὸς τῶν φρενῶν. διὰ τί
δὲ δήποτε τὸ ῥεῦμα ἀναρρήγνυται τὸ μὲν αὐτόματον, τὸ δὲ διὰ πόνους; αὐτόματον μὲν οὖν,
40 ὅταν αὐτόματος ὁ ἀὴρ ἐλθὼν ἐς τὰς φλέβας
στενοχωρίην ποιήσῃ τῇσι τοῦ αἵματος διεξόδοισι·
τότε γὰρ πιεζεύμενον τὸ αἷμα πολὺ γενόμενον
ἀναρρηγνύει τοὺς πόρους, ᾗ ἂν μάλιστα βρίσῃ·
ὅσοι δὲ διὰ πόνων πλῆθος ῃμορράγησαν, καὶ
τούτοις οἱ πόνοι πνεύματος ἐνέπλησαν τὰς
φλέβας· ἀνάγκη γὰρ τὸν πονέοντα τόπον
κατέχειν τὸ πνεῦμα. τὰ δὲ ἄλλα τοῖς εἰρημένοις
48 ὅμοια γίνεται.

XI. Τὰ δὲ ῥήγματα πάντα γίνεται διὰ τόδε·
ὅταν ὑπὸ βίης διαστέωσιν αἱ σάρκες ἀπ᾽ ἀλλήλων,
ἐς δὲ τὴν διάστασιν ὑποδράμῃ πνεῦμα, τοῦτο τὸν
4 πόνον παρέχει.

XII. Ἦν δὲ διὰ τῶν σαρκῶν αἱ φῦσαι
διεξιοῦσαι τοὺς πόρους τοῦ σώματος ἀραιοὺς
ποιέωσιν, ἕπηται δὲ τῇσι φύσῃσιν ὑγρασίη, ἧς[1]
τὴν ὁδὸν ὁ ἀὴρ ὑπειργάσατο, διαβρόχου δὲ
γενομένου τοῦ σώματος, ὑπεκτήκονται μὲν αἱ

[1] The MSS. here present hopeless varieties of readings. For ἕπηται δὲ A has ἐν δὲ and M ἕπεται δὲ. After ὑγρασίη

When therefore the flux has grown used to flowing by this route, and the passages have become channelled, it now spreads even to the chest. Being acrid the phlegm ulcerates the flesh when it strikes it, and bursts open the veins. The extravasated blood rots in course of time and becomes pus, as it can neither ascend nor get away downwards. For a fluid thing cannot easily ascend upwards, and the diaphragm is a barrier to its descent. Why ever then is it that the flux bursts upwards, either spontaneously or through pains? Well, there is a spontaneous flux whenever the air spontaneously enters the veins and makes the channels narrow for the passage of the blood; for on such occasions the blood is compressed because of its volume, and bursts open the passages wherever the pressure is greatest. Whenever excessive pains cause hemorrhage, in these cases also it is wind with which the pains have filled the veins, seeing that any part in pain must retain the wind. Other cases are like those that I have already described.

XI. Lacerations in all cases occur for the following reason. Whenever flesh is violently severed from flesh, and wind slips into the gap, the pain is thereby produced.

XII. If the breaths by passing through the flesh dilate the passages of the body, and these breaths are followed by moisture, the way for which is prepared by the air, then, when the body has become sodden, the flesh melts away and swellings

---

we find *τῇσι* (A), *τοῖσι* (A²), *ἥτις* (M). Nelson conjectures *ἐν δὲ τῇσι φύσῃσι ὑγρασίη ᾖ, τῆς τὴν ὁδὸν κ.τ.λ.*, but surely *τῆς* is impossible.

σάρκες, οἰδήματα δὲ ἐς τὰς κνήμας καταβαίνει· καλεῖται δὲ τὸ τοιοῦτον νόσημα ὕδρωψ.[1] μέγιστον δὲ σημεῖον ὅτι φῦσαι τοῦ νοσήματός εἰσιν αἴτιαι, τόδε ἐστίν· ἤδη τινὲς ὀλεθρίως ἔχοντες ἐκλύσθη-
10 σαν[2] καὶ ἐκενώθησαν τοῦ ὕδατος· παραυτίκα μὲν οὖν τὸ ἐξελθὸν ἐκ τῆς κοιλίης ὕδωρ πολὺ φαίνεται, χρονιζόμενον δὲ ἔλασσον γίνεται. δῆλον οὖν,[3] ὅτι παραυτίκα μὲν τὸ ὕδωρ ἠέρος πλῆρές ἐστιν· ὁ δὲ ἀὴρ ὄγκον παρέχει μέγαν· ἀπιόντος δὲ τοῦ πνεύματος ὑπολείπεται τὸ ὕδωρ αὐτό· διὸ δὴ φαίνεται μὲν ἔλασσον, ἐστὶν δὲ ἴσον. ἄλλο δὲ αὐτῶν τόδε σημεῖον· κενωθείσης γὰρ παντελῶς τῆς κοιλίης, οὐδ' ἐν τρισὶν ἡμέρῃσιν ὕστερον πάλιν πλήρεις γίνονται.[4] τί οὖν ἐστὶ τὸ πληρῶ-
20 σαν ἀλλ' ἢ πνεῦμα; τί γὰρ ἂν οὕτως ἄλλο ταχέως ἐξεπλήρωσεν; οὐ γὰρ δήπου ποτόν γε τοσοῦτον ἐσῆλθεν ἐς τὸ σῶμα· καὶ μὴν οὐδὲ σάρκες ὑπάρχουσιν ἔτι αἱ τηξόμεναι· λείπεται γὰρ ὀστέα καὶ νεῦρα καὶ ῥινός,[5] ἀφ' ὧν οὐδενὸς
25 οὐδεμίῃ δύναιτ' ἂν αὔξησις ὕδατος εἶναι.

XIII. Τοῦ μὲν οὖν τοῦ ὕδρωπος εἴρηται τὸ αἴτιον· αἱ δὲ ἀποπληξίαι γίνονται διὰ τὰς φύσας· ὅταν γὰρ αὗται διαδύνουσαι[6] ἐμφυσήσωσι τὰς σάρκας, ἀναίσθητα ταῦτα γίνεται τοῦ σώματος· ἢν μὲν οὖν ἐν ὅλῳ τῷ σώματι πολλαὶ φῦσαι

[1] As A reads καταβαίνῃ, Nelson conjectures ὑπεκτήκωνται and καταβαίνῃ, and changes δὲ after καλεῖται to δή.

[2] A later hand in A has ἤγουν ἠντλήθησαν (an intelligent gloss), and a note says that there was another reading ἐκαύθησαν, which Littré adopts.

[3] For δῆλον οὖν many MSS. have διὰ τί οὖν γίγνεται καὶ τοῦτο δῆλον.

[4] Nelson has πλήρης γίνεται from the γίνεται of M.

descend to the legs. A disease of this kind is called dropsy. The strongest evidence that breaths cause the disease is the following. Patients already at death's door in some cases are pumped[1] dry of the water. Now the water appears to come copiously from the cavity at first, becoming less plentiful after a time. Now it is plain that at first the water is full of air, and the air makes it of great bulk. But as the wind goes away the water is left by itself, and so it appears to be less, though the quantity is really equal. These patients furnish another proof, in that when the cavity has been completely emptied, not even three days elapse before they are full again. What then filled them except air? What else could fill them up so quickly? Not drink; for surely so much does not enter the body. Not flesh either; as there does not remain flesh to be dissolved. In fact only bones, sinews and skin are left, from none of which could come any increase of water.

XIII. The cause of dropsy then has been set forth; apoplexy, too, is caused by breaths. For when they pass through the flesh and puff it up, the parts of the body affected lose the power of feeling. So if copious breaths rush through the whole body,

[1] An unique use of κλύζω, which accounts for the variant ἐκαύθησαν. I translate the aorists throughout as gnomic, and do not confine their meaning to past instances only.

---

[5] ῥινός A: ἶνες other MSS. Nelson says Erotian also, but ἶνες occurs in *Places in Man* (Littré VI. 284). We must not assume that Erotian read ἶνες here.

[6] αὗται διαδύνουσαι A: αἱ φῦσαι ψυχραὶ οὖσαι καὶ πολλαὶ διαδύνωσι καὶ other MSS. (with slight variations).

διατρέχωσιν, ὅλος ὥνθρωπος ἀπόπληκτος γίνεται. ἢν δὲ ἐν μέρει τινί, τοῦτο τὸ μέρος· καὶ ἢν μὲν ἀπέλθωσιν αὗται, παύεται ἡ νοῦσος· ἢν δὲ
9 παραμείνωσι, παραμένει.[1]

XIV. Δοκεῖ δέ μοι καὶ τὴν ἱερὴν καλεομένην νοῦσον τοῦτο εἶναι τὸ παρεχόμενον· οἷσι δὲ λόγοις ἐμαυτὸν ἔπεισα, τοῖς αὐτοῖσι τούτοισι καὶ τοὺς ἀκούοντας πείθειν πειρήσομαι. ἡγεῦμαι δὲ οὐδὲν ἔμπροσθεν οὐδενὶ εἶναι μᾶλλον τῶν ἐν τῷ σώματι συμβαλλόμενον ἐς φρόνησιν ἢ τὸ αἷμα·[2] τοῦτο δὲ ὅταν μὲν ἐν τῷ καθεστεῶτι μένῃ, μένει καὶ ἡ φρόνησις· ἑτεροιουμένου[3] δὲ τοῦ αἵματος μεταπίπτει καὶ ἡ φρόνησις. ὅτι δὲ ταῦτα οὕτως
10 ἔχει, πολλὰ τὰ μαρτυρέοντα· πρῶτον μέν, ὅπερ ἅπασι ζῴοις κοινόν ἐστιν, ὁ ὕπνος, οὗτος μαρτυρεῖ τοῖς εἰρημένοισιν· ὅταν γὰρ ἐπέλθῃ τῷ σώματι,[4] τὸ αἷμα ψύχεται, φύσει γὰρ ὁ ὕπνος πέφυκεν ψύχειν· ψυχθέντι δὲ τῷ αἵματι νωθρότεραι γίνονται αἱ διέξοδοι. δῆλον δέ· ῥέπει τὰ σώματα καὶ βαρύνεται (πάντα γὰρ τὰ βαρέα πέφυκεν ἐς βυσσὸν φέρεσθαι), καὶ τὰ ὄμματα συγκλείεται, καὶ ἡ φρόνησις ἀλλοιοῦται, δόξαι δὲ ἕτεραί τινες ἐνδιατρίβουσιν, αἳ δὴ ἐνύπνια καλέονται. πάλιν
20 ἐν τῇσι μέθῃσι πλέονος ἐξαίφνης γενομένου τοῦ

[1] After παραμένει M and several other MSS. read ὅτι δὲ ταῦτα οὕτως ἔχει, χασμῶνται συνεχῶς.

[2] ἡγεῦμαι οὐδὲν . . . αἷμα· Nelson. Littré has ἡγεῦμαι δὲ ἔμπροσθεν μηδὲν εἶναι κ.τ.λ. Ermerins transposes ἔμπροσθεν to before ἐμαυτὸν (above). Reinhold has ἔμπρ. μηδενὶ εἶναι μηδὲν ἀλλὸ τῶν . . . ἀλλ' ἢ τὸ αἷμα. The MSS. show a variety of readings, A having the same as the printed text, except that for συμβαλλόμενον (Littré's emendation) it has (with M) ξυμβαλομένων.

the whole patient is affected with apoplexy. If the breaths reach only a part, only that part is affected. If the breaths go away, the disease comes to an end; if they remain, the disease too remains.

XIV. To the same cause I attribute also the disease called sacred. I will try to persuade my hearers[1] by the same arguments as persuaded myself. Now I hold that no constituent of the body in anyone contributes more to intelligence than does blood.[2] So long as the blood remains in its normal condition, intelligence too remains normal; but when the blood alters, the intelligence also changes. There are many testimonies that this is the case. In the first place sleep, which is common to all the animals, witnesses to the truth of my words. When sleep comes upon the body the blood is chilled, as it is of the nature of sleep to cause chill. When the blood is chilled its passages become more sluggish. This is evident; the body grows heavy and sinks (all heavy things naturally fall downwards); the eyes close; the intelligence alters, and certain other fancies linger, which are called dreams. Again, in cases of drunkenness, when the blood has increased in

[1] This word (ἀκούοντας) seems to imply that περὶ φυσῶν was originally a lecture or ἐπίδειξις.

[2] I have followed A and Nelson only because I have nothing better to propose. Although the general meaning is clear, the text is intolerably harsh, both in grammar and in order. If I may hazard a conjecture, the manuscript tradition represents a conflation of simpler readings, one of which worked with ἔμπροσθεν and the other with μᾶλλον.

---

[3] ἑτεροιουμένου A: ἐξαλλάσσοντος M.

[4] After σώματι many MSS. have ὁ ὕπνος τότε.

αἵματος μεταπίπτουσιν αἱ ψυχαὶ καὶ τὰ ἐν τῇσι ψυχῇσι φρονήματα, καὶ γίνονται τῶν μὲν παρεόντων κακῶν ἐπιλήσμονες, τῶν δὲ μελλόντων ἀγαθῶν εὐέλπιδες. ἔχοιμι δ' ἂν πολλὰ τοιαῦτα εἰπεῖν, ἐν οἷσιν αἱ τοῦ αἵματος ἐξαλλαγαὶ τὴν φρόνησιν ἐξαλλάσσουσιν. ἢν μὲν οὖν παντελῶς ἅπαν ἀναταραχθῇ τὸ αἷμα, παντελῶς ἡ φρόνησις ἐξαπόλλυται· τὰ γὰρ μαθήματα καὶ τὰ ἀναγνωρίσματα ἐθίσματά ἐστιν· ὅταν οὖν ἐκ τοῦ
30 εἰωθότος ἔθεος μεταστέωμεν, ἀπόλλυται ἡμῖν ἡ φρόνησις. φημὶ δὲ τὴν ἱερὴν νοῦσον ὧδε γίνεσθαι· ὅταν πνεῦμα πολὺ κατὰ πᾶν τὸ σῶμα παντὶ τῷ αἵματι μιχθῇ, πολλὰ ἐμφράγματα γίνεται πολλαχῇ κατὰ τὰς φλέβας· ἐπειδὰν οὖν ἐς τὰς παχείας καὶ πολυαίμους φλέβας πολὺς ἀὴρ βρίσῃ, βρίσας δὲ μείνῃ, κωλύεται τὸ αἷμα διεξιέναι· τῇ μὲν οὖν ἐνέστηκε, τῇ δὲ νωθρῶς διεξέρχεται, τῇ δὲ θᾶσσον· ἀνομοίης δὲ τῆς πορείης τῷ αἵματι διὰ τοῦ σώματος γενομένης,
40 παντοῖαι αἱ ἀνομοιότητες· πᾶν γὰρ τὸ σῶμα πανταχόθεν ἕλκεται καὶ τετίνακται τὰ μέρεα τοῦ σώματος ὑπηρετέοντα τῷ ταράχῳ καὶ θορύβῳ τοῦ αἵματος, διαστροφαί τε παντοῖαι παντοίως γίνονται· κατὰ δὲ τοῦτον τὸν καιρὸν ἀναίσθητοι πάντων εἰσίν, κωφοί τε τῶν λεγομένων τυφλοί τε τῶν γινομένων, ἀνάλγητοί τε πρὸς τοὺς πόνους· οὕτως ὁ ἀὴρ ταραχθεὶς ἀνετάραξε τὸ αἷμα καὶ ἐμίηνεν. ἀφροὶ δὲ διὰ τοῦ στόματος ἀνατρέχουσιν εἰκότως· διὰ γὰρ τῶν φλεβῶν διαδύνων ὁ ἀήρ,
50 ἀνέρχεται μὲν αὐτός, ἀνάγει δὲ μεθ' ἑωυτοῦ τὸ λεπτότατον τοῦ αἵματος· τὸ δὲ ὑγρὸν τῷ ἠέρι μιγνύμενον λευκαίνεται· διὰ λεπτῶν γὰρ ὑμένων

quantity, the soul and the thoughts in the soul change; the ills of the present are forgotten, but there is confidence that the future will be happy. I could mention many other examples of an alteration in the blood producing an alteration of the intelligence. So if all the blood experience a thorough disturbance, the intelligence is thoroughly destroyed. For learnings and recognitions are matters of habit. So whenever we depart from our wonted habit our intelligence perishes. I hold that the sacred disease is caused in the following way. When much wind has combined throughout the body with all the blood, many barriers arise in many places in the veins. Whenever therefore much air weighs, and continues to weigh, upon the thick, blood-filled veins, the blood is prevented from passing on. So in one place it stops, in another it passes sluggishly, in another more quickly. The progress of the blood through the body proving irregular, all kinds of irregularities occur. The whole body is torn in all directions; the parts of the body are shaken in obedience to the troubling and disturbance of the blood; distortions of every kind occur in every manner. At this time the patients are unconscious of everything—deaf to what is spoken, blind to what is happening, and insensible to pain. So greatly does a disturbance of the air disturb and pollute the blood. Foam naturally rises through the mouth. For the air, passing through the veins, itself rises and brings up with it the thinnest part of the blood. The moisture, mixing with the air, becomes white, for the air being pure is

καθαρὸς ἐὼν ὁ ἀὴρ διαφαίνεται· διὸ δὴ λευκοὶ φαίνονται παντελῶς οἱ ἀφροί. πότε οὖν παύσονται τῆς νούσου καὶ τοῦ παρεόντος χειμῶνος οἱ ὑπὸ τούτου τοῦ νοσήματος ἁλισκόμενοι;[1] ὁπόταν γυμνασθὲν ὑπὸ τῶν πόνων τὸ σῶμα θερμήνῃ τὸ αἷμα· τὸ δὲ διαθερμανθὲν ἐθέρμηνε τὰς φύσας, αὗται δὲ διαθερμανθεῖσαι διαφέρονται καὶ
60 διαλύουσι τὴν σύστασιν τοῦ αἵματος, αἱ μὲν συνεξελθοῦσαι μετὰ τοῦ πνεύματος, αἱ δὲ μετὰ τοῦ φλέγματος· ἀποζέσαντος δὲ τοῦ ἀφροῦ καὶ καταστάντος τοῦ αἵματος καὶ γαλήνης ἐν τῷ
64 σώματι γενομένης πέπαυται τὸ νόσημα.

XV. Φαίνονται τοίνυν αἱ φῦσαι διὰ πάντων τῶν νοσημάτων μάλιστα πολυπραγμονέουσαι· τὰ δ' ἄλλα πάντα συναίτια καὶ μεταίτια· τοῦτο δὴ τὸ αἴτιον τῶν νούσων ἐπιδέδεικταί μοι. ὑπεσχόμην δὲ τῶν νούσων τὸ αἴτιον φράσειν, ἐπέδειξα δὲ τὸ πνεῦμα καὶ ἐν τοῖς ὅλοις[2] πρήγμασι δυναστεῦον καὶ ἐν τοῖσι σώμασι τῶν ζῴων· ἤγαγον δὲ τὸν λόγον ἐπὶ τὰ γνώριμα τῶν ἀρρωστημάτων, ἐν οἷς ἀληθὴς ἡ ὑπόθεσις ἐφάνη·
10 εἰ γὰρ περὶ πάντων τῶν ἀρρωστημάτων λέγοιμι, μακρότερος μὲν ὁ λόγος ἂν γένοιτο, ἀτρεκέστερος
12 δὲ οὐδαμῶς, οὐδὲ πιστότερος.

[1] After ἁλισκόμενοι M adds ἐγὼ φράσω.
[2] ὅλοις A: ἄλλοισι M. Cf. p. 231.

seen through thin membranes. For this reason the foam appears completely white. When then will the victims of this disease rid themselves of their disorder and the storm that attends it? When the body exercised by its exertions has warmed the blood, and the blood thoroughly warmed has warmed the breaths, and these thoroughly warmed are dispersed, breaking up the congestion of the blood, some going out along with the respiration, others with the phlegm. The disease finally ends when the foam has frothed itself away, the blood has re-established itself, and calm has arisen in the body.

XV. So breaths are seen to be the most active agents during all diseases; all other things are but secondary and subordinate causes. This then as the cause of diseases I have now expounded. I promised to declare the cause of diseases, and I have set forth how wind is lord, not only in things as wholes, but also in the bodies of animals. I have led my discourse on to familiar maladies in which the hypothesis has shown itself correct. If indeed I were to speak of all maladies, my discourse, while being longer, would not be in the least more true or more convincing.

LAW

INTRODUCTION

The quaint little piece called *Law* has been strangely neglected by scholars. Yet it presents many fascinating problems, and its style is simple and graceful.

To date it is difficult. Known to Erotian, it is mentioned by no other ancient authority. The internal evidence is very slight, but such as it is it points to Stoic influence. The piece is too short for the historian to base any argument upon general style or subject matter, but the third chapter is so similar to a well-known passage in Diogenes Laertius that it is difficult to believe that they did not both originate in the same school. For the Stoics, of all ancient sects, were the most fond of analogy and imagery,[1] deriving this fondness from the eastern universities in which their earliest teachers were educated.

The passage in Diogenes Laertius is VII. 40: *εἰκάζουσι δὲ ζῴῳ τὴν φιλοσοφίαν, ὀστοῖς μὲν καὶ νεύροις τὸ λογικὸν προσομοιοῦντες· τοῖς δὲ σαρκωδεστέροις τὸ ἠθικόν· τῇ δὲ ψυχῇ τὸ φυσικόν. ἢ πάλιν ᾠῷ· τὰ μὲν γὰρ*

[1] See *e.g.* Sextus Empiricus II. 7: Ζήνων ὁ Κιτιεὺς ἐρωτηθεὶς ὅτῳ διαφέρει διαλεκτικὴ ῥητορικῆς, συστρέψας τὴν χεῖρα καὶ πάλιν ἐξαπλώσας ἔφη τούτῳ κ.τ.λ. and Cicero *Academica* II. 145: (Zeno) cum extensis digitis adversam manum ostenderat, "visum," inquiebat "huiusmodi est" etc. Compare the "parabolic" teaching of the New Testament. Possibly the characteristic was more prominent in Zeno than in other Stoics.

ἐκτὸς εἶναι τὸ λογικόν· τὰ δὲ μετὰ ταῦτα τὸ ἠθικόν· τὰ δ' ἐσωτάτω τὸ φυσικόν· ἢ ἀγρῷ παμφόρῳ· τὸν μὲν περιβεβλημένον φραγμὸν τὸ λογικόν· τὸν δὲ καρπὸν τὸ ἠθικόν· τὴν δὲ γῆν ἢ τὰ δένδρα τὸ φυσικόν.

Chapter III of *Law* reads: ὁκοίη γὰρ τῶν ἐν γῇ φυομένων θεωρίη, τοιήδε καὶ τῆς ἰητρικῆς ἡ μάθησις. ἡ μὲν γὰρ φύσις ἡμέων ὁκοῖον ἡ χώρη· τὰ δὲ δόγματα τῶν διδασκόντων ὁκοῖον τὰ σπέρματα· ἡ δὲ παιδομαθίη, τὸ καθ' ὥρην αὐτὰ πεσεῖν ἐς τὴν ἄρουραν· ὁ δὲ τόπος ἐν ᾧ ἡ μάθησις, ὁκοῖον ἡ ἐκ τοῦ περιέχοντος ἠέρος τροφὴ γιγνομένη τοῖσι φυομένοισι· ἡ δὲ φιλοπονίη, ἐργασίη.

The resemblance may not appear striking, but the similarity of expression makes it probable that *Law* was written by somebody who was under Stoic influence, particularly as there is no positive evidence against the supposition.

It is called "Law" because it gives the essential factors in the education of a good physician.

The last two sentences seem to imply that some physicians were initiated into a craft or guild, but the metaphorical style of the rest of the piece forbids any confident conclusion to be drawn. If, however, we take into account the evidence from *Precepts* and *Decorum*, which I discuss in the introduction to the latter, it seems very probable that some physicians at least joined together in secret societies, with a ritual and a liturgy.

From Chapter IV (ἀνὰ τὰς πόλιας φοιτεῦντας) we see that physicians still wandered like Sophists from city to city.

The most important piece of information in the piece is the assertion, made at the beginning of Chapter I, that there were no penalties to keep erring physicians in order, and that in consequence the profession was in bad repute. So we see that even thus early some men realized the necessity of discipline for practitioners.[1]

We cannot decide whether or not *Law* is a fragment. It is, however, tempting to think that it forms a short address delivered by the head of some medical school to pupils about to begin their professional studies, pointing out to them the necessary conditions of real success.

MSS. and Editions

Law is found in V and M, as well as in several Paris MSS. I have on the whole preferred M to V. The readings I have given show how closely allied V is to the C of Littré.

Littré mentions some twelve editions, the chief of which are those of Coray in his second edition of *Airs Waters Places* (Paris, 1816) and Daremberg (*Hippocrate*, Paris, 1843).

Since Littré's edition there have appeared the editions of Ermerins and Reinhold.

I have myself collated both V and M, as well as Vaticanus Graecus 277.[2] Neither *Oath* nor *Law* appears in Holkhamensis 282, so that it is impossible to compare it and V as far as these two pieces are concerned.

When preparing the text of *Oath* for Volume I was obliged to rely on the critical notes of Ermerins and Littré. It seems convenient to give here such notes on the text of the *Oath* as I should have written if I had seen the manuscripts earlier.

[1] I have treated this question fully in my lecture *Greek Medical Etiquette*.
[2] XIVth century.

My references are to Volume I, pages 298 and 300.

For ὄμνυμι in l. 1 M and V have ὀμνύω; Vat. Gr. 277 has ὄμνυμι.

In l. 2 V has ἅπαντας, and punctuates after ἵστορας; Vat. Gr. 277 has μάρτυς over ἵστορας and συμφωνίαν over ξυγγραφὴν in l. 5.

V has χρέους where M and Vat. Gr. 277 have χρεῶν.

Then occur some most important variants. Though the writing in Vat. Gr. 277 is rather smudged, it seems to have for ἡγήσεσθαι, κοινώσεσθαι and ποιήσεσθαι the aorists ἡγήσασθαι, κοινώσασθαι and ποιήσασθαι. Both M and V clearly have the aorists. When preparing the text I yielded to the authority of certain scholars, and changed the text of Littré to the future, thus securing a uniformity of tense throughout *Oath*. I did not realize at the time how strong the evidence is for the aorist, which I now feel should be adopted. Lower down (l. 13) M and Vat. Gr. 277 have ποιήσασθαι, but V omits all the intervening words from one μετάδοσιν ποιήσασθαι to the other; the eye of the scribe evidently passed from the first occurrence of the phrase to the second. In ll. 20, 21 Vat. Gr. 277 places πεσσὸν after δώσω, but M and V place it before φθόριον. In l. 22 M and V omit both τὸν and τήν, but they appear in Vat. Gr. 277. From this point there seem to be no important variants, but M and V (not Vat. Gr. 277) read ἀνδρείων for ἀνδρῴων, and V (but not M or Vat. Gr. 277) places εἶναι after τοιαῦτα. Vat. Gr. 277 has many notes, both marginal and interlinear, some of which are almost, or quite, illegible. I have noted the glosses μάρτυς and συμφωνίαν. The word *παραγγελίης* also presented difficulty, as it is glossed by a word which seems to be *παράκλησις*. There is a long marginal note on *γενέτησιν* which Littré also quotes from the margin of E (Paris 2255).

The conclusions I have reached are that the vulgate text of *Oath* is approximately correct; that Littré's C (2146) is akin to V, and that E is closely related to Vat. Gr. 277.

ΝΟΜΟΣ

I. Ἰητρικὴ τεχνέων μὲν πασέων ἐστὶν[1] ἐπιφανεστάτη· διὰ δὲ ἀμαθίην τῶν τε χρεωμένων αὐτῇ,[2] καὶ τῶν εἰκῇ τοὺς τοιούσδε κρινόντων, πολύ τι πασέων ἤδη τῶν τεχνέων ἀπολείπεται. ἡ δὲ τῶνδε ἁμαρτὰς μάλιστά μοι δοκεῖ ἔχειν αἰτίην τοιήνδε· πρόστιμον γὰρ ἰητρικῆς μούνης ἐν τῇσι πόλεσιν οὐδὲν ὥρισται, πλὴν ἀδοξίης· αὕτη δὲ οὐ τιτρώσκει τοὺς ἐξ αὐτῆς συγκειμένους. ὁμοιότατοι γάρ εἰσιν[3] οἱ τοιοίδε τοῖσι παρεισαγο-
10 μένοισι προσώποισιν ἐν τῇσι τραγῳδίῃσιν· ὡς γὰρ[4] ἐκεῖνοι σχῆμα μὲν[5] καὶ στολὴν καὶ πρόσωπον ὑποκριτοῦ ἔχουσιν, οὐκ εἰσὶν δὲ ὑποκριταί, οὕτω καὶ οἱ ἰητροί, φήμῃ μὲν πολλοί, ἔργῳ δὲ
14 πάγχυ βαιοί.

II. Χρὴ γάρ, ὅστις μέλλει ἰητρικῆς σύνεσιν ἀτρεκέως ἁρμόζεσθαι, τῶνδέ μιν ἐπήβολον[6] γενέσθαι· φύσιος· διδασκαλίης· τόπου[7] εὐφυέος· παιδομαθίης· φιλοπονίης· χρόνου.[8] πρῶτον μὲν οὖν πάντων δεῖ φύσιος· φύσιος[9] γὰρ ἀντιπρησσούσης κενεὰ πάντα·[10] φύσιος δὲ ἐς τὸ ἄριστον ὁδηγεούσης, διδασκαλίη τέχνης γίνεται· ἣν μετὰ φρονήσιος δεῖ περιποιήσασθαι, παιδομαθέα γενόμενον ἐν τόπῳ ὁκοῖος εὐφυὴς πρὸς μάθησιν ἔσται·
10 ἔτι δὲ φιλοπονίην προσενέγκασθαι ἐς χρόνον

[1] πασῶν ἐστὶν omitted by V. [2] V omits τε and αὐτῇ.
[3] V places εἰσιν after τραγῳδίῃσιν.
[4] V has καὶ γάρ. [5] V omits μέν.

LAW

I. Medicine is the most distinguished of all the arts, but through the ignorance of those who practise it, and of those who casually judge such practitioners, it is now of all the arts by far the least esteemed. The chief reason for this error seems to me to be this: medicine is the only art which our states have made subject to no penalty save that of dishonour, and dishonour does not wound those who are compacted of it. Such men in fact are very like the supernumeraries in tragedies. Just as these have the appearance, dress and mask of an actor without being actors, so too with physicians; many are physicians by repute, very few are such in reality.

II. He who is going truly to acquire an understanding of medicine must enjoy natural ability, teaching, a suitable place, instruction from childhood, diligence, and time. Now first of all natural ability is necessary, for if nature be in opposition everything is in vain. But when nature points the way to what is best, then comes the teaching of the art. This must be acquired intelligently by one who from a child has been instructed in a place naturally suitable for learning. Moreover he must apply diligence

[6] V has μὴν and ἐπήβολος; so apparently Vat. Gr. 277.

[7] For τόπου M has τρόπου. So too below.

[8] The order in V is φύσιος· παιδομαθίης· διδασκαλίης· τόπου εὐφυέος· φιλοπονίης· χρόνου.

[9] V has ταύτης for φύσιος.

[10] V has πάντα κενεά.

πολύν, ὅκως ἡ μάθησις ἐμφυσιωθεῖσα δεξιῶς τε
12 καὶ εὐαλδέως τοὺς καρποὺς ἐξενέγκηται.

III. Ὁκοίη γὰρ τῶν ἐν γῇ φυομένων θεωρίη, τοιήδε καὶ τῆς ἰητρικῆς ἡ μάθησις. ἡ μὲν γὰρ φύσις ἡμέων ὁκοῖον ἡ χώρη· τὰ δὲ δόγματα τῶν διδασκόντων ὁκοῖον τὰ σπέρματα· ἡ δὲ παιδομαθίη, τὸ καθ' ὥρην αὐτὰ πεσεῖν ἐς τὴν ἄρουραν· ὁ δὲ τόπος ἐν ᾧ ἡ μάθησις, ὁκοῖον ἡ ἐκ τοῦ περιέχοντος ἠέρος τροφὴ γιγνομένη τοῖσι φυομένοισιν· ἡ δὲ φιλοπονίη, ἐργασίη· ὁ δὲ χρόνος
9 ταῦτα ἐνισχύει πάντα,† ὡς τραφῆναι τελέως.[1]

IV. Ταῦτα ὦν χρὴ † ἐς τὴν ἰητρικὴν τέχνην ἐσενεγκαμένους, καὶ ἀτρεκέως αὐτῆς γνῶσιν λαβόντας, οὕτως ἀνὰ τὰς πόλιας φοιτεῦντας, μὴ λόγῳ μοῦνον, ἀλλὰ καὶ ἔργῳ ἰητροὺς νομίζεσθαι. ἡ δὲ ἀπειρίη, κακὸς θησαυρὸς καὶ κακὸν κειμήλιον τοῖσιν ἔχουσιν αὐτήν, καὶ ὄναρ καὶ ὕπαρ, εὐθυμίης τε καὶ εὐφροσύνης ἄμοιρος, δειλίης τε καὶ θρασύτητος τιθήνη. δειλίη μὲν γὰρ ἀδυναμίην σημαίνει· θρασύτης δὲ ἀτεχνίην. δύο γάρ,
10 ἐπιστήμη τε καὶ δόξα, ὧν τὸ μὲν ἐπίστασθαι
11 ποιεῖ, τὸ δὲ ἀγνοεῖν.[2]

V. Τὰ δὲ ἱερὰ ἐόντα πρήγματα ἱεροῖσιν ἀνθρώποισι δείκνυται· βεβήλοισι δὲ οὐ θέμις, πρὶν
3 ἢ τελεσθῶσιν ὀργίοισιν ἐπιστήμης.

[1] I reprint Littré, but with no confidence, as both ὡς and ὦν are strange and the reading of M (καὶ τραφῆναι τελέως ταῦτα · ὦν χρή) indicates a deep-seated corruption. V has καὶ τραφῆναι τελείως· ταῦτα ὦν χρεών ἐστιν. This seems to suggest as the correct reading ταῦτα χρεών ἐστιν or perhaps χρὴ οὖν ταῦτα.

[2] After ἀγνοεῖν most MSS (including M) have ἡ μὲν οὖν ἐπιστήμη ποιέει τὸ ἐπίστασθαι, ἡ δόξα τὸ ἀγνοεῖν V has δύο

for a long period, in order that learning, becoming second nature, may reap a fine and abundant harvest.

III. The learning of medicine may be likened to the growth of plants. Our natural ability is the soil. The views of our teachers are as it were the seeds. Learning from childhood is analogous to the seeds' falling betimes upon the prepared ground. The place of instruction is as it were the nutriment that comes from the surrounding air to the things sown. Diligence is the working of the soil. Time strengthens all these things, so that their nurture is perfected.

IV. These are the conditions that we must allow the art of medicine, and we must acquire of it a real knowledge before we travel from city to city and win the reputation of being physicians not only in word but also in deed. Inexperience on the other hand is a cursed treasure and store for those that have it, whether asleep or awake;[1] it is a stranger to confidence and joy, and a nurse of cowardice and of rashness. Cowardice indicates powerlessness; rashness indicates want of art. There are in fact two things, science and opinion; the former begets knowledge, the latter ignorance.

V. Things however that are holy are revealed only to men who are holy. The profane may not learn them until they have been initiated into the mysteries of science.

[1] A proverbial expression meaning "always."

γάρ, ὧν τὸ μὲν (?) ἐπίστασθαι ποιέει, τὸ δὲ μὴ ἐπίστασθαι, ἡ δὲ δόξα τὸ ἀγνοεῖν.

DECORUM

INTRODUCTION

This tract, so far as I can trace, is mentioned by no ancient author.

Strange ideas are current as to its date. The writer in Pauly-Wissowa (*s. v.* "Hippocrates 16") says briefly "Zeit 350 v. Chr." It has even been connected with *Ancient Medicine.*

An examination of its style and language shows that this date is much too early. The broken grammar, strange expressions, and queer turns are too numerous to be explained by the corruptness of the manuscript tradition. They indicate a late date, and probably an imperfect knowledge of Greek. I would in particular call attention to the following unusual expressions, rare compounds and *ἅπαξ λεγόμενα.*

τὰς μηδὲν ἐς χρέος πιπτούσας διαλέξιας.
ἱδρῶτας τίθενται βλέποντες.
νομοθεσίην τίθενται ἀναίρεσιν.
ἀγορὴν ἐργαζόμενοι.
πικροὶ πρὸς τὰς συναντήσιας.

εὔκρητοι	"good-tempered."
ἀνάστασις	"disturbance."
ἀποσίγησις	"silence."
ἐνθυμηματικός	"skilled in argument."
λημματικός	"quick to seize."
ἀποτερματίζεσθαι	"to turn towards."
ἀπαρηγόρητος	"inexorable."

ἀπάντησις "pugnacity."
ἀπεμπόλησις "sale."
ἐγκατάντλησις "washing."
παρέξοδος "traveller's case."
παλαίωσις "a growing old."
προδιαστέλλεσθαι "to give a positive opinion beforehand."
καταστολὴ "moderation."
ἀνακυρίωσις "authoritative affirmation."
ἀταρακτοποιησίη "acting with perfect composure."
ἀδιάπτωτος "infallible."
ἀβλεπτέω "not to see."
ὑπόδεξις "solicitous attention" (as to a guest).

This list by no means exhausts the peculiar words. I would also lay stress upon the late words εἴδησις, εἰδῆσαι, and the constant use of the preposition πρός in a variety of relations.[1]

The general tortuousness of the style is a further indication of late date. The subject matter, again, of the first four chapters is similar to the commonplace moralizing which was the result of Stoicism when it became a rule of life. There is indeed nothing in the tract peculiar to Stoic philosophy, except perhaps the word ἡγημονικός in Chapter IV. But the picture of the true philosopher in Chapter III will, I think, be considered by most readers to

[1] The queerness of the diction of *Decorum* (there is scarcely a sentence which can fairly be called normal) convinces me that we are dealing with an address purposely written in a quaint and obscure manner. It is the language of a secret society, and some parts are completely unintelligible. See pp. 272–276.

be an effort to bring the Stoic "wise man" down to earth as a grave, self-controlled, orderly man of the world. The insistence upon the importance of "nature" (φύσις) is not only not inconsistent with Stoicism, but suggestive of it.

It would be rash to dogmatize about either the date or the authorship of *Decorum*. But perhaps the facts would be accounted for if we suppose that a teacher of medical students, of a later date than 300 B.C., happened to be attracted by Stoic morality, which exerted a wider influence upon the general public than any of the other schools of philosophy, and so displayed forms attenuated to various degrees, "watered down," so to speak, to suit the needs of different types of character. He prepared in writing a lecture on how a physician should conduct himself, in particular how he should be a devotee of true "philosophy."[1] In other words, he gave instruction in etiquette and bed-side manners. Never intended for publication, but for an aid to memory in delivering the lecture, *Decorum* shows all the roughness and irregularities that might be expected in the circumstances.[2] In particular, the first two chapters read as though some unintelligent scribe had tried to make a continuous narrative of rough jottings and alternative expressions.

Whatever its origin, *Decorum* is invaluable to the

[1] The use of σοφία in the sense of ethics, or rather moral conduct, and the description of the φιλόσοφος as the artist in living, are typical of later Greek thought.

[2] I would insist that we must not treat the text of *Decorum* as though it were literature. It is corrupt, but if we could restore the exact words of the writer they would still be in great part a series of ungrammatical notes to remind the lecturer of the heads of his discourse.

historian of medicine. We are told many things which enable us to picture the Greek physician on his rounds, and one chapter gives us the clue to what otherwise would be a mystery, the way in which the Greeks got over the difficulty of nursing serious cases of illness.

How the work came to be included in the Hippocratic collection is not known. Though not in V it is in the V index, and so it must have been in the library of books of which the common ancestor of M and V was composed.

I had written this introduction, and had spent nearly a week in attempting to translate Chapter IV, when the conclusion forced itself upon me that none of my explanations—not even the sum total of them —accounted for the phenomena before me. Let it be granted that M, our most reliable manuscript, shows deep-seated corruption; that the writer wrote a debased Greek; that he was a lecturer who jotted down heads of discourse, and fragments of sentences that he wished particularly to remember, without paying attention to grammar, and without marking the connection between one phrase and another—even though all this is taken for granted the peculiarity of *Decorum* is not fully explained. There is something *unnatural* and fantastic about certain parts of it; one might say that the obscurity was apparently intentional.

While these thoughts were occurring to me I remembered that a similar peculiarity is to be observed in certain parts of *Precepts,* and then it suddenly flashed across my mind that probably the obscurity *was* intentional, and that there were certain formulae and scraps of knowledge which the lecturer conveyed orally, not wishing that his written notes should convey much information to the uninitiated. What if the address was delivered at a meeting of a secret society of physicians, and purposely was intelligible only to those familiar with the formulae and ritual of the society?

We must never forget that secret societies were perfectly familiar to the Greeks from at least the days of Pythagoras. As the vigour of the City-State decayed in the fourth and third centuries B.C., Greek corporate feeling found expression more and more in smaller bodies—in clubs, in friendly societies, and in fraternities generally. That these would have some "secrets" is highly probable if not certain, the great "mysteries" of Eleusis among others setting an example which would very readily be followed.

Physicians too would have a fraternity of their own, probably several fraternities. We must not say that no doctor could practise unless he belonged to such a society, but we may be certain that outsiders would not be looked upon with favour by their fellow-physicians.

Now it is clear that the "secrets" of this society (or societies, if there were several) could not possibly be the ordinary medical knowledge of the age. A moment's thought will show that any attempt to conceal this knowledge would have been futile. The secrets would rather be mystic formulae and maxims of little or no practical value. It is at least curious that Chapter IV of *Decorum* does not become unintelligible until, after a statement of the predominant influence of nature (φύσις), the task of wisdom (σοφία) is mentioned. At once the language becomes dark. Apparently there is also a gap, for the next sentence refers to two λόγοι which have never been mentioned before, at least upon any natural interpretation of the text, and also to two "acts taken together" (πρήγμασι συναμφοτέροισι), these also being mentioned here for the first time. The chapter goes on to speak of a "road traversed by those others," and of rogues "stript bare and then clothing themselves in all manner of badness and disgrace." Shortly after this the chapter becomes comparatively intelligible.

I put it forward as a mere suggestion that the two λόγοι and the two πρήγματα refer to the "secrets," and that at this point in the lecture the λόγοι were spoken and the πρήγματα done. Those clothed in badness and disgrace may be the uninitiated.

If at meetings of medical associations lectures were given to the initiated, we should surely expect them to be on the subjects dealt with in *Precepts* and *Decorum*—professional behaviour, etiquette and so forth. And where, if not in addresses of this type, should we expect to find veiled allusions to the secret formulae and ritual of the society?[1] I believe that *Decorum* and (possibly) *Precepts* are running commentaries on ritualistic observances, and presuppose much knowledge in the hearer. They are φωνάεντα συνετοῖσιν.

A reader may object that all my remarks are pure conjecture. I would point out, however, that this is not so. There is strong evidence that medical

[1] We should also expect in such addresses peculiar words and phrases. A glance at *Decorum* will show that they are common enough. The language in many places is positively grotesque.

secret societies existed, although I confess that I did not appreciate it fully until I saw that it threw light upon the fourth chapter of *Decorum*, which is perhaps the darkest spot in Greek literature. The last sentence of *Law* runs thus:—

> *τὰ δὲ ἱερὰ ἐόντα πρήγματα ἱεροῖσιν ἀνθρώποισι δείκνυται, βεβήλοισι δὲ οὐ θέμις πρὶν ἢ τελεσθῶσιν ὀργίοισιν ἐπιστήμης.*

"Holy things are shown to holy men; to the profane it is not lawful to show them until these have been initiated into the rites of knowledge."

Is it very unnatural to take this language as literal and not metaphorical?

Secondly, in *Precepts V.*, a genuine physician is called ἠδελφισμένος.[1] What can this strange phrase mean except "one made a brother," "initiated into the brotherhood"?

My third passage is taken from *Oath*. The taker of this oath says that only to his own sons, to those of his teacher, and to those pupils who have sworn allegiance νόμῳ ἰητρικῷ, will he impart:—

> *παραγγελίης τε καὶ ἀκροήσιος καὶ τῆς λοιπῆς ἁπάσης μαθήσιος.*

"Precept, oral instruction and all the other teaching."

Note that allusion is made to a νόμος ἰητρικός, and that it is at the end of our Νόμος that the reference to initiation occurs. Moreover, *Precepts* is the title of

[1] The best manuscript of *Precepts*, M, reads in this passage: τίς γὰρ ὦ πρὸς διὸς ἠδελφισμένως ἰητρεύοι πίστει ἢ ἀτεραμνίηι (*sic*). But the correcting hand has written ο over the ω of ἠδελφισμένως; so it is clear that ἰητρὸς has fallen out before ἰητρεύοι.

one of the puzzlingly obscure Hippocratic treatises. Lastly, "Precept, oral instruction and all other teaching," is a curiously verbose expression, and may very well allude, among other things, to mystic λόγοι imparted to initiated members of a physicians' guild.

I trust that the reader will pardon the personal tone of this discussion. I feel that he will be the better able to appreciate and criticize my suggestion if he is told how I came to make it. I would also remark that I leave my notes on Chapters I–V practically as they were before I thought of references to mysterious "secrets."

MSS. and Editions

Decorum is found in seven Paris manuscripts and in M.[1] Foes and Mack note a few readings from manuscripts now lost. Unfortunately there is no manuscript of a superior class which enables us to check M when that manuscript is obviously corrupt.

If parts of *Decorum* were originally rough jottings, it is not surprising that our manuscript tradition is full of errors. It is hopeless to attempt to restore the original text; indeed for a long time I thought the only course to follow was to print M exactly as it is written. Finally I decided to take Littré as a

[1] I have collated this manuscript from excellent photographs sent to me through the kindness of the Librarian of St. Mark's Library, Venice. The collation used by Littré (who calls the manuscript "a") was very accurate. In Chapter VII, however, M reads, not λεχθημονευόμενον as Littré says, but λεσχημονευόμενον. In Chapter XI Littré says that M has ἐσθίης. The photograph, however, shows plainly ἐσίῃς.

basis, and to correct his text wherever I thought the general sense could be made plainer by a simple alteration.[1] I do not pretend, however, that the text I have printed represents the autograph, nor that the English is in many places anything but a rough paraphrase.

I must add that in 1740 *Decorum* was published at Göttingen by G. Matthiae, but I have not seen this work, nor yet *Traités hippocratiques. Préceptes. De la Bienséance.* Traduction par MM. Boyer et Girbal. Montpellier, 1853.

[1] I believe that I have given the reading of M wherever it differs seriously from the printed text.

ΠΕΡΙ ΕΥΣΧΗΜΟΣΥΝΗΣ

I. Οὐκ ἀλόγως οἱ προβαλλόμενοι τὴν σοφίην πρὸς πολλὰ εἶναι χρησίμην, ταύτην δὴ[1] τὴν ἐν τῷ βίῳ. αἱ γὰρ πολλαὶ πρὸς περιεργίην φαίνονται γεγενημέναι· λέγω δέ, αὗται αἱ μηδὲν ἐς[2] χρέος τῶν πρὸς ἃ διαλέγονται· ληφθείη δ' ἂν τουτέων μέρεα ἐς ἐκεῖνο, ὅτι ὅπη[3] οὐκ ἀργίη, οὐδὲ μὴν κακίη· τὸ γὰρ σχολάζον καὶ ἄπρηκτον ζητεῖ ἐς κακίην[4] καὶ ἀφέλκεται·[5] τὸ δ' ἐγρηγορὸς καὶ πρός τι τὴν διάνοιαν ἐντετακὸς ἐφειλκύσατό τι
10 τῶν πρὸς καλλονὴν βίου τεινόντων. ἐῶ δὲ τουτέων[6] τὰς μηδὲν ἐς χρέος πιπτούσας διαλέξιας·[7] χαριεστέρη γὰρ καὶ[8] πρὸς ἕτερόν[9] τι ἐς τέχνην πεποιημένη,[10] τέχνην δὴ[11] πρὸς εὐσχημοσύνην
14 καὶ δόξαν.

II. Πᾶσαι γὰρ αἱ μὴ μετ' αἰσχροκερδείης καὶ ἀσχημοσύνης καλαί, ἧσι[12] μέθοδός τις ἐοῦσα

[1] δὲ M : δὴ Littré. [2] ἐς omitted by M.

[3] ἐς ἐκεῖνα, ἢ ὅτι M : ἐς ἐκεῖνο, ὅτι ὅπη Littré : ἐς ἐκείνην, ὅτι Ermerins.

[4] ζητέει ἐς κακίην M and Littré : ζητέει κακίην Ermerins.

[5] ἀφέλκεται M : ἀφέλκεσθαι Littré : ἐφέλκεται Ermerins.

[6] ἑωυτοῦ· τουτέων τὰς M : ἐῶ δὲ τουτέων τὰς Littré.

[7] διαλέξιας M and Littré : διαλέξιος Ermerins.

[8] καὶ πρὸς M and Ermerins : ἢ πρὸς Littré.

[9] After ἕτερον the MSS. have μέν.

[10] πεποιημένην M.

[11] τέχνην δὲ τὴν πρὸς M and Littré : ταύτην δὴ τὴν πρὸς Ermerins.

[12] κακείνοισι M : καλαὶ ἧσι Littré.

DECORUM

I. Not without reason are those who present as useful for many things wisdom, that is, wisdom applied to life. Most kinds of wisdom, indeed, have manifestly come into being as superfluities; I mean those which confer no advantage upon the objects that they discuss. Parts thereof may be tolerated up to this point, that where idleness is not neither is there evil. Idleness and lack of occupation tend—nay are dragged—towards evil. Alertness, however, and exercise of the intellect, bring with them something that helps to make life beautiful. I leave out of account mere talk that leads to no useful purpose.[1] More gracious is wisdom that even with some other object[2] has been fashioned into an art, provided that it be an art directed towards decorum and good repute.[3]

II. Any wisdom, in fact, wherein works some scientific method, is honourable if it be not tainted

[1] It is hard not to believe that this sentence is a gloss on αὗται . . . διαλέγονται above.

[2] *I. e.* than that of being useful.

[3] The text is so corrupt (or the original was so careless) that one cannot be sure that the version given above is even approximately correct. The general argument seems to be that *σοφία* "keeps a man out of mischief," but that the best kind of *σοφία* is that which has been reduced to an art, and that the art of making life more decorous and honourable—a point of view typical of later Greek thought, particularly of Stoicism.

τεχνικὴ ἐργάζεται· ἀλλ' εἴ γε μή, πρὸς ἀναιδείην δημεύονται.[1] νέοι τε γὰρ αὐτοῖσιν ἐμπίπτουσιν· ἀκμάζοντες δὲ δι' ἐντροπίην ἱδρῶτας τίθενται βλέποντες· πρεσβῦται δὲ διὰ πικρίην νομοθεσίην τίθενται ἀναίρεσιν ἐκ τῶν πόλεων. καὶ γὰρ ἀγορὴν ἐργαζόμενοι οὗτοι,[2] μετὰ βαναυσίης ἀπατέοντες, καὶ ἐν πόλεσιν ἀνακυκλέοντες οἱ
10 αὐτοί.[3] ἴδοι δέ τις[4] καὶ ἐπ' ἐσθῆτος καὶ ἐν τῇσιν ἄλλῃσι περιγραφῇσι· κὴν γὰρ ἔωσιν ὑπερηφανέως κεκοσμημένοι, πολὺ μᾶλλον φευκτέοι καὶ μισητέοι
13 τοῖσι θεωμένοισίν εἰσιν.[5]

III. Τὴν δὲ ἐναντίην χρὴ ὧδε[6] σκοπεῖν· οἷς οὐ διδακτὴ κατασκευή, οὐδὲ περιεργίη· ἔκ τε γὰρ περιβολῆς καὶ τῆς ἐν ταύτῃ εὐσχημοσύνης καὶ ἀφελείης, οὐ πρὸς περιεργίην πεφυκυίης, ἀλλὰ μᾶλλον πρὸς εὐδοξίην, τό τε σύννουν, καὶ τὸ ἐν νῷ πρὸς ἑωυτοὺς διακεῖσθαι, πρός τε τὴν πορείην. οἵ τε ἑκάστῳ σχήματι[7] τοιοῦτοι· ἀδιάχυτοι, ἀπερίεργοι, πικροὶ πρὸς τὰς συναντήσιας, εὔθετοι πρὸς τὰς ἀποκρίσιας, χαλεποὶ πρὸς τὰς ἀντι-
10 πτώσιας, πρὸς τὰς ὁμοιότητας εὔστοχοι καὶ ὁμιλητικοί, εὔκρητοι πρὸς ἅπαντας, πρὸς τὰς ἀναστάσιας σιγητικοί, πρὸς τὰς ἀποσιγήσιας

[1] πρὸς ἀναιτίην δημευταὶ M: πρὸς ἀναιτίην δημευτέαι Littré: πρὸς ἀναιδείην δημεύονται Ermerins (Zwinger, Foes and Mack note a reading δημεύεται). [2] οὗτοι M: οὗτοι Littré.

[3] οἱ αὐτοί is possibly a gloss.

[4] After τις Littré adds ἂν with three Paris MSS. It is not in M. In the Hippocratic writings the optative without ἂν often has the meaning of the optative with it.

[5] φευκτέον καὶ μισητέον τοῖσι θεωμένοισίν ἐστιν M: φευκτέοι καὶ μισητέοι τ. θ. εἰσιν Littré.

[6] χρειῶδες M: χρὴ ὧδε Littré.

[7] οἵ τε ἑκάστῳ σχήματι M: οἷοι ἕκαστοι σχήματι Littré.

with base love of gain and unseemliness. If they be so tainted, such kinds of wisdom become popular only through impudence. Young men fall in with the devotees thereof; when they are grown up they sweat with shame[1] at the sight of them; when they are old, in their spleen they pass laws to banish these devotees from their cities. These are the very men who go around cities, and gather a crowd about them, deceiving it with cheap vulgarity. You should[2] mark them by their dress, and by the rest of their attire; for even if magnificently adorned, they should much more be shunned and hated by those who behold them.[3]

III. The opposite kind of wisdom one should conceive of thus. No studied preparation, and no over-elaboration. Dress decorous and simple, not over-elaborated, but aiming rather at good repute, and adapted for contemplation, introspection and walking.[4] The several characteristics are: to be serious, artless, sharp in encounters, ready to reply, stubborn in opposition, with those who are of like mind quick-witted and affable, good-tempered towards all, silent in face of disturbances, in the

[1] ἐντροπίην is a strange form, and should probably be ἐντροπήν.

[2] Or "may."

[3] The details of this chapter are hopelessly obscured, partly through the corruption of the text, but the general outline is clear. "Quack" philosophers are described, to be compared with genuine philosophers in the next chapter. It is useless to try to rewrite the text so as to make it grammatical and logical. We are dealing with lecture notes, not literature.

[4] So Littré, and the context seems to require such a sense. The construction apparently is: "you may judge of the opposite kind from dress, etc."

ἐνθυμηματικοὶ καὶ καρτερικοί, πρὸς τὸν καιρὸν εὔθετοι καὶ λημματικοί, πρὸς τὰς τροφὰς εὔχρηστοι καὶ αὐτάρκεες, ὑπομονητικοὶ[1] *πρὸς καιροῦ τὴν ὑπομονήν,*[2] *πρὸς λόγους ἀνυστοὺς πᾶν τὸ ὑποδειχθὲν ἐκφέροντες, εὐεπίῃ χρεώμενοι, χάριτι διατιθέμενοι, δόξῃ τῇ ἐκ τούτων διισχυριζόμενοι, ἐς ἀληθείην πρὸς τὸ ὑποδειχθὲν* 20 *ἀποτερματιζόμενοι.*[3]

IV. *Ἡγεμονικώτατον μὲν οὖν τούτων ἁπάντων τῶν προειρημένων ἡ φύσις· καὶ γὰρ οἱ ἐν τέχνῃσιν, ἢν προσῇ*[4] *αὐτοῖσι τοῦτο, διὰ πάντων τούτων πεπόρευνται τῶν προειρημένων. ἀδίδακτον γὰρ τὸ χρέος ἔν τε σοφίῃ καὶ ἐν τῇ τέχνῃ· πρόσθε μὲν ἢ διδαχθῇ,*[5] *ἐς τὸ ἀρχὴν λαβεῖν ἡ φύσις κατερρύη καὶ κέχυται,*[6] *ἡ δὲ σοφίη ἐς τὸ εἰδῆσαι τὰ ἀπ' αὐτῆς τῆς φύσιος ποιεύμενα. καὶ γὰρ ἐν ἀμφοτέροισι τοῖσι λόγοισι πολλοὶ κρατηθέντες* 10 *οὐδαμῇ συναμφοτέροισιν ἐχρήσαντο τοῖσι πρήγμασιν ἐς δεῖξιν· ἐπὴν οὖν τις αὐτῶν ἐξετάζῃ τι*[7] *πρὸς ἀληθείην τῶν ἐν ῥήσει τιθεμένων, οὐδαμῇ*

[1] *ὑπομενητικοὶ* M.

[2] *πρὸς καιρὸν πρὸς ὑπομονήν* M: *πρὸς καιροῦ τὴν ὑπομονήν* Littré.

[3] *ἀποτελματισθῆναι* M: *ἀποτερματιζόμενοι* Coray and Littré.

[4] *προσῆν* M: *προσῇ* Littré.

[5] *προσθεμενη διδαχθῇ* M: *πρόσθε μὲν ἢ διδαχθῇ* Littré: *πρόσθε μὲν ἢ διδαχθῆναι* Ermerins.

[6] *λαβεῖν· ἡ δὲ φύσις κατερρύη καὶ κέχυται τῇ δὲ σοφίῃ* M: *λαβεῖν ἡ φύσις κατερρύη καὶ κέχυται, ἡ δὲ σοφίη* Littré.

[7] *τε* M: *τι* Littré with Van der Linden.

---

[1] I do not believe that a modern can catch the exact associations of these adjectives, many of which are very rare words, if not *ἅπαξ λεγόμενα*. The difficulty is all the greater

face of silence ready to reason and endure, prepared for an opportunity and quick to take it, knowing how to use food and temperate, patient in waiting for an opportunity, setting out in effectual language everything that has been shown forth, graceful in speech, gracious in disposition, strong in the reputation that these qualities bring, turning to the truth when a thing has been shown to be true.[1]

IV. The dominant factor in all the qualities I have mentioned is nature. In fact, if they have natural ability, those engaged in the arts have already made progress in all the qualities mentioned. For in the art, as in wisdom, use is not a thing that can be taught. Before any teaching has taken place nature has rushed down in a flood to make the beginning; it is afterwards that wisdom comes to know the things that are done by nature herself.[2] In fact many, worsted in both words, have in no way used for demonstration both the actual things together.[3] Accordingly, whenever one of them examines in regard to truth something that is being

because the writer works to death his favourite preposition (πρός), using it sometimes in cases which, if a modern may be allowed to judge, make dubious Greek. I find it hard to give ἀνυστός its usual meaning, and may not ὑποδειχθὲν mean "seen as in a glass, darkly"?

[2] The translation of this sentence is largely guess-work. It seems plain, however, that φύσις is contrasted with σοφία; nature comes first and conditions all that wisdom and instruction can accomplish afterwards.

[3] What are ἀμφότεροι οἱ λόγοι? Does λόγοι mean "words" or "respects"? We cannot tell, as the lecturer has in this chapter jotted down merely the heads of his discourse. However λόγοισι seems certainly contrasted with πρήγμασιν. Apparently the meaning is that without natural gifts and training combined no visible achievement can be accomplished.

τὰ πρὸς φύσιν αὐτοῖσι χωρήσει. εὑρίσκονται γοῦν οὗτοι παραπλησίην ὁδὸν ἐκείνοισι πεπορευμένοι.[1] διόπερ ἀπογυμνούμενοι τὴν πᾶσαν ἀμφιέννυνται κακίην καὶ ἀτιμίην. καλὸν γὰρ ἐκ τοῦ διδαχθέντος ἔργου λόγος· πᾶν γὰρ τὸ ποιηθὲν τεχνικῶς ἐκ λόγου ἀνηνέχθη· τὸ δὲ ῥηθὲν τεχνικῶς, μὴ ποιηθὲν δέ, μεθόδου ἀτέχνου
20 δεικτικὸν ἐγενήθη· τὸ γὰρ οἴεσθαι μέν, μὴ πρήσσειν δέ, ἀμαθίης καὶ ἀτεχνίης σημεῖόν ἐστιν· οἴησις γὰρ καὶ μάλιστα ἐν ἰητρικῇ αἰτίην μὲν τοῖσι κεκτημένοισιν,[1] ὄλεθρον δὲ τοῖσι χρεωμένοισιν ἐπιφέρει· καὶ γὰρ ἢν ἑωυτοὺς ἐν λόγοισι πείσαντες οἰηθῶσιν εἰδέναι ἔργον τὸ ἐκ μαθήσιος, καθάπερ χρυσὸς φαῦλος ἐν πυρὶ κριθεὶς τοιούτους αὐτοὺς ἀπέδειξεν. καίτοι γε τοιαύτη ἡ πρόρρησις ἀπαρηγόρητον.[2] ᾗ σύνεσις ὁμογενής ἐστιν, εὐθὺ τὸ πέρας ἐδήλωσε γνῶσις· τῶν δ᾽ ὁ χρόνος τὴν
30 τέχνην † εὐαδέα † [3] κατέστησεν, ἣ τοῖσιν ἐς τὴν

[1] κεκτημένοισιν Coray: κεχρημένοισιν MSS.

[2] ἀπαρηγόρητον ἐς ξύνεσιν ὁμογένεσιν ὡς ἐστιν εὐθὺ τὸ πέρας ἐμήνυσε γνῶσις M: ἀπαρηγόρητος· ᾗ σύνεσις ὁμογενής ἐστιν Littré. I have followed Littré, keeping however ἀπαρηγόρητον. Perhaps ἐμήνυσε is a better reading than ἐδήλωσε.

[3] εὐαδέα M: εὐοδέα Littré. Neither can be right. Perhaps ἐς εὐοδίην.

---

[1] Who are οὗτοι and ἐκεῖνοι? Once more the lecturer's notes are too scanty for us to say, but, unless we are to suppose that he left a gap here to be filled up in his actual delivery of the lecture, ἐκεῖνοι will refer to the "quacks" of Chapter II and οὗτοι to those deficient in natural ability and training.

set out in speech, nature will in no way come to their aid. These are found at any rate to have walked in a path similar to that followed by the others.[1] Wherefore being stripped they clothe themselves with the whole of badness and disgrace. For reasoning[2] that comes as the result of work that has been taught is a good thing; for everything that has been done artistically has been performed as the result of reasoning. But when a thing is not done, but only expressed artistically, it indicates method divorced from art.[3] For to hold opinions, without putting them into action, is a sign of want of education and of want of art.[4] For mere opining brings, in medicine most particularly, blame upon those who hold opinions and ruin upon those who make use of them.[5] In fact, if they persuade themselves by word,[6] and opine that they know the work that is the result of education, they show themselves up like gold proved by fire to be dross. And yet such a forecast is something inexorable. Where understanding is on a par with action, knowledge at once makes plain the end. In some cases time has put the art on the right track, or has made clear

[2] Apparently λόγος here means "theory," "hypothesis" (so Littré), although the usual contrast, "word" as opposed to "deed," is not lost sight of.

[3] Here the lecturer, having mentioned the necessity of theory, passes on to the mistake of words being allowed to take the place of deeds.

[4] We must remember when we translate τέχνη "art," that it includes both what we call art and what we call science. The importance of uniting both these aspects of τέχναι seems to be the subject of part of this difficult chapter.

[5] This seems adapted from *Breaths*, p. 226.

[6] Possibly, "by reasoning."

παραπλησίην οἶμον ἐμπίπτουσι τὰς ἀφορμὰς
32 δήλους ἐποίησε.

V. Διὸ δὴ[1] ἀναλαμβάνοντα τούτων τῶν προειρημένων ἕκαστα, μετάγειν τὴν σοφίην ἐς τὴν ἰητρικὴν καὶ τὴν ἰητρικὴν ἐς τὴν σοφίην. ἰητρὸς γὰρ φιλόσοφος ἰσόθεος· οὐ[2] πολλὴ γὰρ διαφορὴ ἐπὶ τὰ ἕτερα· καὶ γὰρ ἔνι τὰ πρὸς σοφίην ἐν ἰητρικῇ πάντα, ἀφιλαργυρίη, ἐντροπή, ἐρυθρίησις, καταστολή, δόξα, κρίσις, ἡσυχίη, ἀπάντησις, καθαριότης, γνωμολογίη, εἴδησις τῶν πρὸς βίον χρηστῶν καὶ ἀναγκαίων, καθάρσιος[3]
10 ἀπεμπόλησις, ἀδεισιδαιμονίη, ὑπεροχὴ θείη.[4] ἔχουσι γὰρ ἃ ἔχουσι πρὸς ἀκολασίην, πρὸς βαναυσίην, πρὸς ἀπληστίην, πρὸς ἐπιθυμίην, πρὸς ἀφαίρεσιν, πρὸς ἀναιδείην.[5] αὕτη γὰρ[6] γνῶσις τῶν προσιόντων καὶ χρῆσις τῶν πρὸς φιλίην, καὶ ὡς καὶ ὁκοίως τὰ[7] πρὸς τέκνα, πρὸς χρήματα. ταύτῃ μὲν οὖν ἐπικοινωνὸς σοφίη

[1] δὴ M: δεῖ Littré.

[2] οὐ one MS., and also mentioned in Zwinger and Foes. So Littré. M omits.

[3] καθαρσίης M: ἀκαθαρσίης Littré: καθάρσιος my conjecture.

[4] M has θεῖα and Littré reads θεία. I suspect a gap in the text at this place. See note 6 of the translation.

[5] ἐνιδεῖν M: ἀναιδείην Littré.

[6] Before γνῶσις Littré with one MS. has ἡ.

[7] τὰ Littré with one MS.: τε M.

---

[1] Nature and education; practice and theory; fact and reasoning; deed and word—such seem to be the complementary correlatives insisted upon in this chapter. The last sentence means that long experience sometimes makes up for deficient education. See, however, the *Introduction*, p. 273.

[2] So Littré; but the Greek can hardly bear that meaning,

the means of approach to those who have chanced upon the like route.[1]

V. Wherefore resume each of the points mentioned, and transplant wisdom into medicine and medicine into wisdom. For a physician who is a lover of wisdom is the equal of a god. Between wisdom and medicine there is no gulf fixed;[2] in fact medicine possesses all the qualities that make for wisdom. It has disinterestedness, shamefastness, modesty, reserve,[3] sound opinion, judgment, quiet, pugnacity,[4] purity, sententious speech, knowledge of the things good and necessary for life, selling of that which cleanses,[5] freedom from superstition, pre-excellence divine. What they have, they have in opposition to[6] intemperance, vulgarity, greed, concupiscence, robbery, shamelessness. This is knowledge of one's income, use of what conduces to friendship, the way and manner to be adopted towards one's children and money.[7] Now with medicine

even the debased Greek of *Decorum*, and the omission of οὐ in M and many other MSS. points to corruption.

[3] Possibly (as Littré) modesty in dress.

[4] The word in the text (ἀπάντησις) must mean "power to stand up against opponents."

[5] Littré's "rejet de l'impureté" merely repeats καθαριότης above, and gives an impossible sense to ἀπεμπόλησις. My emendation is simple, and suggests that as the physician cleanses the sick body, so wisdom cleanses the sick mind. "Dispensation" would perhaps be a better word than "selling."

[6] The author's favourite word is πρός, and here he uses it in a sense exactly opposite to that in which he employs it scores of times—in fact in the very next sentence (πρὸς φιλίην). Surely there is a gap in the text, the filling of which would give a suitable subject to ἔχουσι.

[7] This sentence is strangely out of place, and most obscurely expressed.

τις, ὅτι καὶ ταῦτα καὶ[1] τὰ πλεῖστα ὁ ἰητρὸς ἔχει.

VI. Καὶ γὰρ μάλιστα ἡ περὶ θεῶν εἴδησις ἐν νόῳ αὐτῇ[2] ἐμπλέκεται· ἐν γὰρ τοῖσιν ἄλλοισι πάθεσι καὶ ἐν συμπτώμασιν εὑρίσκεται τὰ πολλὰ πρὸς θεῶν ἐντίμως κειμένη ἡ ἰητρική. οἱ δὲ ἰητροὶ θεοῖσι παρακεχωρήκασιν· οὐ γὰρ ἔνι περιττὸν ἐν αὐτῇ τὸ δυναστεῦον. καὶ γὰρ οὗτοί πολλὰ μὲν μεταχειρέονται, πολλὰ δὲ καὶ κεκράτηται αὐτοῖσι δι' ἑωυτῶν. †ἃ δὲ καταπλεονεκτεῖ νῦν ἡ ἰητρική, ἐντεῦθεν παρέξει. τίς γὰρ ὁδὸς τῆς ἐν σοφίῃ ὧδε· καὶ γὰρ αὐτέοισιν ἐκείνοισιν· οὕτω δ' οὐκ οἴονται ὁμολογέουσιν ὧδε τὰ περὶ σώματα παραγινόμεναι,[3]† ἃ δὴ διὰ πάσης αὐτῆς πεπόρευται, μετασχηματιζόμενα ἢ μεταποιούμενα, ἃ δὲ μετὰ χειρουργίης ἰώμενα, ἃ δὲ βοηθεόμενα, θεραπευόμενα ἢ διαιτώμενα. τὸ δὲ κεφαλαιωδέστατον ἔστω ἐς τὴν τούτων εἴδησιν.

[1] After ταῦτα M has καί. It is omitted by Littré.

[2] αὐτῇ M: αὐτὴ Littré with one MS.: αὐτῷ Ermerins (conjectured also by Foes).

[3] τίς γὰρ ὁδὸς τῆς ἐν σοφίῃ ὧδε· καὶ γὰρ αὐτέοισιν ἐκείνοισιν· οὕτω δ' οὐκ οἴονται ὁμολογέουσιν ὧδε τὰ περὶ σώματα παραγινόμεναι M. A hopelessly corrupt passage. The restoration of Littré is almost as obscure as the MS. ἔστι γὰρ ὁδός τις ἐν σοφίῃ ὧδε καὶ αὐτέοισιν ἐκείνοισιν· οὕτω δ' οὐκ οἴονται, ὁμολογέουσι δὲ τὰ περὶ σώματα παραγενόμενα.

---

[1] The words ὅτι to ἔχει read like a gloss.

[2] Surely not "symptoms," as Littré translates it.

[3] Littré says "la médecine est, dans la plupart des cas, pleine de révérence à l'égard des dieux." This is an impossible rendering of πρὸς θεῶν ἐντίμως κειμένη.

[4] I take the general sense of this chapter to be that though physicians may be the means, the gods are the cause, of cures in medicine and surgery. The gods confer this honour on medicine, and medical men must realize that the gods are their masters. Unfortunately the middle of the chapter is

a kind of wisdom is an associate, seeing that the physician has both these things and indeed most things.[1]

VI. In fact it is especially knowledge of the gods that by medicine is woven into the stuff of the mind. For in affections generally, and especially in accidents,[2] medicine is found mostly to be held in honour by the gods.[3] Physicians have given place to the gods. For in medicine that which is powerful is not in excess. In fact, though physicians take many things in hand, many diseases are also overcome for them spontaneously. † All that medicine has now mastered it will supply thence. The gods are the real physicians, though people do not think so. But the truth of this statement is shown by the phenomena of disease,† which are co-extensive with the whole of medicine, changing in form or in quality, sometimes being cured by surgery, sometimes being relieved, either through treatment or through regimen. The information I have given on these matters must serve as a summary.[4]

the most corrupt passage in the *Corpus*, and I have been compelled to print the reading of M, faulty as it is, between daggers. Littré makes οὗτοι μεταχειρέονται to refer to quack doctors, as though only charlatans would take the credit of their cures. I would note that μεταχειρέονται and καταπλεονεκτεῖ appear to be ἅπαξ λεγόμενα, while παρέξει in M is written with the -έ- altered, as though the scribe were uncertain what to write. It is at least curious that we again have a passage where, if the writer in his address referred to the mystical formulae of a secret fraternity, he would be likely to write words conveying no meaning to the uninitiated. We should expect these formulae to contain references to the action of the gods in healing diseases. Be this as it may, the exact meaning of the chapter seems lost to us. It is most unfortunate, as it would have been an interesting development of the thesis worked out in *Airs Waters Places* and *The Sacred Disease*, that all diseases are equally divine and equally natural.

VII. Ὄντων[1] οὖν τοιούτων τῶν προειρημένων ἁπάντων, χρὴ τὸν ἰητρὸν ἔχειν τινὰ εὐτραπελίην παρακειμένην· τὸ γὰρ αὐστηρὸν δυσπρόσιτον καὶ τοῖσιν ὑγιαίνουσι καὶ τοῖσι νοσέουσιν. τηρεῖν δὲ χρὴ ἑωυτὸν ὅτι μάλιστα, μὴ πολλὰ φαίνοντα τῶν τοῦ σώματος μερέων, μηδὲ πολλὰ λεσχηνευόμενον τοῖσιν ἰδιώτῃσιν, ἀλλὰ τἀναγκαῖα· † νομίζει γὰρ τοῦτο βίη εἶναι ἐς πρόσκλησιν θεραπηίης.†[2] ποιεῖν δὲ κάρτα μηδὲν περιέργως αὐτῶν, μηδὲ
10 μετὰ φαντασίης· ἐσκέφθω δὲ ταῦτα πάντα, ὅκως ᾖ σοι προκατηρτισμένα ἐς τὴν εὐπορίην, ὡς δέοι· εἰ δὲ μή, ἐπὶ τοῦ χρέους ἀπορεῖν αἰεὶ
13 δεῖ.[3]

VIII. Μελετᾶν δὲ χρὴ ἐν ἰητρικῇ ταῦτα μετὰ πάσης καταστολῆς, περὶ ψηλαφίης, καὶ ἐγχρίσιος, καὶ ἐγκαταντλήσιος, πρὸς τὴν εὐρυθμίην τῶν χειρῶν, περὶ τιλμάτων, περὶ σπληνῶν, περὶ ἐπιδέσμων, περὶ τῶν ἐκ καταστάσιος, περὶ φαρμάκων, ἐς τραύματα καὶ ὀφθαλμικά, καὶ τούτων τὰ πρὸς τὰ γένεα, ἵν' ᾖ σοι προκατηρτισμένα ὄργανά τε καὶ μηχαναὶ καὶ σίδηρος καὶ τὰ ἑξῆς· ἡ γὰρ ἐν τούτοισιν ἀπορίη ἀμηχανίη καὶ
10 βλάβη. ἔστω δέ σοι ἑτέρη παρέξοδος ἡ λιτοτέρη πρὸς τὰς ἀποδημίας[4] ἡ διὰ χειρῶν· ἡ δ' εὐχερε-

[1] ὄντων M: perhaps ἐόντων or ἀληθευόντων.

[2] The sentence within daggers is as it appears in M, and shows obvious signs of corruption. Littré emends to νομίζειν γὰρ τωὐτὸ βίῃ εἶναι ἐς πρόκλησιν θεραπηΐης. M writes ἐς πρόσκλησιν as one word. The sense seems to be that gossip may cause criticism of the treatment proposed by the doctor. It would perhaps be given by reading:

νομίζει γὰρ οὕτως ἰητρὸν ἰέναι ἐς ἔγκλησιν θεραπείης.

Possibly νομίζεται γὰρ οὕτως ἰητρὸς κ.τ.λ. would be even better.

VII. As all I have said is true, the physician must have at his command a certain ready wit, as dourness is repulsive both to the healthy and to the sick. He must also keep a most careful watch over himself, and neither expose much of his person nor gossip to laymen, but say only what is absolutely necessary. For he realizes that gossip may cause criticism of his treatment. He will do none at all of these things in a way that savours of fuss or of show. Let all these things be thought out, so that they may be ready beforehand for use as required. Otherwise there must always be lack when need arises.

VIII. You must practise these things in medicine with all reserve, in the matter of palpation, anointing, washing, to ensure elegance in moving the hands, in the matter of lint, compresses, bandages, ventilation, purges, for wounds and eye-troubles, and with regard to the various kinds of these things, in order that you may have ready beforehand instruments, appliances, knives and so forth. For lack in these matters means helplessness and harm. See that you have a second physician's case, of simpler make, that you can carry in your hands when on a journey. The most convenient is

[3] *ἀπορίη αἰεὶ δεῖ* M: *ἀπορίη ἀηδής* Littré: *ἀπορεῖν αἰεὶ δεῖ* my emendation. Ermerins omits *δεῖ* (dittography).

[4] M has *επιδημίας*.

στάτη διὰ μεθόδων·[1] *οὐ γὰρ οἷόν τε διέρχεσθαι*[2]
13 *πάντα τὸν ἰητρόν.*

IX. *Ἔστω δέ σοι εὐμνημόνευτα φάρμακά τε καὶ δυνάμιες ἁπλαῖ καὶ ἀναγεγραμμέναι, εἴπερ ἄρα ἐστὶν ἐν νόῳ καὶ τὰ περὶ νούσων ἰήσιος, καὶ οἱ τούτων τρόποι, καὶ ὁσαχῶς καὶ ὃν τρόπον περὶ ἑκάστων ἔχουσιν· αὕτη γὰρ ἀρχὴ ἐν ἰητρικῇ καὶ*
6 *μέσα καὶ τέλος.*

X. *Προκατασκευάσθω*[3] *δέ σοι καὶ μαλαγμάτων γένεα πρὸς τὰς ἑκάστων χρήσιας, ποτήματα τέμνειν δυνάμενα ἐξ ἀναγραφῆς ἐσκευασμένα πρὸς τὰ γένεα. προητοιμάσθω δὲ καὶ τὰ πρὸς φαρμακίην ἐς τὰς καθάρσιας, εἰλημμένα ἀπὸ τόπων τῶν καθηκόντων, ἐσκευασμένα ἐς ὃν δεῖ τρόπον, πρὸς τὰ γένεα καὶ τὰ μεγέθεα ἐς παλαίωσιν μεμελετημένα, τὰ δὲ πρόσφατα ὑπὸ*
9 *τὸν καιρόν, καὶ τἄλλα κατὰ λόγον.*

[1] Should we not read διὰ μεθοδίων?

[2] διέρχεσθαι Littré (without stating authority): περιέρχεσθαι M.

[3] In M προσκατασκευάσθω was written first and then the σ of προσ- was smudged out.

---

[1] I retain the reading of Littré without confidence, for διὰ μεθόδων is very curious Greek for "methodically," and M reads plainly περιέρχεσθαι. Hesychius has a gloss μεθόδιον = ἐφόδιον, and I suspect that we should read here διὰ μεθοδίων, and περιέρχεσθαι with M. The μεθόδια would be packets or compartments, filled with small quantities of the chief medical necessaries, with convenient instruments of a portable size, and so on, so that the physician, on arriving at his destination, would not be obliged "to go round everywhere" to get what he wanted. The article before λιτοτέρη is strange, and suggests that ἡ λιτοτέρη and perhaps ἡ διὰ χειρῶν are glosses.

one methodically arranged, for the physician cannot possibly go through everything.[1]

IX. Keep well in your memory drugs and their properties, both simple and compound,[2] seeing that after all it is in the mind that are also the cures of diseases;[3] remember their modes, and their number and variety in the several cases. This in medicine is beginning, middle and end.

X. You must have prepared in advance emollients classified according to their various uses, and get ready powerful[4] draughts prepared according to formula after their various kinds. You must make ready beforehand purgative medicines also,[5] taken from suitable localities, prepared in the proper manner, after their various kinds and sizes, some preserved so as to last a long time, others fresh to be used at the time, and similarly with the rest.

[2] Literally, "written down," because compounded according to a written formula.

[3] Littré says, "si déjà sont dans l'esprit les notions sur le traitement." This is an impossible translation of εἴπερ ἄρα κ.τ.λ. Apparently Littré did not see that the εἴπερ clause is a parenthesis, and that καὶ οἱ τούτων continues the first clause. The general sense is, "carry your knowledge in your head, not on paper, seeing that it is with your mind that you must work a cure."

[4] Littré takes τέμνειν δυνάμενα = "breuvages incisifs," whatever this may mean, adding that some critics suggest ἀνύειν for τέμνειν. It is more likely that τέμνειν is an imperatival infinitive, and that it has its usual meaning of "cutting simples." But δυνάμενα is strange, unless it means "having the appropriate δυνάμεις." Cf. Chapter IX (beginning).

[5] Littré brackets ἐς τὰς καθάρσιας as a gloss, and he may be right. But *Decorum* is alternately over-concise and verbose, and ἐς τὰς καθάρσιας may have been added for the sake of clearness.

XI. Ἐπὴν δὲ ἐσίῃς πρὸς τὸν νοσέοντα, τούτων σοι ἀπηρτισμένων, ἵνα μὴ ἀπορῇς, εὐθέτως ἔχων ἕκαστα πρὸς τὸ ποιησόμενον, ἴσθι γινώσκων ὃ χρὴ ποιεῖν πρὶν ἢ ἐσελθεῖν· πολλὰ γὰρ οὐδὲ συλλογισμοῦ, ἀλλὰ βοηθείης δεῖται τῶν πρηγμάτων. προδιαστέλλεσθαι[1] οὖν χρὴ τὸ ἐκβησόμενον ἐκ τῆς ἐμπειρίης· ἔνδοξον γὰρ καὶ εὐμαθές.

7

XII. Ἐν δὲ τῇ ἐσόδῳ μεμνῆσθαι καὶ καθέδρης, καὶ καταστολῆς, περιστολῆς, ἀνακυριώσιος, βραχυλογίης, ἀταρακτοποιησίης, προσεδρίης, ἐπιμελείης, ἀντιλέξιος πρὸς τὰ ἀπαντώμενα, πρὸς τοὺς ὄχλους τοὺς ἐπιγινομένους εὐσταθείης τῆς ἐν ἑωυτῷ, πρὸς τοὺς θορύβους ἐπιπλήξιος, πρὸς τὰς ὑπουργίας ἑτοιμασίης. ἐπὶ τούτοισι μέμνησο παρασκευῆς τῆς πρώτης· εἰ δὲ μή, † τὰ κατ' ἄλλα ἀδιάπτωτον, ἐξ ὧν παραγγέλλεται ἐς ἑτοιμασίην.†

10

XIII. Ἐσόδῳ χρέο πυκνῶς, ἐπισκέπτεο ἐπιμελέστερον, τοῖσιν ἀπατεωμένοισιν κατὰ τὰς μεταβολὰς ἀπαντῶν·[2] ῥᾷον γὰρ εἴσῃ, ἅμα δὲ καὶ εὐμαρέστερος ἔσῃ· ἄστατα γὰρ τὰ ἐν ὑγροῖσι· διὸ καὶ εὐμεταποίητα ὑπὸ φύσιος καὶ ὑπὸ τύχης· ἀβλεπτηθέντα γὰρ τὰ κατὰ τὸν καιρὸν τῆς ὑπουργίης ἔφθασαν[3] ὁρμήσαντα καὶ ἀνελόντα·

[1] M reads προσδιαστέλλεσθαι.

[2] ἀπάντων M: ἀπαντῶν Littré without comment. He probably followed some Paris MS.

[3] Query, ἔφθασεν.

---

[1] I agree with Littré that the text cannot be right, but I should hesitate to restore it confidently. I believe that here, too, we have the lecturer's rough, ungrammatical notes. The quaintness, the apparently purposed strangeness of the

XI. When you enter a sick man's room, having made these arrangements, that you may not be at a loss, and having everything in order for what is to be done, know what you must do before going in. For many cases need, not reasoning, but practical help. So you must from your experience forecast what the issue will be. To do so adds to one's reputation, and the learning thereof is easy.

XII. On entering bear in mind your manner of sitting, reserve, arrangement of dress, decisive utterance, brevity of speech, composure, bedside manners, care, replies to objections, calm self-control to meet the troubles that occur, rebuke of disturbance, readiness to do what has to be done. In addition to these things be careful of your first preparation. Failing this, make no further mistake in the matters wherefrom instructions are given for readiness.[1]

XIII. Make frequent visits; be especially careful in your examinations, counteracting the things wherein you have been deceived at the changes.[2] Thus you will know the case more easily, and at the same time you will also be more at your ease.[3] For instability is characteristic of the humours, and so they may also be easily altered by nature and by chance. For failure to observe the proper season for help gives the disease a start and kills the patient, as there was nothing to relieve him.

diction of this chapter makes me more than ever convinced that we have in *Decorum* the language of ritual and not of every-day life. In this particular case the sense is quite plain.

[2] Apparently the "changes" shown by a disease in passing from one phase to another.

[3] I can find no parallel for *εὐμαρῆς* in this sense, but the context makes it necessary to interpret it as I have done.

οὐ γὰρ ἦν τὸ ἐπικουρῆσον. πολλὰ γὰρ ἅμα τὰ ποιέοντά[1] τι χαλεπόν· τὸ[2] γὰρ καθ' ἓν κατ'
10 ἐπακολούθησιν εὐθετώτερον καὶ ἐμπειρότερον.

XIV. Ἐπιτηρεῖν δὲ χρὴ καὶ τὰς ἁμαρτίας τῶν καμνόντων, δι' ὧν πολλάκις[3] διεψεύσαντο ἐν τοῖσι προσάρμασι τῶν προσφερομένων· ἐπεὶ[4] τὰ μισητὰ ποτήματα οὐ[5] λαμβάνοντες, ἢ φαρμακευόμενοι ἢ θεραπευόμενοι, ἀνῃρέθησαν· καὶ αὐτῶν μὲν οὐ πρὸς ὁμολογίην τρέπεται τὸ ποιηθέν, τῷ
7 δὲ ἰητρῷ τὴν αἰτίην προσῆψαν.

XV. Ἐσκέφθαι δὲ χρὴ καὶ τὰ περὶ ἀνακλίσεων, ἃ μὲν αὐτῶν πρὸς τὴν ὥρην, ἃ δὲ πρὸς τὰ γένεα· οἱ μὲν γὰρ αὐτῶν ἐς εὐπνόους, οἱ δὲ ἐς καταγείους καὶ σκεπινοὺς τόπους·[6] τά τε ἀπὸ ψόφων καὶ ὀσμῶν, μάλιστα δ' ἀπὸ οἴνου, χειροτέρη[7] γὰρ
6 αὕτη, φυγεῖν δὲ καὶ μετατιθέναι.

XVI. Πρήσσειν δ' ἅπαντα ταῦτα ἡσύχως, εὐσταλέως, μεθ' ὑπουργίης τὰ πολλὰ τὸν νοσέοντα ὑποκρυπτόμενον· ἃ δὲ[8] χρή, παρακελεύοντα ἱλαρῶς καὶ εὐδιεινῶς, σφέτερα δὲ ἀποτρεπόμενον, ἅμα μὲν ἐπιπλήσσειν μετὰ πικρίης καὶ ἐντάσεων, ἅμα δὲ[9] παραμυθεῖσθαι μετ' ἐπιστροφῆς καὶ

[1] ποιέοντα M: προσιόντα Littré. I see no reason for the change.

[2] τὸ Littré, apparently following some MSS.: τῶν M.

[3] Before πολλάκις Littré has πολλοί.

[4] For ἐπεὶ M reads ἐπί.

[5] The MSS. omit οὐ before λαμβάνοντες. Apparently it was added by Calvus.

[6] οἱ μὲν γὰρ αὐτέων ἐς πόνους, οἱ δ' ἐς καταγείους καὶ σκεπινοὺς τόπους M: οἱ μὲν γὰρ αὐτέων ἐς ὑψηλούς, οἱ δὲ ἐς μὴ ὑψηλούς, οἱ δὲ ἐς καταγείους καὶ σκοτεινοὺς τόπους Littré. Ermerins has εὐπνόους for πόνους. I have kept as closely to the reading of M as is possible, merely changing πόνους to εὐπνόους with Ermerins, who adopted this reading from a note of Foes.

For when many things together produce a result there is difficulty. Sequences of single phenomena are more manageable, and are more easily learnt by experience.[1]

XIV. Keep a watch also on the faults of the patients, which often make them lie about the taking of things prescribed. For through not taking disagreeable drinks, purgative or other, they sometimes die. What they have done never results in a confession, but the blame is thrown upon the physician.

XV. The bed also must be considered. The season and the kind of illness[2] will make a difference. Some patients are put into breezy spots, others into covered places or underground. Consider also noises and smells, especially the smell of wine. This is distinctly bad, and you must shun it or change it.[3]

XVI. Perform all this calmly and adroitly, concealing most things from the patient while you are attending to him. Give necessary orders[4] with cheerfulness and serenity, turning his attention away from what is being done to him; sometimes reprove sharply and emphatically, and sometimes comfort

[1] Such must be the meaning, but the Greek is strange.

[2] Littré takes γένεα to refer to different kinds of bed.

[3] I suppose by eating something with a strong and pleasant odour.

[4] Perhaps, "give encouragement to the patient to allow himself to be treated."

---

[7] M has χειριστοτέρη, apparently a "portmanteau" of χειρίστη and χειροτέρη.

[8] ὧδε M: ἃ δὲ Matthiae.

[9] For ἅμα δὲ M has ἃ δέ.

ὑποδέξιος, μηδὲν ἐπιδείκνυντα τῶν ἐσομένων ἢ ἐνεστώτων αὐτοῖσι· πολλοὶ γὰρ δι' αἰτίην ταύτην ἐφ' ἕτερα[1] ἀπεώσθησαν, διὰ τὴν πρόρρησιν τὴν
10 προειρημένην τῶν ἐνεστώτων ἢ ἐπεσομένων.

XVII. Τῶν δὲ μανθανόντων ἔστω τις ὁ ἐφεστὼς ὅκως τοῖσι παραγγέλμασιν οὐ πικρῶς[2] χρήσεται,[3] ποιήσει δὲ ὑπουργίην τὸ προσταχθέν·[4] ἐκλέγεσθαι δὲ αὐτῶν τοὺς ἤδη[5] ἐς τὰ τῆς τέχνης εἰλημμένους, προσδοῦναί τι τῶν ἐς τὸ χρέος, ἢ ἀσφαλέως προσενεγκεῖν· ὅκως τε ἐν διαστήμασι μηδὲν λανθάνῃ σε· ἐπιτροπὴν δὲ τοῖσιν ἰδιώτῃσι μηδέποτε διδοὺς περὶ μηδενός· εἰ δὲ μή, τὸ κακῶς πρηχθὲν ἐς σὲ χωρῆσαι τὸν ψόγον ἐᾷ·[6] μήποτ'
10 ἀμφιβόλως ἔχῃ, ἐξ ὧν τὸ μεθοδευθὲν χωρήσει, καὶ οὐ σοὶ τὸν ψόγον περιάψει,[7] τευχθὲν δὲ πρὸς τὸ γάνος[8] ἔσται· πρόλεγε οὖν ταῦτα πάντα ἐπὶ
13 τῶν ποιευμένων, οἷς καὶ τὸ ἐπεγνῶσθαι πρόκειται.

XVIII. Τούτων οὖν ἐόντων τῶν πρὸς εὐδοξίην καὶ εὐσχημοσύνην τῶν ἐν τῇ σοφίῃ καὶ ἰητρικῇ καὶ ἐν τῇσιν ἄλλῃσι τέχνῃσι, χρὴ τὸν ἰητρὸν

[1] ἕτερα M : ἑκάτερα Littré (with other MSS.).
[2] Littré reads οὐκ ἀκαίρως for οὐ πικρῶς.
[3] M has χρήσηται, which Littré emends to the future.
[4] τὸ προσταχθέν I take to be a gloss on ὑπουργίην. It is just possible that ποιήσει ὑπουργίην is a compound expression governing τὸ προσταχθέν in the accusative. Cf. Chapter II νομοθεσίην τίθενται ἀναίρεσιν.
[5] I have transposed ἤδη, which in the MSS. is after αὐτῶν.
[6] τοῦ ψόγου ἐὰν M. The text is Littré's.
[7] περιάψει Littré with one Paris MS. : περιάψειεν M.
[8] γένος M : κλέος Littré's emendation I think the writer used the poetic word γάνος.

with solicitude and attention, revealing nothing of the patient's future or present[1] condition. For many patients through this cause have taken a turn for the worse, I mean by the declaration I have mentioned of what is present,[1] or by a forecast of what is to come.

XVII. Let one of your pupils be left in charge, to carry out instructions without unpleasantness, and to administer the treatment. Choose out those who have been already admitted into the mysteries of the art, so as to add anything necessary, and to give treatment with safety. He is there also to prevent those things escaping notice that happen in the intervals between visits. Never put a layman in charge of anything, otherwise if a mischance occur the blame will fall on you.[2] Let there never be any doubt about the points which will secure the success of your plan,[3] and no blame will attach to you, but achievement will bring you pride.[4] So say beforehand all this at the time the things are done,[5] to those whose business it is to have fuller knowledge.[6]

XVIII. Such being the things that make for good reputation and decorum, in wisdom, in medicine, and in the arts generally, the physician must mark

[1] I am in doubt whether or not *ἐνεστὼς* in these two cases means "imminent." But *ἐσομένων* and *ἐπεσομένων* seem to suggest the meaning "present."

[2] I make no attempt to correct the broken grammar, holding that the remarks are a lecturer's notes.

[3] The meaning is very obscure.

[4] The *γένος* of M points to the reading *γάνος*, "brightness," perhaps here "glory."

[5] The meaning of *ἐπὶ τῶν ποιεομένων* is very uncertain.

[6] Apparently *ἐπιγιγνώσκω* here means "to know in addition."

διειληφότα τὰ μέρεα περὶ ὧν εἰρήκαμεν, περιεννύμενον πάντοτε τὴν ἑτέρην διατηρέοντα φυλάσσειν, καὶ παραδιδόντα ποιεῖσθαι· εὐκλεᾶ γὰρ ἐόντα πᾶσιν ἀνθρώποισι διαφυλάσσεται· οἵ τε δι' αὐτῶν ὁδεύσαντες δοξαστοὶ πρὸς γονέων καὶ τέκνων· κἤν τινες αὐτῶν μὴ πολλὰ γινώσκωσιν,
10 *ὑπ' αὐτῶν τῶν πρηγμάτων ἐς σύνεσιν καθ-*
11 *ίστανται.*

---

[1] Probably a reference to Chapter I, ληφθείη δ' ἂν τούτων μέρεα.

[2] What is τὴν ἑτέρην? I must once more revert to my suggestion that *Decorum*, with its stilted and often unnatural language, is full of the secret formulae of a medical fraternity, the most "holy" phrases being omitted or disguised. I think τὴν ἑτέρην is one of these phrases. Surely at the

off the parts[1] about which I have spoken, wrap himself round always with the other,[2] watch it and keep it, perform it and pass it on. For things that are glorious are closely guarded among all men. And those who have made their way through them are held in honour by parents and children; and if any of them do not know many things, they are brought to understanding by the facts of actual experience.

end of an address to "the brethren" (ἠδελφισμένος ἰητρός, *Precepts V.*) we should expect references to the mysteries of the craft. And this last chapter seems full of them. How else can we explain διατηρέοντα φυλάσσειν, παραδιδόντα (handing on the pass-words), εὐκλεᾶ διαφυλάσσεται, δι' αὐτῶν ὁδεύσαντες? The word σύνεσις, too, seems to be a word of this class.

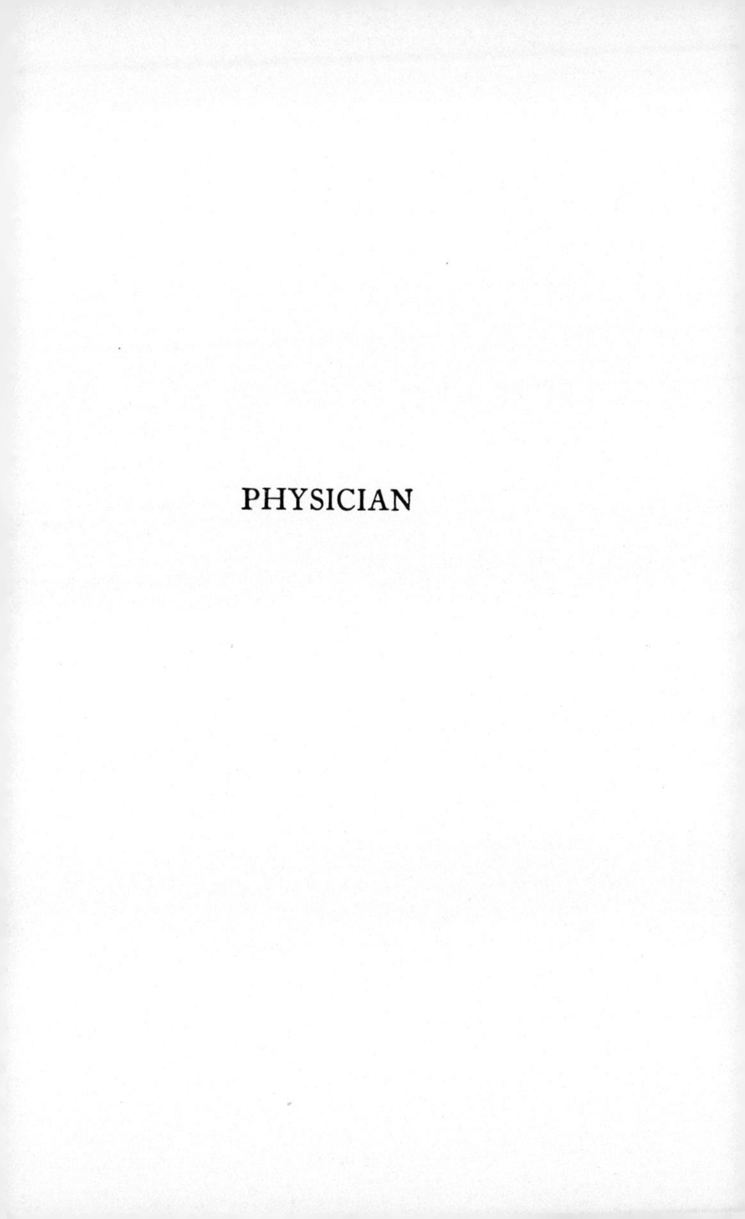

PHYSICIAN

INTRODUCTION

CHAPTER I

In order to give a fairly complete account of what was anciently considered good manners and good behaviour for doctors I must add to *Law, Oath, Precepts* and *Decorum* the first chapter of the work *Physician.*

Very little is known about the position of *Physician* in the history of medicine. "Cet opuscule," says Littré,"[1] "n'est mentionné par aucun des anciens critiques." And later on; "Dans le silence des anciens commentateurs il n'est pas possible de se faire une idée sur l'origine de l'opuscule du Médecin."[2]

After the first chapter the piece goes on to discuss the arrangement of the surgery, the preparation of bandages, instruments, and so forth. Then follows a short discussion of tumours and sores, and the book finishes with a recommendation to a student to attach himself to mercenary troops in order to have practice in surgery[3]—a fairly sure indication of a date later than 400 B.C.

[1] I. 412. [2] I. 414.

[3] Ἐν τῇσι κατὰ πόλιν διατριβῇσι βραχεῖά τίς ἐστι τούτων ἡ χρῆσις· ὀλιγάκις γὰρ ἐν παντὶ τῷ χρόνῳ γίνονται πολιτικαὶ στρατιαὶ καὶ πολεμικαί· συμβαίνει δὲ τὰ τοιαῦτα πλειστάκις καὶ συνεχέστατα περὶ τὰς ξενικὰς στρατιὰς γίνεσθαι. τὸν μὲν οὖν μέλλοντα χειρουργεῖν στρατεύεσθαι δεῖ καὶ παρηκολουθηκέναι στρατεύμασι ξενικοῖς· οὕτω γὰρ ἂν εἴη γεγυμνασμένος πρὸς ταύτην τὴν χρείαν. Chapter XIV.

INTRODUCTION

In Chapter IV an interesting passage occurs in which the surgeon is advised to avoid showiness and ostentation in manipulating bandages, as all such conduct savours of vulgarity and charlatanism.[1]

Dr. J. F. Bensel[2] holds that *Physician* is closely connected with the treatises *Precepts* and *Decorum*. It is most important to come to some conclusion as to whether there is a real connection, or whether there are merely resemblances.

Bensel's monograph (it is really an edition of *Physician*) is very instructive, and compares well with the somewhat arid discussions to be found in most similar works. The author sees that all three books are intended for young beginners; he points out that the artifices we associate with the style of Isocrates are to be seen in *Physician,* and in particular that in some cases there are verbal parallels. These tend to indicate that the date of *Physician* is 350–300 B.C.

Up to this point it is easy to agree with Bensel. But when he goes on to assert that *Physician* is contemporary with *Precepts* and *Decorum,* and that the last shows Epicurean tendencies, it is difficult to follow his argument. *Physician* is comparatively simple, and the Greek is rarely strange or obscure. There are none of the signs of late date. *Precepts* and *Decorum,* on the other hand, are not only strange but even fantastic. No extant Greek prose

[1] εὐρύθμους δὲ ἐπιδέσιας (surely this is the right accentuation and not ἐπιδεσίας with Littré) καὶ θεητρικὰς μηδὲν ὠφελεούσας ἀπογινώσκειν· φορτικὸν γὰρ τὸ τοιοῦτον καὶ παντελῶς ἀλαζονικόν, πολλάκις τε βλαβὴν οἶσον τῷ θεραπευομένῳ· ζητεῖ δὲ ὁ νοσέων οὐ καλλωπισμόν, ἀλλὰ τὸ συμφέρον.

[2] See *Philologus* for 1922, LXXVIII. 88–130.

shows such peculiar vagaries in diction. The signs of late date are many and insistent. Finally, the supposed Epicureanism of *Decorum* cannot possibly be reconciled with the assertion made in that work that physicians give way before the gods, and know that their art is under the direction of a higher power. Surely this is Stoic rather than Epicurean doctrine. The truth seems to be that what Bensel takes to be Epicureanism is really the received ethical teaching of later Alexandrine times, which is in part common to both schools of thought.

The likeness, then, between *Physician* and the other two works is a similarity of subject. All are addresses to young men at the beginning of their medical course, and lay down the rules of conduct and practice that such students must follow. In the face of the evidence it is illegitimate to go further, and to assert that all were written at the same time. On the contrary, there is every reason to think that *Physician* is considerably earlier than the other two.

Littré, having pointed out parallel passages to parts of *Physician* in *Surgery*, *Ancient Medicine* and several other Hippocratic works, concludes his *Argument* with a paragraph so admirable that I quote it in full.

"A l'aide de ces renseignements on entrevoit comment un étudiant faisait son éducation. Il était, ainsi que l'indique le *Serment*, d'ordinaire de famille médicale ; sinon, il s'agrégeait à une de ces familles ; il commençait de bonne heure ; on le plaçait dans *l'iatrion* ou officine, et là il s'exerçait au maniement des instruments, à l'application des bandages, et à tous les débuts de l'art ; puis il voyait les malades avec son maître, se familiarisait avec les maladies, apprenait à reconnaître les *temps opportuns* et à user des *remèdes.* De la sorte il devenait un praticien, et, si son zèle et ses dispositions le favorisaient, un praticien habile. Dans tout cela il n'est question ni d'anatomie ni de physiologie ; c'est qu'en effet ces choses-là n'existaient qu'à l'état de rudiment, et dès lors ne servaient pas de fondement à une éducation. Un médecin pouvait, comme celui dont parle Hippocrate, croire que l'apophyse styloïde du cubitus et l'apophyse de l'humérus, qui est dans le pli du coude, appartenaient à un même os (*des Fractures,* § 3), ou, comme un autre dont il se raille aussi, prendre les apophyses épineuses du rachis pour le corps même des vertèbres (*des Articulations,* § 46) ; ceux-là, on le voit, n'avaient pas la moindre notion, je ne dirai pas d'anatomie, mais de l'ostéologie la plus élémentaire. Les hippocratiques, sans avoir une vue distincte des rapports de l'anatomie avec la médecine, nous montrent les premiers essais pour sortir de l'empirisme primitif, obligé nécessairement de se passer d'anatomie et de physiologie. Hippocrate avait une connaissance très-précise des os. Passé cela, son école n'avait plus rien de précis ; des notions, en gros, sur les principaux viscères, des efforts infructueux pour débrouiller la marche des vaisseaux sanguins, une méconnaissance complète des nerfs proprement dits, confondus sous le nom de νεῦρα avec toutes les parties blanches, et, pour me servir du langage hippocratique, la mention *de deux cavités qui reçoivent et expulsent les matières alimentaires, et de beaucoup d'autres cavités que connaissent ceux qui s'occupent de ces objets* (*de l'Art,* § 10). Les choses étant ainsi à l'état rudimentaire, on ne s'étonnera pas que toute la partie théorique roule essentiellement sur les quatre humeurs et leurs modifications; la spéculation ne pouvait se généraliser qu'à l'aide de ces éléments qui avaient assez de réalité apparente pour permettre quelques tentatives de théorie. Mais ce point de vue suffit pour faire apprécier, sans plus de détail, ce qu'étaient ces systèmes primitifs qu'on a si longtemps surfaits, et qui ne peuvent pas mieux valoir que les bases qui les supportent."

MSS. and Editions

Physician is found in V, C, E and Holkhamensis 282. It has been edited by J. F. Bensel in *Philologus* LXXVIII. (1922), pp. 88–130.

I have collated V and Holkhamensis 282. The hand of V does not appear to be the same as that of this manuscript in *Dentition*, though possibly the same scribe adopted another style of writing. It is finer and somewhat neater, while λ and α are written with long strokes that slope downwards from left to right. Iota subscript is not written, so that as δικαιοσύνη is the reading towards the end of Chapter I, the dative is almost certainly correct. V agrees very nearly with the vulgate.

ΠΕΡΙ ΙΗΤΡΟΥ

Ἰητροῦ μέν ἐστι προστασίη[1] ὁρᾶν εὔχρως τε καὶ εὔσαρκος πρὸς τὴν ὑπάρχουσαν αὐτῷ φύσιν· ἀξιοῦνται γὰρ ὑπὸ τῶν πολλῶν οἱ μὴ εὖ διακείμενοι τὸ σῶμα οὕτως[2] οὐδ' ἂν ἑτέρων ἐπιμεληθῆναι καλῶς· ἔπειτα τὰ περὶ αὐτὸν καθαρίως[3] ἔχειν, ἐσθῆτι[4] χρηστῇ καὶ χρίσμασιν εὐόδμοις, ὀδμὴν ἔχουσιν ἀνυπόπτως πρὸς ἅπαντα· τοῦτο γὰρ ἡδέως ἔχειν συμβαίνει τοὺς νοσέοντας.[5] δεῖ δὲ σκοπεῖν τάδε περὶ τὴν ψυχὴν τὸν σώφρονα,[6] μὴ 10 μόνον τὸ σιγᾶν, ἀλλὰ καὶ περὶ τὸν βίον πάνυ εὔτακτον, μέγιστα γὰρ ἔχει πρὸς δόξαν ἀγαθά, τὸ δὲ ἦθος εἶναι καλὸν καὶ ἀγαθόν, τοιοῦτον δ' ὄντα[7] πᾶσι καὶ σεμνὸν καὶ φιλάνθρωπον· τὸ γὰρ

[1] εἶναι προστασίην with ἔσται after εὔσαρκος MSS.: ἐστι προστασίη, with ἔσται omitted, Ermerins: ἰητροῦ μὲν προστασίη ὁρᾶν ὡς εὔχρως τε καὶ εὔσαρκος ἔσται Bensel.

[2] οὕτως ὡς MSS.: οὕτως Littré: αὐτοὶ Ermerins.

[3] V has ἔπειτα περὶ αὐτῶν καθαίρειν ὡς. πρέπει (for τὰ περὶ) Ermerins. Bensel reads καθαρείως.

[4] After ἐσθῆτι Ermerins adds τε.

[5] I think that εὐόδμοις is a gloss on ὀδμὴν ἔχουσιν ἀνυπόπτως πρὸς ἅπαντα, and that τοῦτο . . . νοσέοντας is a gloss on the whole preceding sentence. It should be noticed that the grammar of the second gloss is faulty, and perhaps τοῖς νοσέουσι should be read.

[6] περὶ τὴν ψυχὴν σώφρονα V, which has also τοῦτο before

THE PHYSICIAN

CHAPTER I

THE dignity of a physician requires that he should look healthy, and as plump as nature intended him to be; for the common crowd consider those who are not of this[1] excellent bodily condition to be unable to take care of others. Then he must be clean in person, well dressed, and anointed with sweet-smelling unguents that are not in any way suspicious. This, in fact, is pleasing to patients. The prudent man must also be careful of certain moral considerations[2]—not only to be silent, but also of a great regularity of life,[3] since thereby his reputation will be greatly enhanced; he must be a gentleman in character, and being this he must be grave and kind to all. For an over-forward

[1] The οὕτως of this sentence is not otiose: "those who are not well off in these respects" (*i.e.* of a healthy complexion and not too thin). Ermerins emendation to αὐτοί is therefore not necessary, though it is ingenious.

[2] Bensel's reading will mean "the following are important characteristics of a prudent soul."

[3] It is easy to understand εἶναι with εὔτακτον from the εἶναι in the clause after the parenthesis. This understanding of a word or phrase in a first clause, which is actually used in a second clause, being unknown in modern English, is often a cause of obscurity.

σκοπεῖν. Ermerins reads and punctuates τὴν ψυχήν· σώφρονα . . . εὔτακτον εἶναι. Bensel has τὴν ψυχὴν τὴν σώφρονα.

[7] Perhaps ἐόντα.

προπετὲς καὶ τὸ πρόχειρον καταφρονεῖται, κῆν πάνυ χρήσιμον ᾖ· σκεπτέον[1] δὲ ἐπὶ τῆς ἐξουσίης· τὰ γὰρ αὐτὰ παρὰ τοῖς αὐτοῖς σπανίως ἔχοντα[2] ἀγαπᾶται. σχήμασι δὲ ἀπὸ μὲν προσώπου σύννουν μὴ πικρῶς· αὐθάδης[3] γὰρ δοκεῖ εἶναι καὶ μισάνθρωπος, ὁ δὲ ἐς γέλωτα ἀνιέμενος καὶ
20 λίην ἱλαρὸς φορτικὸς ὑπολαμβάνεται· φυλακτέον δὲ τὸ τοιοῦτον οὐχ ἥκιστα. δίκαιον δὲ πρὸς πᾶσαν ὁμιλίην εἶναι· χρὴ γὰρ πολλὰ ἐπικουρεῖν δικαιοσύνην·[4] πρὸς δὲ ἰητρὸν οὐ μικρὰ συναλλάγματα τοῖσι νοσέουσίν[5] ἐστιν· καὶ γὰρ αὐτοὺς[6] ὑποχειρίους ποιέουσι τοῖς ἰητροῖς, καὶ πᾶσαν ὥρην ἐντυγχάνουσι γυναιξί, παρθένοις, καὶ[7] τοῖς ἀξίοις πλείστου κτήμασιν· ἐγκρατέως οὖν δεῖ πρὸς ἅπαντα ἔχειν ταῦτα. τὴν μὲν οὖν
29 ψυχὴν καὶ τὸ σῶμα οὕτω διακεῖσθαι.

[1] Bensel with V reads σκοπὸν for σκεπτέον.

[2] σπανίως ἔχουσιν MSS. : σπανίως ἔχοντα Littré.

[3] αὐθάδης . . . μισάνθρωπος MSS. : Ermerins has neuters.

[4] δικαιοσύνην Holkhamensis 282, and apparently E : δικαιοσύνη V and C : δικαιοσύνῃ Bensel.

[5] νοσοῦσίν V : ἀρρωστέουσιν Ermerins.

[6] αὐτοὺς MSS. : αὑτοὺς Zwinger, Linden : ἑωυτοὺς Ermerins.

[7] Ermerins omits καὶ after παρθένοις.

obtrusiveness is despised, even though it may be very useful. Let him look to the liberty of action that is his; for when the same things are rarely presented to the same persons there is content.[1] In appearance, let him be of a serious but not harsh countenance; for harshness is taken to mean arrogance and unkindness, while a man of uncontrolled laughter and excessive gaiety is considered vulgar, and vulgarity especially must be avoided. In every social relation he will be fair, for fairness must be of great service.[2] The intimacy also between physician and patient is close. Patients in fact put themselves into the hands of their physician, and at every moment he meets women, maidens and possessions very precious indeed. So towards all these self-control must be used. Such then should the physician be, both in body and in soul.

[1] So Littré. But it is more than doubtful if the Greek will bear this meaning. The reading of V (*σκοπὸν*) points to corruption of the text, as does the *σπαυίως ἔχουσιν* of the MSS.

[2] Bensel's emendation to the dative is very attractive, and is probably right: "for on many occasions one must come to the help of fairness."

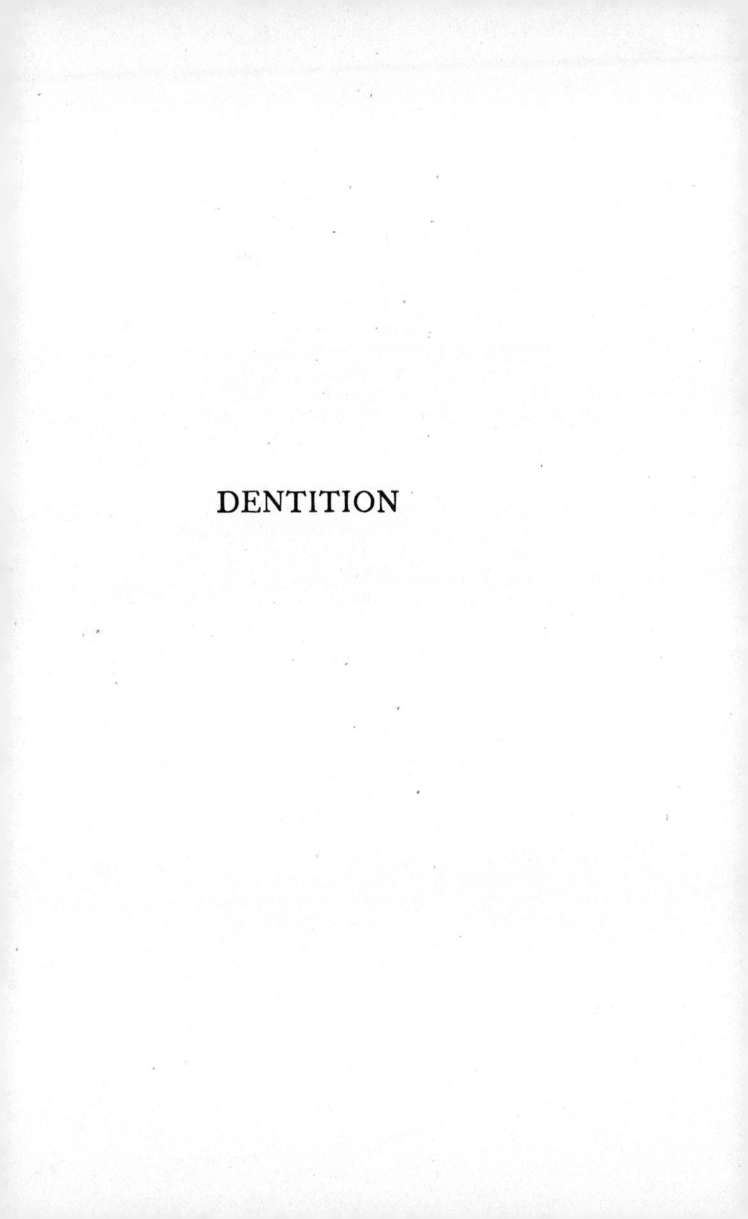

DENTITION

INTRODUCTION

Of this short piece Littré[1] says: "Ce très court fragment n'est cité par aucun ancien commentateur, rien ne peut nous faire deviner de qui il est, ni où il a été pris." In his *Argument* he begins: "Cet opuscule est rédigé dans la forme aphoristique, et, tout court qu'il est, il témoigne que l'auteur avait étudié, non sans fruit, l'état des enfants à la mamelle et leurs maladies."[2]

Adams'[3] remarks are very similar: "This little tract is destitute of any competent evidence of its authenticity. Some of the observations contained in it bespeak a familiar acquaintance with the diseases of infancy."

The account in Pauly-Wissowa is even scantier in its information: "ein Blättchen über das Zahnen der Kinder, wie das vorige weder von Galen noch Erotian erwähnt."

In spite of these rather discouraging remarks *Dentition* is a work of no little interest. In the first place it is written in aphorisms, and like most medical aphorisms deals with prognosis rather than treatment. Then again it is curiously short and abrupt, and the reader wonders why it was written in the present form. The answer to this puzzle may

[1] I. p. 415. [2] VIII. p. 542.
[3] Vol. I. p. 124.

perhaps become plainer after a discussion of the subject matter of *Dentition.*

It is obvious to any medical man that the tract is divided into two parts, both of which contain propositions apparently irrelevant to the main subject. Roughly speaking, however, one may put the matter thus:

(1) Propositions I.—XVII. deal with dentition (ὀδοντοφυΐα), and incidentally with the suckling and weaning of infants.

(2) Propositions XVIII.—XXXII. deal with ulceration of the tonsils (παρίσθμια), uvula and throat.

Teething and ulcerated throats are not connected, and it may be asked why they are here placed side by side. A short work dealing with both dentition and ulcerated throats is indeed a strange mixture.

It is remarkable that the key-word to most or the first part is ὀδοντοφυΐα, while of the second part it is παρίσθμια. This suggests that *Dentition* is an extract from a larger collection of aphorisms, which were arranged in a kind of alphabetical order. Ir the tract consisted only of propositions VI.—XII. and XVIII., XX.—XXVII., XXX.—XXXII., no doubt would be possible; every proposition would contain one or the other of the key-words. But there remain:—

(*a*) I.—V., with the key-words γάλα and θηλάζω.

(*b*) XIII.—XVII., with the key-words οὐρεῖσθαι, παράκειται (?), παρεσθίω, παρηθῶ, leading on to παρίσθμια in XVIII.

(*c*) XIX., the key-word of which is doubtful.

(*d*) XXVIII., XXIX., the key-words of which are doubtful.

Now surely ὀδοντ-, οὐρ-, παρα-, παρε-, παρη-, παρι-, must be intentionally set in alphabetical order, and I suggest that a scribe, copying a larger collection of aphorisms, omitted accidentally ὀδοντοφυΐα to παρίσθμια. This larger collection was arranged alphabetically, and probably dealt with diseases of childhood. When the scribe found out his mistake, he wrote out the omitted portion at the end, and added to it a few other propositions that he had missed. A later scribe, misinterpreting the facts, regarded the appendix as a fresh work, and gave it the not unnatural name *Dentition*. These remarks may be condemned as speculative guesses, but they are guesses to which an interesting parallel is to be found in the Paris manuscript 2255(E). At the end of this manuscript is a piece called περὶ προγνώσεως ἐτῶν. On examining it we find that it is a fragment of *Airs Waters Places*, which some scribe omitted, placed at the end of his volume, and so added a fresh treatise to the Hippocratic collection!

It is not at all unlikely that there are other similar fragments in the Hippocratic collection. Possibly, too, longer works contain fragments inserted by scribes who thought that they had found a suitable place for them. One or two passages, for instance, in *Epidemics I.* strongly suggest by their irrelevance an origin such as I have described.

The language of *Dentition* is in some respects unusual.

Proposition II. βορός. A poetic word(?). See Aristophanes *Peace* 38. ἕλκω, "I drink," seems poetic. See Euripides *Phoen.* 987 (ἕλκειν μαστόν).

Proposition III. ἐπιναύσιος is apparently a late word.

Proposition IV. πολλὴ φέρεται ἡ κοιλίη. εὐπεπτῶ is very rare.

Proposition XII. χειμῶνας ἔχει, if this reading be correct.

Proposition XIV. παρηθῶ, of the bowels being moved.

Proposition XV. ἀναλαμβάνω, of eating.

Proposition XVII. παρηθῶ.

Proposition XXV. ἀσμενίζω. This is apparently a late word.

Proposition XXVIII. ἀναλαμβάνω, of taking food or drink.

Proposition XXIX. εὐτροφὴς (if the reading be correct). It is apparently ἅπαξ λεγόμενον.

The number of strange expressions in so short a piece points to a late date. If *Dentition* be late, it forms an exception to my general statement that the aphoristic style ceased to prevail among medical writers after 400 B.C.

MSS. and Editions

The manuscripts containing *Dentition* are V, C, E, and Holkhamensis 282.

I have collated V and Holkhamensis 282. In this treatise the two are not strikingly alike; in fact, the close correspondence between the two manuscripts seems to end where they no longer correspond in the order of the treatises, namely after *Eight Months' Child*.

On the other hand, if I may judge from Littré's *apparatus criticus*, V and C (Paris 2146) are almost identical, and they also contain the treatises in the same order. It seems quite certain that C is a mere copy of V.

V reads πολὺ in Proposition III. and in others, but πουλὺ in V and in XXVIII. (τὰ πουλὺ γάλα κ.τ.λ.), although later in the same sentence πολὺ occurs.

The pronominal forms in ὁπ- are the almost universal rule, but in XIX. and XXII. ὁκ- is found.

The scribe regularly omits iota subscript, but in one place (XXX.) iota is written subscript between the -η- and -σ- of τῇσιν ἄλλῃσιν ὥρῃσι.

Sometimes, instead of dividing a word between one line and the next, the scribe preferred to write part of the word with a mark of abbreviation. Thus χειμῶνας appears as χειμῶν√, δυναμένων as δυναμέν͡ and θηλάζειν as θηλάζ⁄⁄. It is quite likely that corruptions have sometimes been caused by systems of abbreviation and contraction.

Examination of *Dentition* as it appears in V confirms my belief that no confidence can be placed in the spelling of even our best manuscripts in the matter of such points as ὁπ- and ὁκ-.

In places the text of *Dentition* is very corrupt. Accordingly, instead of attempting to restore hopeless passages, I have printed the text of Littré between daggers. In the footnotes emendations are mentioned, and in some cases discussed.

I know of no separate editions of the piece, although it is included in the editions of Littré and Ermerins.

ΠΕΡΙ ΟΔΟΝΤΟΦΥΙΗΣ

I. Τὰ φύσει εὔτροφα τῶν παιδίων οὐκ ἀνάλογον τῆς σαρκώσεως [1] καὶ τὸ γάλα θηλάζει.

II. Τὰ βορὰ καὶ πολὺ ἕλκοντα γάλα οὐ πρὸς λόγον σαρκοῦται.

III. Τὰ πολὺ διουρέοντα τῶν θηλαζόντων ἥκιστα ἐπιναύσια.[2]

IV. Οἷσι [3] πολλὴ φέρεται ἡ κοιλίη καὶ εὐπεπτοῦσιν,[4] ὑγιεινότερα· ὁπόσοισιν ὀλίγη, βοροῖσιν ἐοῦσι καὶ μὴ ἀνάλογον τρεφομένοισιν,[5] ἐπίνοσα.

V. Ὁπόσοισι [6] δὲ πολὺ γαλακτῶδες ἀπεμεῖται, κοιλίη συνίσταται.

VI. Ὁπόσοισιν ἐν ὀδοντοφυΐῃ ἡ κοιλίη πλείω ὑπάγει ἧσσον σπᾶται ἢ ὅτῳ ὀλιγάκις.

VII. Ὁπόσοισιν ἐπὶ ὀδοντοφυΐῃ πυρετὸς ὀξὺς ἐπιγίγνεται ὀλιγάκις σπῶνται.

VIII. Ὁπόσα ὀδοντοφυεῦντα εὔτροφα μένει καταφορικὰ ἐόντα κίνδυνος σπασμὸν [7] ἐπιλαβεῖν.

IX. Τὰ ἐν χειμῶνι ὀδοντοφυεῦντα, τῶν ἄλλων ὁμοίων ἐόντων, βέλτιον ἀπαλλάσσει.

[1] σαρκώσεως MSS. : σαρκώσιος Mack.

[2] ἐπιναύσια V, Holk. 282, C: ἐνιαύσια vulgate: ναυσίᾳ Ermerins.

[3] οἷσι V : Holk. 282 has ὁκόσοισι in the margin, but οἷσι in the text.

[4] The form of εὐπεπτοῦσιν arouses suspicion.

DENTITION

I. CHILDREN who are naturally well-nourished do not suck milk in proportion to their fleshiness.

II. Children with voracious appetites, and who suck much milk do not put on flesh in proportion.

III. Of sucking children those that pass much urine are the least subject to vomiting.

IV. Children that pass copious stools and have good digestion are the more healthy; those that pass stools scantily, and with voracious appetites are not nourished in proportion, are unhealthy.[1]

V. Those that vomit copiously milky matters suffer from constipation.

VI. Those who while teething have their bowels moved often are less subject to convulsions than those who have them moved seldom.

VII. Those who while teething are attacked by acute fever seldom suffer from convulsions.

VIII. Those who while teething are lethargic while remaining well-nourished run a risk of being seized with convulsions.

IX. Those who teethe in winter, other things being equal, come off better.

[1] Or, "subject to illness."

---

[5] ὑγιεινότερα . . . τρεφομένοισιν omitted by Holk. 282.

[6] ὁπόσοισι V: Holk. 282 reads ὁκόσοισι with π written over the κ, and so also in other places.

[7] σπασμὸς V and C; σπασμὸν Littré.

X. Οὐ πάντα τὰ ἐπὶ ὀδοῦσι σπασθέντα τελευτᾷ· πολλὰ δὲ καὶ διασῴζεται.

XI. Τὰ μετὰ βηχὸς ὀδοντοφυεῦντα χρονίζει·[1] ἐν δὲ τῇ διακεντήσει ἰσχναίνεται μᾶλλον.

XII. Ὁπόσα ἐν τῷ ὀδοντοφυεῖν χειμῶνας ἔχει, ταῦτα καὶ[2] προσεχόντως ἠγμένα ῥᾷον φέρει ὀδοντοφυΐαν.

XIII. Τὰ διουρεῦντα πλέον ἢ διαχωρεῦντα πρὸς λόγον εὐτροφώτερα.

XIV. Ὁπόσοισιν οὐρεῖται μὴ πρὸς λόγον, κοιλίη δὲ πυκνῶς ὠμὸν ἐκ παιδίων παρηθεῖ, ἐπίνοσα.

XV. Τὰ εὔυπνα καὶ εὔτροφα πολὺ †ἀναλαμβάνειν[3] καὶ παράκειται οὐχ ἱκανῶς διῳκημένον.†[4]

XVI. Τὰ παρεσθίοντα ἐν τῷ θηλάζειν ῥᾷον φέρει ἀπογαλακτισμόν.

XVII. Τὰ πολλάκις παρηθεῦντα[5] δίαιμον καὶ ἄπεπτον κατὰ κοιλίην πλεῖστα τῶν ἐν πυρετῷ ὑπνώδεα.

[1] χρονίζει Littré: χρονίζειν V and C.
[2] ταῦτα καὶ is omitted by Ermerins.
[3] ἀναλαμβάνει Foes: ἀναλαμβάνειν MSS.
[4] It is hard to decide whether Holk. 282 has διωκημένον or διωκειμένον.
[5] παρηθεῦντα Foes: παριθεῦντα or παρυθεῦντα MSS.

[1] For this sense of χειμὼν see *e.g. Breaths* XIV. τῆς νούσου καὶ τοῦ παρεόντος χειμῶνος, and also Ermerins' note on this passage. The meaning seems to be that during teething stormy "tantrums" on the part of the child are a better sign than a subdued, semi-comatose state.

[2] Perhaps πρὸς λόγον goes with διαχωρεῦντα, though the order of words is against this. The sense, however, would be improved. "Those who, in proportion, pass more urine than faeces are better nourished." So Littré.

X. Not all children die that are seized with convulsions while teething; many recover.

XI. Teething is protracted when complicated with a cough, and emaciation in such cases is excessive while the teeth are coming through.

XII. Children who have a troublesome time while teething, if they are suitably attended to, bear up more easily against teething.[1]

XIII. Those that pass more urine than faeces are proportionately better nourished.[2]

XIV. Those who do not pass urine in proportion, but from babyhood discharge undigested food frequently, are unhealthy.[3]

XV. Children who sleep well, and are well-nourished, may take a great deal of food, even though it is placed before them insufficiently prepared for digestion.[4]

XVI. Those that eat solid food while being suckled bear weaning more easily.

XVII. Those that often pass stools of undigested food mixed with blood, the great majority of them when feverish are drowsy.[5]

[3] Or, "subject to illness."

[4] It is fairly certain that the general sense of this proposition is to the effect that children who have healthy constitutions may without harm put a strain upon their digestive organs. But the exact reading is more than uncertain. παράκειται is strange, and cannot mean πάρεστι, as Littré thinks. But παράκειται seems to be the key-word (παρα-, with παρε- in the next proposition), and so is probably right. Perhaps εἰ has fallen out after καὶ (the scribe may have thought that οὐχ was wrong after εἰ), but I can find no parallel to this sense of διῳκημένον.

[5] Here too the Greek is strange, and I am not satisfied with the text, though I can offer no better reading. Possibly τῶν should be τούτων or ἐόντα; possibly it should be omitted.

XVIII. Τὰ ἐν παρισθμίοις ἕλκεα ἄνευ πυρετῶν γιγνόμενα ἀσφαλέστερα.

XIX. Ὁπόσοισιν ἐν τῷ θηλάζειν τῶν νηπίων[1] βὴξ προσίσταται, σταφυλὴν εἴωθε μείζονα ἔχειν.

XX. Ὁπόσοισι ταχέως ἐν παρισθμίοις νομαὶ ἐφίστανται, τῶν πυρετῶν μενόντων καὶ βηχίων, κίνδυνος πάλιν γενέσθαι ἕλκεα.[2]

XXI. Τὰ παλινδρομήσαντα ἐν ἰσθμίοις ἕλκεα †τοῖς ὁμοίοισι†[3] κινδυνώδεα.

XXII. †Τοῖσι παιδίοισιν ἀξιολόγοις ἕλκεσιν[4] ἐν παρισθμίοισι, καταπινομένων,[5] σωτηρίας[6] ἐστίν, ὁπόσα[7] δὴ[8] μᾶλλον τῶν πρότερον μὴ δυναμένων καταπίνειν.†[9]

XXIII. Ἐν παρισθμίοις ἕλκεσι, πολὺ[10] τὸ χολῶδες ἀνεμεῖσθαι ἢ κατὰ κοιλίην ἔρχεσθαι,[11] κινδυνῶδες.

XXIV. Ἐν τοῖσιν ἐν παρισθμίοισιν ἕλκεσιν ἀραχνιῶδές[12] τι ἐὸν οὐκ ἀγαθόν.

XXV. Ἐν τοῖσιν ἐν παρισθμίοισιν ἕλκεσι

[1] Ermerins places τῶν νηπίων after ὁπόσοισι.

[2] Ermerins omits ἕλκεα.

[3] ὁμοίοισι (or ὁμοίωσι) MSS. : ὡμοῖσι Calvus : νηπίοισι Cornarius and Ermerins.

[4] ἀξιόλογοις ἕλκεσιν MSS. : ἀξιόλογα ἕλκεα Ermerins.

[5] καταπινομένων MSS. : καταπίνειν δυναμένων Ermerins after Linden.

[6] σωτηρίας ἐστίν MSS. : σωτήριά ἐστι Ermerins.

[7] The MSS. punctuate before ὁπόσα and after χολῶδες in the next proposition. Littré suggested the punctuation in the text and he is followed by Ermerins.

[8] δὴ MSS. : δὲ Ermerins.

[9] Ermerins punctuates after πρότερον and marks an hiatus after καταπίνειν.

[10] Holk. 282 has τὸ πολύ.

[11] ἔρχεσθαι MSS. : διέρχεσθαι Ermerins.

XVIII. Ulcers on the tonsils that come without fever are less dangerous.

XIX. Babies that are attacked by a cough while being suckled usually have an enlarged uvula.

XX. When corroding sores form quickly on the tonsils, the fevers and coughs remaining, there is a danger of ulcerations occurring again.

XXI. Ulcerations that recur on the tonsils are dangerous.[1]

XXII. When children have considerable ulceration of the tonsils, if they can drink, it is a sign that they may recover, the more so if they could not drink before.[2]

XXIII. In cases of ulcerated tonsils, to vomit bilious matters, or to evacuate them by stools, is attended with danger.

XXIV. In cases of ulcerated tonsils, the formation of a membrane like a spider's web is not a good sign.[3]

XXV. In cases of ulcerated tonsils, after the first

[1] The conjecture of Cornarius ("of babies") is most ingenious and may be right. I suspect, however, that *τοῖς ὁμοίοισι* is part of a corrupted gloss on *ἰσθμίοις*, which some scholiast saw was used in the same sense as (*ὁμοίως*) *παρισθμίοις*.

[2] The most corrupt proposition in *Dentition*. It seems impossible to restore the exact text of the original. One suspects, however, that Ermerins is right in reading *ἀξιόλογα ἕλκεα* and *σωτήριά ἐστι*, and that Linden correctly changed *καταπινομένων* to *καταπίνειν δυναμένων*. The sense of *ὁπόσα . . . καταπίνειν* is fairly certain, but the Greek to represent it could be written in several ways.

[3] It would be interesting if we could interpret this proposition correctly.

[12] Holkamensis 282 omits *ἀνεμεῖσθαι . . . ἀραχνιῶδες*, the eye of the scribe passing from -*ῶδες* to -*ῶδες*.

μετὰ τοὺς πρώτους χρόνους διαρρεῖν φλέγμα διὰ τοῦ στόματος, πρότερον οὐκ ὄν,[1] χρήσιμον, ὅμως ἀνακτέον· ἢν δὲ ἄρξηται συνδιδόναι,[2] πάντως ἀσμενιστέον· τὸ δὲ μὴ οὕτως διαρρέον εὐλαβητέον.

XXVI. Ῥευματιζομένοις παρίσθμια κοιλίη κατενεχθεῖσα πλείω [3] λύει τὰς ξηρὰς βῆχας· παιδίοισιν ἀνενεχθέν τι [4] πεπεμμένον πλείω λύει.

XXVII. Τὰ πολὺν χρόνον ἐν παρισθμίοις ἕλκεα ἀναυξῆ μένοντα ἀκίνδυνα πρὸ τῶν πέντε ἢ ἓξ ἡμερέων.

XXVIII. Τὰ πολὺ γάλα τῶν θηλαζόντων ἀναλαμβάνοντα ὡς τὸ πολὺ ὑπνώδεα.

XXIX. Τὰ μὴ †εὐτροφέα† [5] τῶν θηλαζόντων ἄτροφα καὶ δυσανάληπτα.

XXX. Ἕλκεα ἐν θέρει γιγνόμενα ἐν παρισθμίοις χείρονα τῶν ἐν τῇσιν ἄλλῃσιν ὥρῃσι· τάχιον γὰρ νέμεται.

XXXI. Τὰ περὶ σταφυλὴν νεμόμενα ἕλκεα ἐν παρισθμίοισιν, σῳζομένοισι [6] τὴν φωνὴν ἀλλοιοῖ.

XXXII. Τὰ περὶ φάρυγγα νεμόμενα ἕλκεα χαλεπώτερα καὶ ὀξύτερα ὡς ἐπιπολὺ δύσπνοιαν ἐπιφέρει.

[1] ὂν MSS. : ἰὸν Ermerins. Perhaps ἐόν.

[2] ἄρξηται ξυνδιδῷ MSS. : ἄρξηται καὶ ξυνδιδῷ Mack : ἄρξηται ἢν ξυνδιδῷ Ermerins : ἄρξηται ξυνδιδόναι Littré : ἄρξηται συνδιδόν would be nearer the MSS.

[3] Ermerins omits πλείω.

[4] Ermerins omits παιδίοισιν and reads ἀνενεχθὲν δέ τι. V has τί.

[5] εὐτροφέα MSS. : εὔτροφα Ermerins.

[6] Before σῳζομένοισι V, Holk. 282 and C have γῆν, but read παρισθμίοις, not παρισθμίοισιν. Possibly γῆν has arisen from the -ιν.

periods it is useful for phlegm to flow from the mouth, which before did not do so; nevertheless it must be brought up. If the symptoms begin to disappear, it is altogether a welcome sign. If the phlegm does not flow in this way, you must be careful.[1]

XXVI. When there is a discharge on the tonsils, in most cases dry coughs are resolved by evacuation through the bowels; with children most cases are resolved by the vomiting of concocted matters.

XXVII. Ulcerations on the tonsils, that remain for a long time without increasing, are not attended with danger before five or six days.[2]

XXVIII. Children at the breast that take much milk are generally drowsy.

XXIX. Children at the breast that are ill nourished[3] also pick up strength with difficulty.

XXX. Ulcerated tonsils that occur in summer are worse than those that occur at other seasons, for they spread more rapidly.

XXXI. Ulcers on the tonsils that spread over the uvula alter the voice of those who recover.

XXXII. Ulcers that spread about the throat are more serious and acute, as they generally bring on difficulty of breathing.

[1] The readings *ὃν* and *συνδιδόναι* are uncertain, but the sense is quite clear.

[2] Littré points out that it is difficult to fit in *πολὺν χρόνον* with *πρὸ τῶν πέντε ἢ ἓξ ἡμερέων.* I agree with him, and believe that the first phrase is a gloss on the second.

[3] The word *εὐτροφέα* can scarcely be right; it should be *εὐτραφέα* or *εὔτροφα.* But even when it is corrected it is otiose with *ἄτροφα.* I suspect that there were once two readings (the Hippocratic collection has hundreds of such slight variations), namely, *τὰ μὴ εὐτραφέα τῶν θηλαζόντων καὶ δυσανάληπτα* and *τὰ ἄτροφα τῶν θηλαζόντων καὶ δυσανάληπτα.* At some time these two versions were combined into one.

POSTSCRIPT

(1) Objections may be raised to the use of "abscession" to translate ἀπόστασις. It is certainly not used in modern English, but neither are the ideas associated with ἀπόστασις accepted by modern science. The only alternative to the use of the term "abscession" would be to transliterate the Greek word with a footnote giving its meaning.

(2) *Regimen in Acute Diseases*, XIX. p. 78, ll. 11 foll. I am in doubt whether the sentence ἢν δὲ μὴ ὑπεληλύθῃ ὁ παλαιότερος σῖτος νεοβρῶτι ἐόντι, κ.τ.λ. refers or not to the former part of the chapter (ἐπισχεῖν τὴν δόσιν τοῦ ῥυφήματος, ἔστ' ἂν οἴηται κεχωρηκέναι ἐς τὸ κάτω μέρος τοῦ ἐντέρου τὸ σιτίον). My translation so takes it, identifying ὑπεληλύθῃ and κεχωρηκέναι ἐς τὸ κάτω μέρος τοῦ ἐντέρου. It is possible, however, that a new case is introduced. The patient has recently eaten food, but his bowels were even *before* this (παλαιότερος) full of unevacuated food. In such cases the doctor is recommended to use an enema or a suppository.

(3) *Regimen in Acute Diseases*, XXXVII. p. 92, l. 27. Though all the MSS. read μὴ I feel inclined to delete it. Possibly it may be retained as a pleonastic or redundant μή, but it would be difficult if not impossible to find a parallel. This pleonastic μή, so far as I know, is not found with an infinitive depending upon θαυμαστὸν εἶναι or its equivalent. It

might easily be a repetition (in uncials) of the last syllable of εἶναι.

(4) *Regimen in Acute Diseases,* XLVIII. p. 104. I feel that the whole of this chapter, and perhaps the next, is an interpolation. The sentence ὅτι . . . ἐς τὰ ῥυφήματα μεταβάλλουσιν ἐκ τῆς κενεαγγίης is either corrupt or a rather inane truism. The next sentence, ἔπειτα οὐ χρηστέον ῥυφήμασιν, πρὶν ἡ νοῦσος πεπανθῇ ἢ ἄλλο τι σημεῖον φανῇ ἢ κατὰ ἔντερον, κενεαγγικὸν ἢ ἐριστικόν, ἢ κατὰ τὰ ὑποχόνδρια, is directly contrary to the whole teaching of *Regimen in Acute Diseases,* and in particular cannot possibly be reconciled with Chapters XII–XIV. Chapter XLIX is perhaps not an interpolation, but a parenthesis which in a modern book would take the form of a footnote.

(5) *Regimen in Acute Diseases,* LXV. p. 120, l. 12. All the MSS. read προσκαταχεῖσθαι. But it is immediately followed by μετακαταχεῖσθαι. As προσ- and προ- are constantly confused by scribes, it is just possible that we should read προκαταχεῖσθαι. "Water should be poured over the body both *before* and *after* it is rubbed with soap."

(6) *Sacred Disease,* IV. p. 146, ll. 9–11. I am dissatisfied not only with the editors' emendations, but also with my own conjecture. The more I study the passage the more I am convinced that the words οὔτε εἴργεσθαι down to αὐτοῖς εἰσίν; are a gloss or glosses. The variants in the MSS. (besides those given on p. 146, M has δεινοὶ ἄρ' αὐτοῖς εἰσίν, and θ has δεινοὶ αὐτοῖς ἐῶσιν) point in the same direction. Moreover, οὔτε before εἴργεσθαι should be οὐδέ.

Both (*a*) οὔτε εἴργεσθαι ἂν οὐδενὸς τῶν ἐσχάτων, and
(*b*) ποιέοντες ἕνεκά γε· πῶς οὐ δεινοί (M) or
ποιέοντες ὡς οὐ δεινοί (θ)

look like rather childish glosses on σελήνην καθαιρήσει καὶ ἥλιον ἀφανιεῖ κ.τ.λ. It should be remembered that no Greek writings were so likely to become corrupted by glosses as were the medical works. If the two phrases I have indicated are taken away the text runs: θεοὺς οὔτε εἶναι νομίζειν οὔτε ἰσχύειν οὐδέν· εἰ γὰρ ἄνθρωπος μαγεύων καὶ θύων σελήνην καθαιρήσει . . . οὐκ ἂν ἔγωγέ τι θεῖον νομίσαιμι, which is both good grammar and good logic.

Sacred Disease, XIX. p. 178. In θ the passage from l. 5 to l. 10 appears thus (I do not correct mistakes):

οἱ δ' ὀφθαλμοὶ καὶ τὰ ὦτα· καὶ ἡ γλῶσσα· καὶ αἱ χεῖρες· καὶ οἱ πόδες· οἱ ἂν ὁ ἐγκέφαλος γινώσκη. τοιαύτα πρήσσουσι· γίνεται γὰρ ἐν ἅπαντι τῶ σώματι τῆς φρονήσιος τε ὡς ἂν μετέχηι τοῦ ἠέρος· ἐς δὲ τὴν ξύνεσιν ὁ ἐγκέφαλός ἐστιν ὁ διαγγέλλων.

In M we have:

οἱ δε ὀφθαλμοὶ καὶ τὰ οὔατα καὶ ἡ γλῶσσα καὶ αἱ χεῖρες καὶ οἱ πόδες, οἷα ἂν ὁ ἐγκέφαλος γινώσκηι, τοιαῦτα ὑπηρετοῦσι· γίνεται γὰρ παντὶ τῶ σώματι τῆς φρονήσιός τε. ὡς ἂν μετέχη τοῦ ἠέρος· ἐς δὲ τὴν σύνεσιν. ὁ ἐγκέφαλός ἐστιν ὁ διαγγέλλων.

The reading ὑπηρετοῦσι ("the limbs are the servants of the decisions of the brain") is attractive, and may be right. But the form is suspicious, and in spite of its attractiveness the word probably arose out of πρήσσουσι spelt πρήττουσι.

But the second sentence is ungrammatical, and Littré's text, which I have printed between daggers, is little, if any, better than the manuscripts. It is easy to rewrite something grammatical with the required sense, *e.g.*:

> *γίνεται γὰρ ἅπαντι τῷ σώματι ἡ φρόνησις ἕως ἂν μετέχῃ τοῦ ἠέρος,*

or

> *γίνεται γὰρ ἅπαντι τῷ σώματι τῆς φρόνησιός τι, ὡς ἂν μετέχῃ τοῦ ἠέρος.*

Even when the grammar is corrected other difficulties remain. The writer indeed is not very careful in his use of psychological terms, but it is quite impossible to reconcile this attribution of *φρόνησις* to *all* the body with the statement (Chapter XX):

> *τῆς μέντοι φρονήσιος οὐδετέρῳ μέτεστιν,*

i.e. neither heart nor midriff participate in *φρόνησις*. They have *αἴσθησις* only.

When we consider the ease with which glosses, and stupid glosses, would find their way into the Hippocratic texts,[1] it is difficult not to believe that we have here an unintelligent note. If the sentence be deleted the text runs:

> *οἷα ἂν ὁ ἐγκέφαλος γινώσκῃ, τοιαῦτα πρήσσουσι· ἐς δὲ τὴν σύνεσιν ὁ ἐγκέφαλός ἐστιν ὁ διαγγέλλων.*

The brain tells the limbs how to act, and is the messenger to consciousness, telling it what is happening.

(7) Secret Societies and the Hippocratic Writings

I suggest in my introduction to *Decorum* that this work represents an address delivered before a secret

[1] See pp. xlvii., xlviii.

society of physicians. It will be well briefly to review the evidence.

(1) *Decorum* is written in fantastic Greek of such a peculiar nature that no hypothesis, except that the author was in parts intentionally quaint and in others intentionally obscure, will account for the facts.

It is well known that the liturgies of secret societies affect strange words and expressions.

(2) The obscurity is greatest when the writer is speaking of *σοφία,* the gods, and the necessity of guarding and preserving certain knowledge. These are just the places where "secrets" would be mentioned.

(3) The taker of the Hippocratic *Oath* promises to impart *παραγγελίη, ἀκρόησις,* and *ἡ λοίπη ἅπασα μάθησις* only to (*a*) his sons, (*b*) his teacher's sons, and (*c*) indentured (*συγγεγραμμένοι*) pupils who have adopted the *νόμος ἰητρικός.*

(4) *Law* is a short address delivered to medical students before the beginning of their medical course. After stating the conditions without which a medical course cannot be a success, the writer concludes thus :—

τὰ δὲ ἱερὰ ἐόντα πρήγματα ἱεροῖσιν ἀνθρώποισι δείκνυται· βεβήλοισι δὲ οὐ θέμις, πρὶν ἢ τελεσθῶσιν ὀργίοισιν ἐπιστήμης.

(5) In *Precepts* (Chapter V) a genuine physician of sound principles is called *ἠδελφισμένος ἰητρός,* "a physician who has been made a brother."

On the other hand there are the following objections.

POSTSCRIPT

(1) All the ancient θίασοι had a distinctly religious association with some deity, and there is no trace of such a special cult in either *Precepts* or *Decorum*. In fact the absence of superstition is the most striking characteristic of all the Hippocratic writings, and proves their independence of the priest-physicians superintending the temples of Asclepius. Nevertheless *Decorum* is unique in insisting on the function of the gods in curing diseases.

(2) The Asclepiadae could not have been a θίασος, as the form of the word is against such a view. The proper style of a θίασος under the titular protection of Asclepius would have been Asclepiastae. There are as a matter of fact many references in inscriptions to such θίασοι of Asclepiastae.

Like nearly all the questions arising out of a study of the Hippocratic writings, this one of secret societies must be left in uncertainty and doubt. Further research may in the future throw light upon a dark problem, but for the present the following conclusions seem as positive as the facts warrant:—

(1) Among the hundreds of θίασοι and similar organizations in ancient Greece, particularly in Alexandrine and post-Alexandrine times, it is most unlikely that none would be limited to medical men.

(2) Such societies would have their ritual and liturgy, full of quaint expressions and unusual words.

(3) These words and expressions would be found,

if anywhere, in treatises of the type of *Decorum*.

(4) Our documents use language which, on a literal interpretation, do imply the existence of "mysteries," "initiation" and "brotherhood."